Geoinformatics for Sustainable Urban Development

This book provides compelling new insights into how cities are attempting to address sustainability challenges via major applications of geospatial technology in an urban area. It elucidates the role of geospatial techniques such as GIS and GNSS, including remote sensing in urban management, and covers the theory and practice of urban sustainability transitions. It provides case studies and contextualised tools for the governance of urban transitions to present various applications of geospatial techniques in an urban environment.

Features:

- Covers hands-on approaches on quantitative measures of urban analytics
- Focuses on sustainability issues in urban planning and development
- Includes pertinent global case studies for implementation of urban planning practices
- Reviews the inter-relationship between smart cities and sustainable development

This book is aimed at graduate students, researchers, and professionals in GIS, urban sciences, and geography.

Geoinformatics for Sustainable Urban Development

Edited by
Sulochana Shekhar and Deepak Kumar

CRC Press
Taylor & Francis Group
Boca Raton London New York

CRC Press is an imprint of the
Taylor & Francis Group, an **informa** business

First edition published 2024
by CRC Press
6000 Broken Sound Parkway NW, Suite 300, Boca Raton, FL 33487-2742

and by CRC Press
4 Park Square, Milton Park, Abingdon, Oxon, OX14 4RN

CRC Press is an imprint of Taylor & Francis Group, LLC

© 2024 selection and editorial matter, Sulochana Shekhar and Deepak Kumar; individual chapters, the contributors

ISBN: 9781032362564 (hbk)
ISBN: 9781032362571 (pbk)
ISBN: 9781003331001 (ebk)

DOI: 10.1201/9781003331001

Typeset in Times
by Newgen Publishing UK

Contents

Editor Biographies

Sulochana Shekhar is Dean of the School of Earth Sciences and Head of the Department of Geography at the Central University of Tamil Nadu, Thiruvarur, India. She is the proud recipient of the Erasmus Mundus (European Commission) Fellowship, Commonwealth Academic Fellowship (United Kingdom), and Endeavour Research Fellowship (Australia). Her research interests include urban geography and the application of remote sensing and geospatial techniques in urban environments. She is presently teaching urban geography and the fundamentals and applications of Geoinformatics at the post-graduation level. She has worked on cellular automata-based urban growth models and urban sprawl assessment using entropy for her doctoral research. Object-based image analysis was the main theme of her post-doctoral research at ITC, Netherlands. She has previously worked on projects involving spatial decision support systems for slums (UCL, UK) and extraction of slums and urban green space using object-based image analysis (UTAS, Australia). She has completed major funded projects on housing the urban poor (HUDCO), and the environmental impact of urbanisation (UGC-Major) using geospatial techniques. She has collaborated with Cambridge University, UK, and completed the UKIERI research project on the interactive spatial decision support system for managing public health. She recently wrapped up a DST project about urban heat island studies using multi-source remote sensing data. Under this project, she has developed an urban spectral library for selected urban built-up surface materials. Currently, she is working on building sustainable and resilient cities.

Deepak Kumar is an academic research professional with over 10 years of experience. Dr Kumar has been working as a Research Scientist since August 2022 at the Center of Excellence in Weather and Climate Analytics, Atmospheric Science Research Center (ASRC), State University of New York (SUNY), University at Albany (UAlbany), New York (USA). He has been associated with Amity University Uttar Pradesh, Noida (India), as a full-time Assistant Professor from 2016. He has been revisiting a wide range of issues associated with trad-itional research activities in the intersection area of Remote Sensing and Geoinformatics, Information Science, Energy Sciences, Climate Sciences, Sustainable Urban Studies, Environment and Climate Change, Urban Weather and Climate Modelling, Database Systems, Data Analysis and Visualization. He has completed two government-sponsored research projects namely "Hybrid Urban Landscape Analysis for Green Smart Cities through Geospatial Technology" sponsored by the Science and Engineering Research Board, Department of Science and Technology, Government of India, and "Meta-sensing of the urban footprint from airborne synthetic aperture radar (ASAR) data" sponsored from Space Applications Centre (ISRO), Ahmedabad, Gujarat, India as a sole principal investigator. Dr Kumar has led a cross-functional research team for project development, management, laboratory setups, and supervision, along with academic research training for undergraduate, postgraduate, and PhD students and fellow researchers. He has published 45+ research articles in high-impact web of science/Scopus-indexed international journals by Elsevier, Springer, Taylor & Francis, Emerald, SAGE, and Wiley publications. He has reviewed more than 150+ research articles as an ad-hoc/adjunct/invited reviewer or editorial board member for high-impact factor journals published by IEEE, Elsevier, Springer Nature, Taylor & Francis Group, Wiley, SAGE, Emerald, MDPI. He has supervised four PhD students and 29 postgraduate students. He has conducted 15 academic, research, and industrial visits for undergraduate students. He encourages researchers to develop their skills, knowledge, and experience through conference appearances, outreach activities, and the contribution of professional membership of learned bodies. He is a lifetime member of various prestigious professional bodies like IEEE Geoscience and Remote Sensing Society, American Geophysical Union (AGU), European Geosciences Union (EGU), Member of American Meteorological Society (AMS), International Association for Urban Climate (IAUC), American Association of Geographers (AAG), International Geographical Union (IGU), International Society for Photogrammetry and Remote Sensing etc.

Contributors

Abel Abebe
Department of Geology, College of
  Natural Sciences, Arba Minch
  University, Ethiopia

Pritha Acharya
National Institute of Disaster
  Management, India

Akanksha
Department of Civil Engineering, Amity
  University, Uttar Pradesh, Noida,
  India

Wasim Ayub Bagwan
School of Rural Development, Tata
  Institute of Social Sciences, Tuljapur,
  Maharashtra, India

K. Balaji
Vellore Institute of Technology, Vellore,
  Tamil Nādu, India

Deepali Bansal
Wildlife Institute of India, Chandrabani,
  Dehradun, India

Venkatesh Baskaran
Government College of Engineering,
  Tirunelveli, Tamil Nadu, India

Kamal Bisht
Shaheed Bhagat Singh College,
  University of Delhi, Delhi, India

Meseret Desalegn
Department of Geology, College of
  Natural Sciences, Arba Minch
  University, Ethiopia

Amit M. Deshmukh
Visvesvaraya National Institute of
  Technology Nagpur, Nagpur, India

Renu Dhupper
Amity Institute of Environmental
  Sciences, Amity University, Uttar
  Pradesh, Noida, India

Adebayo Oluwole Eludoyin
Obafemi Awolowo University, Ile-Ife,
  Nigeria

Meet Fatewar
Department of Regional Planning,
  School of Planning and Architecture,
  New Delhi.
School of Architecture, Planning and
  Design, DIT University, Dehradun,
  Uttarakhand

Ephrem Getahun
Department of Geology, College of
  Natural Sciences, Arba Minch
  University, Ethiopia

Sanjay Kumar Ghosh
Indian Institute of Technology Roorkee,
  India

Jagadeshan Gunalan
Department of Geology, College of
  Natural Sciences, Arba Minch
  University, Ethiopia

Anil K. Gupta
National Institute of Disaster
  Management, New Delhi, India

Niruti Gupta
Malaviya National Institute of
  Technology, Jaipur, India

Harshita Jain
Amity Institute of Environmental
Sciences, Amity University, Uttar
Pradesh, Noida, India

Prerna Jasuja
Malaviya National Institute of
Technology, Jaipur, India

Colins Johnny Jesudhas
University VOC College of Engineering,
Thoothukudi, India

Nupur Joshi
Amity Institute of Environmental
Sciences, Amity University, Uttar
Pradesh, Noida, India

Muralitharan Jothimani
Department of Geology, College of
Natural Sciences, Arba Minch
University, Ethiopia

A. Krishna Kumar
Government College of Engineering,
Tirunelveli, Tamil Nadu, India

Deepak Kumar
Amity Institute of Geoinformatics &
Remote Sensing, Amity University,
Uttar Pradesh, Noida, India

Dharmendra Kumar
Indian Institute of Remote Sensing,
ISRO, Dehradun, India

Sumit Kumar
Punjab Remote Sensing Centre,
Ludhiana, India

Maya Kumari
Amity School of Natural Resources
& Sustainable Development, Amity
University, Uttar Pradesh, Noida,
India

R. Pushpa Lakshmi
PSNA College of Engineering and
Technology, Dindigul, India

Ashish Mani
Amity School of Natural Resources
& Sustainable Development, Amity
University, Uttar Pradesh, Noida,
India

Kalpana Markandey
Osmania University, Hyderabad, India

Gaurav Kumar Mishra
Visvesvaraya National Institute of
Technology Nagpur, Nagpur, India

Oladeji Quazeem Muhammed
Obafemi Awolowo University, Ile-Ife,
Nigeria

R. Nagalakshmi
Department of Civil Engineering, SRM
Institute of Science & Technology,
Kattankulathur, Tamil Nadu, India

A.R. Narayani
School of Architecture and Interior
Design, SRM Institute of Science &
Technology, Kattankulathur, Tamil
Nadu, India

Pranjal Pandey
Department of Civil Engineering, Amity
University, Uttar Pradesh, Noida,
India

Brijendra Pateriya
Punjab Remote Sensing Centre,
Ludhiana, India

E. Aswin Raj
Government College of Engineering,
Tirunelveli, Tamil Nadu, India

C. Rakesh
Vellore Institute of Technology, Vellore,
Tamil Nadu, India

Meghna Rout
Punjab Remote Sensing Centre,
Ludhiana, India

Shubham Kumar Sanu
Delhi School of Economics, University
of Delhi, Delhi, India

Ambrina Sardar Khan
Amity Institute of Environmental
Sciences, Amity University, Uttar
Pradesh, Noida, India

Hanuth Saxena
Savitribai Phule Pune University, Pune,
India

Priya Sharma
Gautam Buddha University, Greater
Noida, Uttar Pradesh, India

Reenu Sharma
Punjab Remote Sensing Centre,
Ludhiana, India

Vishwa Raj Sharma
Shaheed Bhagat Singh College,
University of Delhi, Delhi, India

Sulochana Shekhar
Central University of Tamil Nadu, Tamil
Nadu, India

Aman Singh Rajput
IPE Global Ltd., New Delhi, India

Chetna Singh
School of Planning and Architecture,
New Delhi, India

M.A.M. Mannar Thippu Sulthan
Government College of Engineering,
Tirunelveli, Tamil Nadu, India

Rina Surana
Malaviya National Institute of
Technology, Jaipur, India

C. Jeswin Titus
EIC-ADM, ESRI India Technologies,
India

T. Vivek
Vellore Institute of Technology, Vellore,
Tamil Nādu, India

Vinita Yadav
Department of Regional Planning,
School of Planning and Architecture,
New Delhi

1 Subaltern Urbanisation in the State of Haryana, India

Meet Fatewar and Vinita Yadav

CONTENTS

1.1 INTRODUCTION

Subaltern urbanisation refers to the growth of settlement agglomerations, which neither depends on large settlements (metropolis) nor is it planned in a systematic manner (planned cities). Such settlements are autonomous in nature and independent of metropolises. Subaltern urbanisation is generated by the market and historical forces (Denis et al. 2012a). It is an outcome of the limited possibility of metro-centric development and metropolis-based urbanisation. It helps to understand the importance of small and medium-sized settlements within the framework of metropolis-based urbanisation (Denis and Zérah, 2017). In India, the urban system is a combination

DOI: 10.1201/9781003331001-1

1

of diversified towns and cities. The rate of urbanisation is slower as compared to the
other developing countries of the world (Planning Commission, 2011). However, the
definition of urban settlements differs across countries based on single or multiple
parameters. For example, Austria follows a single-fold classification using population as the only parameter, whereas India uses a four-fold classification method by
considering the criteria of total population, population density, male working population engaged in non-agricultural pursuits, and civic status of the governing institu-
tion. The value of a single parameter such as population varies from 1000 in Austria
to 50,000 in the Republic of Korea (Aijaz, 2017). Therefore, it is difficult to compare
the level of urbanisation between different countries using national criteria. This non-
comparability is mainly due to the variations in the definition of "urban settlement"
across countries (Fatewar and Yadav, 2021).

The urban population of India increased from 27.82% (286.1 million) in 2001
to 31.14% (377.1 million) in 2011 (Census of India, 2011a). In absolute terms,
the increase in the urban population (91 million) was greater than the increase in
the rural population (90.5 million) in 2011, for the first time since independence
(Bhagat, 2011). The Census of India recognises two types of urban settlements, i.e.
statutory towns (STs), and census towns (CTs). Out of the total urban population,
84.46% (318.5 million) and 15.54% (58.6 million) of the urban population reside in
4041 STs and 3892 CTs, respectively. The absolute number of STs increased only
by 6.37% (242) from 3799 in 2001 to 4041 in 2011. However, the number of CTs
increased from 1362 to 3892, with a decadal growth rate of 185.76% from 2001 to
2011 (MoUD, 2015). The reclassification of rural settlements led to urban population
growth of 32.80% (29.9 million). Approximately 2647 new CTs were expected to be
added by the next census year of 2021 (Pradhan, 2017). Therefore, the unanticipated
increase in CTs needs to be analysed in terms of the demographic changes in small
urban settlements, which is one of the major contributors to subaltern urbanisation.

The chapter is structured into six sections. Section 1.1 includes an introduction
to the chapter. Section 1.2 describes the research methodology, particularly the data
collection, tools, and methods of data analysis along with the scope and limitations of
the research. Section 1.3 explains the concept of subaltern urbanisation and its classification axis. The difference between subaltern and non-subaltern urbanisation is
also explained in this section. Section 1.4 critically reviews the trend of urbanisation
and contribution of STs and CTs in the urbanisation level for the state of Haryana.
Section 1.5 illustrates the framework to identify the subaltern towns and analyses
the contribution of subaltern urbanisation at the district and state levels. The state
and district levels analyses have been carried out to understand the spatial pattern of
identified subaltern towns. Section 1.6 concludes the research and suggests future
research areas.

1.2   RESEARCH DESIGN

A spatial-statistical analysis has been carried out using ArcGIS software to identify
the subaltern towns and estimate the level of subaltern urbanisation for the state of
Haryana, India. This analysis is mainly based on secondary data published by the
"Office of the Registrar General & Census Commissioner" on the "Census of India"

website. The census data have been processed, tabulated, analysed, and mapped using spatial-statistical analysis. At the state level, the transformation in the level of urbanisation has been studied by examining the data from 1961 to 2011, which are later compared with the nation's urbanisation level. The growth rate of CTs has been calculated by evaluating the data of the last census decade, i.e. 2001–11. Finally, the subaltern towns are identified based on the census data for the year 2011. For the identification of subaltern towns, only the urban settlements are taken into the account. All the class-I towns, having a population of more than 1 lakh, are considered as large urban settlements. Since the scope of the research is limited to Haryana state, the impact of nearby states and regions beyond states has not been taken into account for the identification of subaltern towns. Towns or cities have multiple characteristics but are considered only under a single category based on their predominant function to avoid duplicity in the data analysis.

1.3   SUBALTERN URBANISATION

"Subaltern urbanisation" refers to the growth of settlement agglomerations, which is independent of large settlements and planned cities. Faridabad (Haryana), with a population of more than 1 million, is defined as a metropolis under the category of large settlements. The planned cities consist of (a) new or planned towns such as Panchkula (Haryana) and Bhubaneshwar (Odisha), and (b) industrial towns such as Manesar (Haryana) and Mithapur (Gujarat). The subaltern urbanisation is generated by the market and historical forces, due to the autonomous growth of settlement agglomeration (Denis et al. 2012a). In India, the urbanisation process of many small towns is driven by their own economic dynamism and not by the big city located in proximity. The sense of not being dependent on the economic system of big cities, lack of visibility and voice, drive the process of "subaltern urbanisation." However, the Census of India neither recognises nor provides the definition of subaltern urbanisation.

Non-subaltern urbanisation is exactly the opposite of subaltern urbanisation. Therefore, non-subaltern urbanisation is dependent on large settlements and planned cities. The urbanisation occurring in large settlements, such as metropolitan towns, is called metropolitan urbanisation. The urbanisation occurring in planned settlements is divided into administrative urbanisation and corporative urbanisation. Administrative urbanisation takes place in new towns, whereas corporative urbanisation takes place in industrial towns (Figure 1.1). Hence, urbanisation, arising beyond metropolitan, administrative, and corporative urbanisation, is defined as subaltern urbanisation (Denis et al. 2012b).

Subaltern urbanisation is categorised into two axes, i.e., spatial proximity and administrative recognition. The spatial proximity axis deals with the location of one settlement with respect to another. It is sub-divided into two categories, viz., peripheral and non-peripheral settlements. The peripheral settlements are located on the periphery of the large settlements, whereas the remaining settlements are classified as non-peripheral settlements. The administrative recognition axis deals with the civic status of the settlements, which are described as invisible, denied, recognised, and contested settlements. The invisible settlements are not recognised as urban settlements, whereas denied settlements are classified as census towns. The recognised

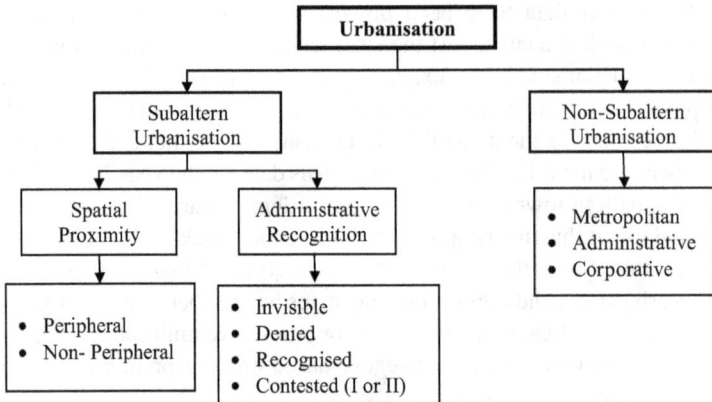

FIGURE 1.1   Categorisation of urbanisation.

Source: Authors' interpretation based on data from Denis et al., 2012a and Denis et al., 2012b.

settlements are classified as statutory towns, whereas contested settlements are those which are dissatisfied with their civic status. The contestation is amongst those rural settlements which want to be urban (Contested-I) and those classified as urban but that want to be rural (Contested-II) (Denis et al. 2012a).

1.4   HARYANA'S URBANISATION

As per the Census of India (2011), the total population of Haryana state is 25.3 million, which resides in 154 urban and 6841 rural settlements. The state's urbanisation level is 34.88% (8.8 million). Despite this, the major proportion of the population resides in rural settlements, i.e. 65.12% (16.5 million). The level of urbanisation in Haryana has increased from 17.23% (1.3 million) in 1961 to 34.88% (8.8 million) in 2011. The low level of urbanisation in 1961 until 2001 was primarily due to the agrarian-based population residing in rural settlements. For the first time, the state's level of urbanisation (28.92%) surpassed the national level of urbanisation (27.82%) in 2001. The share of urban population residing in urban settlements of the state was 34.88%, which was more than the national average of 31.14% in 2011 (Figure 1.2). this reveals that a similar trend of a higher level of urbanisation than the national average continued in 2011. The rapid increase in the level of urbanisation is the result of the development of the primary sector (agriculture and allied activities) which led to the establishment of Mandi Towns in different regions of the state during the green revolution (Kumar, Singh and Kalotra, 2013; Census of India, 2011a).

Haryana state has 22 districts, covering a geographical area of 44,212 square kilometres. Amongst these districts, Faridabad is the most urbanized, with an urbanisation level of 79.51% (1.4 million) followed by Gurgaon and Panchkula with 68.82% (1.0 million) and 55.81% (0.3 million), respectively. The proximity to the nation's capital (NCT of Delhi) and the establishment of a large number of industries are the

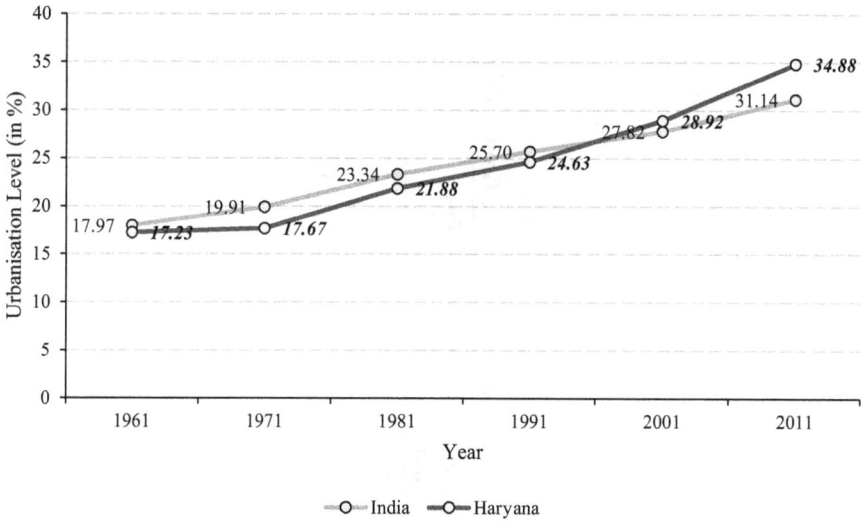

FIGURE 1.2   Trend of urbanisation from 1961 to 2011.

Source: Authors' calculations using data from Census of India, 2001 and 2011.

causes of the high level of urbanisation in Faridabad district. Faridabad city is the only metropolis in the state, and is also a district headquarter, and the oldest industrial hub of Haryana. The urbanisation level of Gurgaon district increased from 35.58% in 2001 to 68.82% in 2011. It is the only district of the state which has observed a phenomenal increase in the urbanisation level during the last census decade because of its proximity to the national capital and good connectivity with IGI Airport, New Delhi. It also hosts Gurgaon city, which is one of the fastest growing settlements, being a financial and technological hub. Panchkula district is developing because of its proximity to Chandigarh, which is the capital of the states of Punjab and Haryana. The planned city of Panchkula has a multi-functional character, including as an administrative, educational, and service centre (Sangwan and Mahima, 2019).

1.4.1   CONTRIBUTION OF CENSUS TOWNS

The growth rate of Haryana state declined from 50.52% in 2001 to 44.59% in 2011. The number of STs decreased from 84 to 80, whereas the number of CTs increased from 22 to 74 during 2001–11. Statutory towns have observed a negative growth rate of 4.76%, while census towns have a positive growth rate of 236.36%. Out of the total urban population, the share of the population residing in CTs was low (4.71%) till 2001, with the numbers growing gradually (Kumar et al., 2013; Census of India, 2011; Census of India, 2001). The population residing in CTs increased abruptly in 2011 (Table 1.1). During the last census decade (2001–11), the burgeoning number of CTs was one of the major factors for the complex development of small urban settlements. It also shifted the focus from metropolitan-based urbanisation to subaltern urbanisation.

TABLE 1.1
Urban Growth Rate of Haryana, 2001–11

S. No.	Civic Status	2001			2011			Increment		Growth Rate	
		Towns (no.)	Population (000s)	Share (%)	Towns (no.)	Population (in 000s)	Share (%)	Towns (no.)	Population (000s)	Towns (%)	Population (%)
1.	CT	22	288.2	4.71	74	913.8	10.33	52	625.7	236.36	217.13
2.	ST	84	5827.1	95.29	80	7928.3	89.67	–4	2101.1	–4.76	36.06
3.	Total	106	6115.3	100.00	154	8842.1	100.00	48	2726.8	45.28	44.59

Source: Authors, calculations using data from Census of India, 2011 and Census of India, 2001.

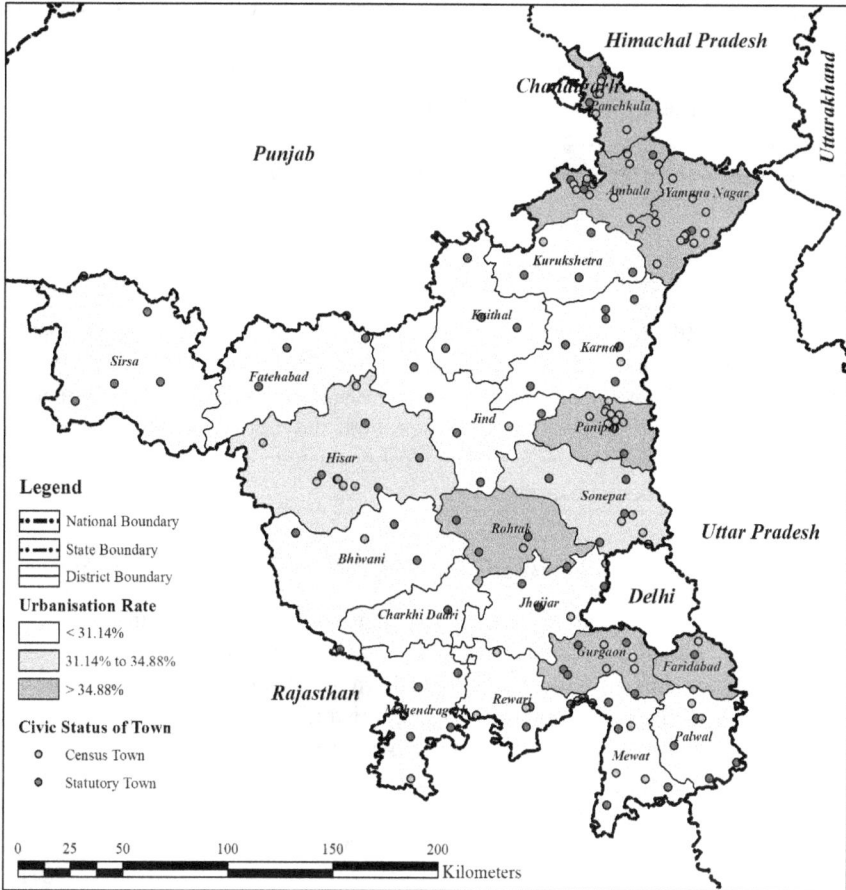

FIGURE 1.3 Relationship between urbanisation level and census towns of Haryana.

Source: Authors' interpretation using data from Census of India, 2011 and Charkhi Dadri, 2016. Note: Charkhi Dadri district was carved out from Bhiwani district on 1 December 2016.

1.4.2 DISTRICT LEVEL ANALYSIS: LEVEL OF URBANISATION AND CTs

As per the spatial-statistical analysis for the state of Haryana, the level of urbanisation in a district depends on two major factors, i.e. proximity to the Union Territory (UT), and the number of CTs within the district (Figure 1.3). Faridabad district has the highest level of urbanisation with 79.51% (1.4 million) followed by Gurgaon and Panchkula districts with 68.82% (1.0 million), and 55.81% (0.3 million), respectively. The majority of urbanised districts are located either near or adjacent to the UTs. Panchkula (55.81%), Ambala (44.38%), and Yamuna Nagar (38.94%) districts are located near to Chandigarh, while Faridabad (79.51%), Gurgaon (68.82%), Panipat (46.05%), and Rohtak (42.04%) districts are situated close to the National Capital Territory (NCT) of Delhi.

Furthermore, the districts with higher urbanisation levels, as compared to the national average (31.14%), have a greater number of CTs. Ambala district (11) followed by Panipat (10) and Yamuna Nagar (10) have the highest number of CTs and urbanisation levels of 44.38%, 46.05%, and 38.94%, respectively. The urbanisation levels of Charkhi Dadri, Kaithal, and Sirsa districts are 11.22%, 21.97%, and 24.65%, respectively, which are less than the national average. None of these districts have any CTs.

1.5   HARYANA'S SUBALTERN TOWNS

In order to identify the subaltern towns of a state, it is important to first identify the non-subaltern towns. Statistically, there is no direct way to identify the subaltern towns of a state. Therefore, an indirect method has been applied to identify them, in which, first, the non-subaltern towns are identified, and then the remaining towns are categorised as "Subaltern towns". Hence, it is important to understand the criteria for the identification of non-subaltern towns.

1.5.1   NON-SUBALTERN URBANISATION

It is essential to understand the characteristics of non-subaltern urbanisation as it is the opposite of subaltern urbanisation. The towns having non-subaltern urbanisation are known as non-subaltern towns. Non-subaltern urbanisation is classified into three categories, i.e. metropolitan urbanisation, administrative urbanisation, and corporate urbanisation. The urbanisation occurring in metropolitan towns is defined as metropolitan urbanisation, such as Faridabad city in Haryana. The urbanisation taking place in planned or new towns, such as Panchkula city of Haryana, is identified as administrative urbanisation. The urbanisation happening in industrial towns such as Manesar city of Haryana is termed as corporative urbanisation (Denis et al., 2012b). As per the Cantonments Act, 2006, the cantonments areas are directly under the control of the Ministry of Defence (MoD) for development-related activities (MoL&J, 2006). For this reason, cantonment areas are also considered under the category of non-subaltern town. Hence, metropolitan towns, industrial towns, planned cities, and containment areas are categorized under this category. The settlements under the shadow of metropolitan towns are also within this category. Thus, there is a need to understand the demographic changes occurring outside the metropolitan, planned, and industrial towns in the state of Haryana. Based on an understanding of the literature review, the authors have classified the following urban settlements under the category of non-subaltern towns.

1.5.1.1   Large Settlements

The development of large settlements is partially influenced by the population size and majorly developed either by the state or central government. Large settlements usually change dynamically and develop due to natural growth. Many medium-sized and small settlements depend on the large size settlement for development and economic growth. As compared to small and medium-sized settlements, large settlements do not depend on the small settlements for development. Hence, large settlements are

TABLE 1.2
Population Contribution of Large Urban Settlements

		Towns		Population	
S. No.	Type of Towns	(no.)	(%)	(000s)	(%)
1	*Large settlements*	*20*	*12.99*	*6014.7*	*68.02*
1.1	1 to 5 lakh population	18	11.69	3714.2	42.01
1.2	5 to 10 lakh population	1	0.65	886.5	10.03
1.3	More than 10 lakh population	1	0.65	1414.1	15.99
2	*Small settlements (less than 1 lakh population)*	*134*	*87.01*	*28.3*	*31.98*
3	*Total*	*154*	*100.00*	*8842.1*	*100.00*

Source: Authors' calculations using data from Census of India, 2011.

classified as non-subaltern towns. In Haryana, the variation in the population size of urban settlements is very high. Even in the class-I category, the population of the urban settlement varies from 1 lakh to 15 lakh. Hence, for the purpose of identification of subaltern towns, urban settlements having a population of more than one lakh (also classified as Class-I towns by the Census of India) are also categorised into a single category of large settlements.

In Haryana, there are 154 urban settlements. Faridabad is the only metropolitan city in the state. Approximately, 15.99% of the urban population resides in just one metropolitan city. It has been observed that the government provides more funds for the development of metropolitan cities as there is a significant number of people residing in such settlements. The urbanisation occurring in metropolitan towns is classified as metropolitan urbanisation. Further, there is only one town in the state which has a population of 5–10 lakhs, i.e. Gurgaon. This has the second largest share of the urban population, with 10.03%. These settlements (having populations of 5–10 lakhs) are comparatively smaller in size as compared to the metropolitan towns. However, the development of such towns occurs due to their high population. Such towns are treated as the upcoming metropolitan towns by the local or state government in the absence of recent census data. Approximately 42.01% of the urban population resides in 18 urban settlements, which have a population size of 1–5 lakh (Table 1.2). Due to the high population threshold, such towns cannot be defined as smaller towns and are treated as the upcoming large settlements of the state. Interestingly, more than two-thirds of the urban population resides in just 20 large urban settlements (Class-I towns) of the state, whereas the remaining one-third of the urban population resides in 134 urban settlements.

1.5.1.2 Industrial Development

There are approximately 15 industrial towns in Haryana, which either developed as an industrial town or were transformed into industrial towns over a period of time. Only 5.74% of the urban population lives in these industrial towns of the state. The urbanisation occurring in industrial towns is classified as corporative urbanisation. Haryana State Industrial and Infrastructure Development Corporation Limited (HSIIDC) is responsible for the development of industrial infrastructure in the state. In 2022, there were around 24 Industrial Estates (IEs), seven Industrial Model Townships (IMTs), and one Information Technology (IT) Park in Haryana (HSIIDC 2022).

1.5.1.3   Planned Development

The cities which are planned and/or developed by the state government are considered under this category. In Haryana, there are two towns which are treated as planned towns of the state. The share of these two towns in the overall urbanisation level is only 0.82%. Panchkula is a planned town with a population of more than 2 lakhs, but it is already counted in the category of large settlements. Therefore, it is not counted under this category to avoid duplication. A special case of a planned CT is observed in the Panipat district of Haryana. "Sector 11 & 12 Part II" is developed as a planned area.

1.5.1.4   Cantonment Board

As per the Cantonments Act, 2006, development-related activities of the cantonment areas are directly under the control of the Ministry of Defence (MoD) (MoL&J, 2006). There are 62 notified containment boards in India. Of these, only one is present in the state of Haryana, i.e. Ambala Cantonment Board (DGDE, 2022). The Ambala Cantonment Board was established in 1843 as a statutory body under the Cantonments Act, 2006. It is a Category-I cantonment board with a population of 55,370 (including troops) in 2011 (Ambala Cantonment Board, 2022). The Ambala Cantonment Board is classified as a non-subaltern settlement as it is entirely regulated by the MoD.

1.5.1.5   Towns Developed under the Shadow of Large Settlements

These towns, which are dependent on the large settlements (including metropolis) for their growth and development, are classified as shadow towns. The towns falling within a specific radial distance from a large settlement are considered to be developed due to the impact of such settlements. However, each town has its own shape. Hence, a radius calculated for one town may not be applicable to another town of similar size. Also, a single radius may not be suitable even for towns located in a hilly region within the same class. To address this issue, different buffers have been used for different towns based on their population size. For the purpose of detailed analysis, the towns with a minimum population of 1 lakh and above are divided into three categories, i.e. 1–5 lakh population (Category I), 5–10 lakh population (Category II), and 10–40 lakh population (Category III). In order to identify the areas under the shadow of large settlements, the radii of 0–10 km, 10–15 km, and 15–20 km are considered for category I, category II, and category III, respectively (Pradhan, 2017). The spatial-statistical analysis in ArcGIS reveals that 37 towns are developed under the shadow area of large urban settlements and they accommodate approximately 5.70% of the total urban population (Table 1.3).

TABLE 1.3
Towns under the Shadow of Large Urban Settlements

S. No.	Town Categories (Buffer radii in km)	Towns		Population	
		(no.)	(%)	(000s)	(%)
1	Category I (0–10 km)	32	20.78	438.6	4.96
2	Category II (10–15 km)	3	1.95	40.9	0.46
3	Category III (15–20 km)	2	1.30	24.8	0.28
4	*Total*	*37*	*24.03*	*504.3*	*5.70*

Source: Authors' calculations using data from Census of India, 2011.

1.5.2   SUBALTERN URBANISATION

The subaltern towns are calculated by a reverse calculation. As mentioned in Section 1.5.1, first the non-subaltern towns are calculated, and then, the remaining towns are classified as subaltern towns. As per the authors' estimate, of 154 towns, 79 are categorised as subaltern towns in Haryana state. In 2011, the share of subaltern urbanisation in the urbanisation level of the state was estimated to be around 19.08% (Table 1.4). The subaltern towns are further classified into two categories, i.e. ancient towns and remaining towns. The towns with over 2,000 years of historical importance and/or with certain cultural and religious backgrounds are considered as ancient towns. Of the 79 subaltern towns, 29 are categorised as ancient towns which have historical importance at the state or national level. The remaining 53 towns do not have any historical importance in the ancient history of the state or nation. The subaltern towns are distributed unevenly throughout the districts of Haryana state. On the contrary, the non-subaltern towns are concentrated close to the large urban settlements within the district and follow a metro-centric development (Figure 1.4).

1.5.2.1   Spatial Proximity

Based on the spatial proximity axis, the subaltern towns are categorised into peripheral towns and non-peripheral towns. These towns, which are located on the periphery of the urban settlements, are known as peripheral towns, whereas the remaining towns are known as non-peripheral towns. Of the 79 subaltern towns in Haryana, only four are classified as peripheral towns. They constitute approximately 7.41% of the total subaltern population. The maximum percentage (92.59%) of the population resides in 70 non-peripheral towns (Table 1.5).

1.5.2.2   Administrative Recognition

Based on the administrative recognition, subaltern towns are categorized into five sub-categories, i.e. (i) Invisible Towns, which are not recognised as urban settlements; (ii) Denied Towns, which are Census Towns; (iii) Recognised Towns, which are Statutory Towns; (iv) Contested (I) which refers to a situation where the settlement

TABLE 1.4
Identification of Subaltern Towns in Haryana

S. No.	Type of Towns	Towns (no.)	Towns (%)	Population (000s)	Population (%)
A	**Non-subaltern towns**	**75**	**48.70**	**7154.8**	**80.92**
1	*Large settlements*	*20*	*12.99*	*6014.7*	*68.02*
1.1	1 to 5 lakh population	18	11.69	3714.2	42.01
1.2	5 to 10 lakh population	1	0.65	886.5	10.03
1.3	More than 10 lakh population	1	0.65	1414.1	15.99
2	*Under shadow*	*37*	*24.03*	*504.3*	*5.70*
2.1	1 to 5 lakh population	32	20.78	438.6	4.96
2.2	5 to 10 lakh population	3	1.95	40.9	0.46
2.3	More than 10 lakh population	2	1.30	24.8	0.28
3	*Cantonment board*	*1*	*0.65*	*55.4*	*0.63*
4	*Industrial development*	*15*	*9.74*	*507.6*	*5.74*
5	*Planned development*	*2*	*1.30*	*72.7*	*0.82*
B	**Subaltern towns**	**79**	**51.30**	**1687.3**	**19.08**
1	*Ancient towns*	*26*	*16.88*	*779.4*	*8.81*
2	*Remaining towns*	*53*	*34.42*	*908.0*	*10.27*
C	*Total*	*154*	*100.00*	*8842.1*	*100.00*

Source: Authors' calculations using data from Census of India, 2011.

is administratively defined as rural but it wants to be urban; and Contested (II) which refers to a situation where the settlement is classified as urban but wants to be rural. In the case of Haryana, only the urban settlements are considered and analysed. Therefore, no town has been identified under the category of invisible towns and contested (I) towns. In the remaining three categories, the maximum share of the population resides in the recognised towns with 82.18% (1.3 million), followed by denied towns with 17.26% (0.3 million). Under the category of contested (II), only three towns accommodating just 0.56% of the total subaltern population are situated (Table 1.6).

1.5.3 DISTRICT LEVEL ANALYSIS

In 2011, the share of subaltern urbanisation in the state was estimated to be around 6.66% (Table 1.7). As per the spatial-statistical analysis, it is observed that the maximum number (eight) of subaltern towns are located in the Mewat district of Haryana, which is the second least urbanised (11.39%) district of the state.

Additionally, no subaltern towns are identified in Faridabad and Panipat districts. Interestingly, Faridabad is the most urbanised district of the state with 79.51%, and Panipat is the fourth most urbanised district with 46.05%. All the towns in the Charkhi Dadri and Mewat districts are identified as subaltern towns, and remarkably,

FIGURE 1.4 Identification of subaltern towns in Haryana.

Source: Authors' interpretation using data from Census of India, 2011.

TABLE 1.5
Categorisation Based on Spatial Proximity Axis

		Towns		Population	
S. No.	Type of Towns	(no.)	(%)	(000s)	(%)
1	Peripheral towns	9	11.39	124.98	7.41
2	Non-peripheral towns	70	88.61	1562.3	92.59
3	*Total*	*79*	*100.00*	*1687.3*	*100.00*

Source: Authors' calculations using data from Census of India, 2011.

TABLE 1.6
Categorisation Based on Administrative Recognition Axis

S. No.	Type of Towns	Towns		Population	
		(no.)	(%)	(000s)	(%)
1	Invisible towns (not recognised as urban)	N.A.	N.A.	N.A.	N.A.
2	Denied towns (census town)	28	35.44	291.3	17.26
3	Recognised (statutory town)	48	60.76	1386.6	82.18
4	Contested (I)	N.A.	N.A.	N.A.	N.A.
5	Contested (II)	3	3.80	9.4	0.56
6	*Total*	*79*	*100.00*	*1687.3*	*100.00*

Source: Authors' calculations using data from Census of India, 2011.

both these districts are the least urbanised, with urbanisation levels of 11.22% and 11.39%, respectively. Therefore, an inverse relationship has been observed between the district's urbanisation level and the number of subaltern towns in the districts of the state. This clearly illustrates that non-subaltern urbanisation is dominated by metropolitan urbanisation in a district.

The highest number (eight) of subaltern towns are observed in Mewat district, followed by Hisar and Karnal districts with seven and six, respectively. All three districts are located on the Haryana border. Mewat district is located in the south-east, and Hisar district is located in the south-west. Both districts share a border with Rajasthan state. Karnal district is located in the north-east and shares a border with Uttar Pradesh state. Ironically, all the districts with a high number of subaltern towns are located distant from the state capital (Chandigarh) as well as the national capital (NCT of Delhi). The districts with a lower number of subaltern towns are located close to the state capital (Panchkula and Panipat) and national capital (Faridabad and Sonepat).

The district's urbanisation level has a negative correlation with the number of subaltern towns, level of subaltern urbanisation, and contribution of subaltern urbanisation to the district's urbanisation level. The urbanisation level has a weak negative correlation (−0.332) with subaltern towns and a moderately stronger negative correlation (−0.621) with the level of subaltern urbanisation. A strong negative correlation (−0.718) is observed between the urbanisation level and the contribution of subaltern urbanisation to the district's urbanisation level (Table 1.8). This means that most towns in the highly urbanised districts are developed due to the impact of large settlements. This is exemplified in the case of Faridabad district, which is developed as a case of metro-centric development. The towns in the low urbanised districts of Mewat and Charkhi Dadri have developed on their own. Thus, the urbanisation level has an inverse relation with subaltern urbanisation. It is also proved that it plays a significant role in understanding the contribution of subaltern urbanisation.

TABLE 1.7
District-wise Contribution of Subaltern Urbanisation, 2011

S. No.	Districts	Urbanisation Level (%)	Total Towns (no.)	Subaltern Towns (no.)	Subaltern Towns (%)	Subaltern Population (000s)	Subaltern Urbanisation (%)	Subaltern Urbanisation Contribution to Urbanisation Level (%)
1	Faridabad	79.51	3	0	0.00	0.0	0.00	0.00
2	Gurgaon	68.82	9	3	33.33	54.8	3.62	5.26
3	Panchkula	55.81	8	2	25.00	44.9	8.01	14.35
4	Panipat	46.05	12	0	0.00	0.0	0.00	0.00
5	Ambala	44.38	15	5	33.33	58.5	5.19	11.69
6	Rohtak	42.04	5	3	60.00	64.4	6.07	14.43
7	Yamuna Nagar	38.94	12	5	41.67	59.8	4.93	12.65
8	Hisar	31.74	11	7	63.64	204.1	11.70	36.88
9	Sonipat	31.27	8	2	25.00	72.6	5.01	16.02
10	Karnal	30.21	8	6	75.00	143.7	9.55	31.61
11	Kurukshetra	28.95	5	4	80.00	124.1	12.86	44.43
12	Rewari	25.93	9	5	55.56	37.3	4.15	15.99
13	Jhajjar	25.39	5	4	80.00	72.6	7.57	29.82
14	Sirsa	24.65	5	3	60.00	83.8	6.47	26.26
15	Bhiwani	23.41	5	4	80.00	68.9	6.09	26.01
16	Jind	22.90	6	4	66.67	75.9	5.69	24.84
17	Palwal	22.69	6	3	50.00	76.1	7.30	32.19
18	Kaithal	21.97	4	3	75.00	91.1	8.48	38.60
19	Fatehabad	19.06	4	3	75.00	115.7	12.28	64.43
20	Mahendragarh	14.41	5	4	80.00	58.3	6.32	43.86
21	Mewat	11.39	8	8	100.00	124.1	11.39	100.00
22	Charkhi Dadri	11.22	1	1	100.00	56.3	11.22	100.00
23	*Haryana*	*34.88*	*154*	*79*	*51.30*	*1687.33*	*6.66*	*19.08*

Source: Authors' calculations using data from Census of India, 2011.

TABLE 1.8
Correlation between Urbanisation Level and Subaltern Urbanisation

Particulars		Urbanisation Level	Subaltern Towns	Subaltern Urbanisation	Subaltern Urbanisation Contribution to Urbanisation Level
Urbanisation level	Pearson Correlation	1	−0.332	−0.621*	−0.718*
	Sig. (two-tailed)		0.131	0.002	0.000
	N	22	22	22	22

Source: Authors' calculations using data from Census of India, 2011.
Note: * Correlation is significant at the 0.01 level (two-tailed).

1.6 CONCLUSION

The share of India's urban population of its total population is approximately 31.14% (377.1 million). The population resides in 4041 STs (318.5 million) and 3892 CTs (58.6 million). During the last census (2011), in absolute terms, the increase in the urban population (91 million) was more than the increase in the rural population (90.5 million) of India for the first time since independence. The number of CTs has also increased, with a growth rate of 185.76% (2530), which accounts for 32.80% of the urban growth. The abrupt increase in the number of CTs has necessitated the need to critically analyse the demographic changes in these settlements, leading to a shift in the focus from metropolitan-based urbanisation to subaltern urbanisation. Thus, the concept of subaltern urbanisation focuses on settlements which are developed due to the market and historical forces.

Haryana's urbanisation level (34.88%) is more than the nation's average urbanisation level (31.14%). In Haryana, the growth rate in the absolute number of CTs is 217.13% (52), which is more than the national level. This, in turn, contributes to subaltern urbanisation. A state level spatial-statistical analysis was carried out. It was observed that Haryana's level of subaltern urbanisation is around 6.66% only. The contribution of subaltern urbanisation in Haryana's level of urbanisation is about 19.89%. At the district level, the urbanisation level has an inverse correlation with the number of subaltern towns, level of subaltern urbanisation, and contribution of subaltern urbanisation in the district's urbanisation level. The non-subaltern towns are located and developed near large urban settlements. Subaltern towns are distributed unevenly throughout the state. The districts with the highest number of subaltern towns are located away from the state capital and national capital border but close to the state border.

This research attempts to explain the basic concept of subaltern urbanisation and the process to identify the subaltern settlements through spatial-statistical analysis by using the last decadal data from the "Census of India." However, for the identification of subaltern towns, factors such as regional impact, effect of large urban settlements, and employment opportunities could not be considered during this analysis. This research gap can be addressed in future research by academicians, scholars, and researchers.

ACKNOWLEDGEMENTS

The authors are grateful to Mrs. Priya Bhardwaj, Urban Planner (AMRUT), Urban Development Directorate, Dehradun, Uttarakhand, for her detailed comments and suggestions on the first draft of the chapter. The authors are also thankful to Mr. Ashok, Assistant Town Planner, Department of Town & Country Planning, Government of Haryana, for providing help in data collection.

REFERENCES

Aijaz, Rumi. 2017. "Measuring Urbanisation in India." *ORF Issue Brief*, no. 218 (December): 1–14.
Ambala Cantonment Board. 2022. "About Us." Accessed April 23, 2022. https://ambala.cantt.gov.in/about-us/.
Bhagat, R.B. 2011. "Emerging Pattern of Urbanisation in India." *Economic & Political Weekly* 46, no. 34: 10–12.
Census of India. 2001. "Tables." Town Directory: Haryana. Accessed April 15, 2022. https://censusindia.gov.in/DigitalLibrary/MFTableSeries.aspx.
Census of India. 2011a. *Primary Census Abstract: Data Highlights India*. New Delhi: Registrar General and Census Commissioner, India.
Census of India. 2011b. "District Census Hand Book." Town Amenities: Haryana. Accessed April 15, 2022. https://censusindia.gov.in/2011census/dchb/DCHB.html.
Charkhi Dadri. 2016. "About District." District Profile. Accessed April 15, 2022. https://charkhidadri.gov.in/district-profile/.
Denis, Eric, and Marie-Héléne Zérah, eds. 2017. *Subaltern Urbanisation in India: An Introduction to the Dynamics of Ordinary Towns*. New Delhi, India: Springer.
Denis, Eric, Partha Mukhopadhyay, and Marie-Héléne Zérah. 2012a. "Subaltern Urbanisation in India." *Economic & Political Weekly* 47, no. 30 (July): 52–62.
Denis, Eric, Partha Mukhopadhyay, and Marie-Héléne Zérah. 2012b. "Subaltern Urbanisation in India." HAL archives-ouvertes.fr (October): 1–36. https://halshs.archives-ouvertes.fr/halshs-00743051/file/Subaltern_Urbanisation_in_India_7_July_2012_EPW.pdf.
DGDE: Directorate General Defence Estates. "Cantonments." Accessed April 23, 2022. www.dgde.gov.in/content/cantonments.
Fatewar, Meet, and Vinita Yadav. 2021. "Assessment of Urbanisation's Trends and Patterns: A Global Perspective." *Nagarlok* 53, issue 4 (December): 1–15.
HSIIDC: Haryana State Industrial and Infrastructure Development Corporation Limited. 2022. "Activities and Services." Infrastructure Development. Accessed April 24, 2022. https://hsiidc.org.in/industrial-estates.
Kumar, Satish, Ranbir Singh, and Dr. Gaurav Kalotra. 2013. "Urban Population: Distribution and its Socio-Economic Attributes in Haryana (2011)." *Global Research Analysis* 2, issue 12 (December): 79–84.
MoL&J: Ministry of Law & Justice. 2006. *The Cantonments Act, 2006*. New Delhi, India: Government of India.

MoUD: Ministry of Urban Development. 2015. *Urban and Regional Development Plans Formulation and Implementation (URDPFI) Guidelines (Volume I).* https://smartnet. niua.org/content/d19f4f87-aaa1-4e9a-9651-534cb28ddd3c.

Planning Commission. 2011. *Mid-Term Appraisal Eleventh Five Year Plan 2007–2012.* New Delhi, India: Oxford University Press.

Pradhan, Kanhu Charan. 2017. "Unacknowledged Urbanisation: The New Census Towns in India." In *Subaltern Urbanisation in India: An Introduction to the Dynamics of Ordinary Towns,* edited by Eric Denis, and Marie-Héléne Zérah, 39–66. New Delhi, India: Springer.

Sangwan, Himanshu, and Ms. Mahima. 2019. "Growth of Urban Population in Haryana: A Spatio-Temporal Analysis." *International Journal of Research and Analytical Reviews* 6, issue 1: 752–756.

2 Urban Sprawl Modelling Using the CA-MARKOV Model for Thoothukudi City

C. Jeswin Titus and Colins Johnny Jesudhas

CONTENTS

2.1 INTRODUCTION

Rapid and intensive population explosion is an alarming factor for the economy of developing countries. Rapid urbanisation in developing countries is predominantly because of the movement of people from rural areas to urban areas for better opportunities, education, etc. By 2025, about 60% of the population is expected to live in urban areas (Lambin et al. 2001). This causes the development of the urban fringe, a very complex phenomenon referred to as urban sprawl (USp). USp is uncontrollable and uncoordinated growth existing along the periphery of the existing urban landscapes to accommodate migrant populations, as a result the density of urban areas decreases, and the areal distribution of built-up area is exploded. In general, USp is in fragmented patches (Padmanaban et al. 2017) because of land availability and land procurement costs. These fragmented patches will have multi-nuclei, and remote sensing technologies can be used to map and model this dynamic phenomenon (Kumar, Garg, and

Khare 2008). Studies reveal that since the 1990s the pattern of landscapes has drastically altered (Liping, Yujun, and Saeed 2018), because of an increase in the urban landscape, a decrease in agricultural land (Feng et al. 2016) and forest cover (Stow et al. 2004), and considerable deterioration of water resources leading to drinking water shortages and degradation (Matlhodi et al. 2021) in the urban environment. USp is also linked with an increase in exposure to bare soil because of vacant fallow lands in urban fringes (Alkaradaghi et al. 2018).

In this study, urban sprawl is modelled using the Cellular Automata (CA)-Markov model using the datasets obtained from the Landsat series (Ebrahimipour, Saadat, and Farshchin 2016) for Thoothukudi city in the same seasons to avoid variations due to vegetation phenology for 2009, 2013, and 2017. For estimating the LULC changes in 2009, 2013, and 2017, the maximum likelihood classification (MLC) algorithm was used as it is more reliable (Jadawala, Shukla, and Tiwari 2021). For modelling the sprawl, the CA-Markov model (Markov chain model and a CA filter) was used because of good results when compared with other models (Hamad, Balzter, and Kolo 2018). Initially, LULC maps of 2009, 2013, and 2017 were prepared using the MLC algorithm that was chosen as the model's input to evaluate the conversion areas and conversion probabilities using chain analysis. To simulate the LULC for 2017, the LULC map from 2009 was used as a base map, along with matrix conversion probabilities from 2009–2013 and conditional probability data. Comparison and validation of the anticipated and measured LULCs for the year 2017 and the LULC for the year 2021 were predicted using the model after validation.

2.2 STUDY AREA AND DATA USED

Thoothukudi is an industrial city with the status of a Municipal Corporation. The city has an area of 90.66 km^2, is in Tamil Nādu, India (Figure 2.1), and has a population of approximately four lakhs. A 10-km buffer of the city covering an area of 314.20 km^2 is considered for this study. The city is well connected through all transport routes, namely roadways, railways, airways, and waterways. The town is located at longitude 78°13' E and latitude 8°45' N. The ground is flat, sloping from west to east, and the average elevation is 5 meters above mean sea level. It has a semi-arid climate with hot summers, and receives rainfall mainly from the northeast monsoon. Most of the people of the city are employed in saltpans, sea-borne trading, fishing, and tourism. Using temporal remote sensing data for the years 2009 and 2017, urban dynamics were examined. The satellite data were downloaded from the USGS website's (http://earthexplorer.usgs.gov) Landsat Thematic Mapper (TM) and Operational Land Imager (OLI) sensors. These satellite images were captured using the conventional projection method and geometrically adjusted (UTM 43N, WGS84 datum). The boundary of Thoothukudi city was downloaded from the Tamil Nādu Geographic Information System website (www.tngis.tn.gov.in).

2.3 METHODOLOGY

In the current study, land use/land cover features were mapped, and urban expansion was assessed using satellite data. Both TM and OLI data for the years 2013 and

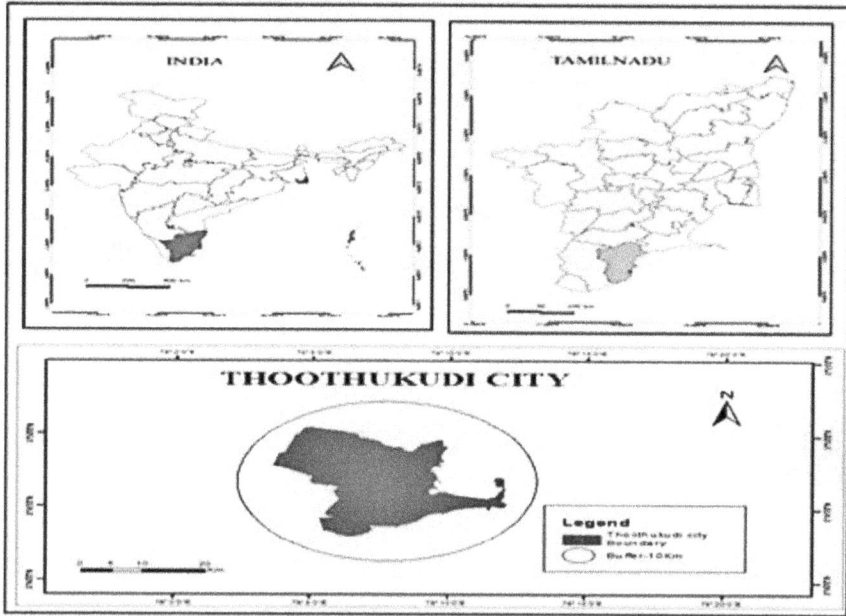

FIGURE 2.1   Representation of Thoothukudi city and its location in Tamil Nādu, India.

TABLE 2.1
Data Specifications

Satellite	Date of acquisition	Spatial resolution	Path/row
Landsat 5 TM	21/01/2009	30 m	143/54
Landsat 8 OLI	10/01/2013	30 m	143/54
Landsat 8 OLI	29/12/2017	30 m	143/54

2017 were downloaded (Table 2.1). A schematic representation of the methodology is shown in Figure 2.2. The Landsat satellite data of the study area were extracted by clip operation in QGIS software. The classification process provides a quantitative depiction of each land use class of interest in the image. The USp model was developed by importing the classified layers into IDRISI software for USp prediction (Liping, Yujun, and Saeed 2018).

2.3.1   CLASSIFICATION

According to their digital values, pixels are sorted into a limited number of distinct classes through the process of classification. The land use/land cover map was created using the supervised image classification approach known as maximum likelihood classification (Feng et al. 2016). The QGIS software's semi-automatic classification

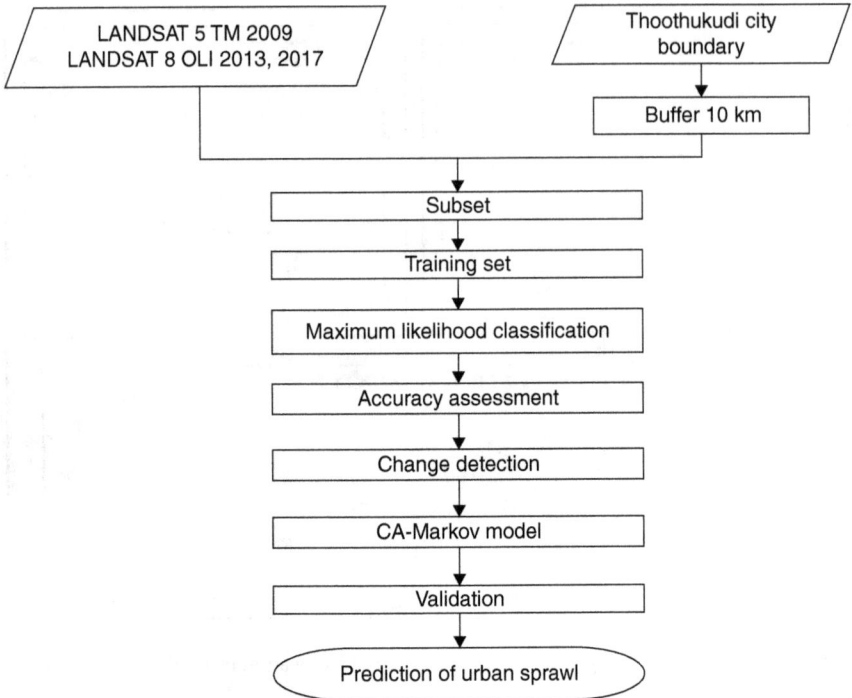

FIGURE 2.2   Schematic representation of the methodology.

plug-in (SCP) toolbox was used to classify the data. Specifying training sites is a process for locating spectrally comparable regions on an image. Training sites must be homogeneous to avoid false classification and must be known targets from where the spectral signatures are extracted. Training site data that are used to train the learning algorithm should be of good quality because of the fact that it exemplifies only a small portion of the entire image. The learning algorithm can be applied to those sites to recognise similar pixel characteristics in the imagery. The selection of multiple training sites distributed throughout the image can significantly improve classification accuracy.

2.3.2   MAXIMUM LIKELIHOOD SUPERVISION

A signature file was created from the training dataset and multivariate classification was adopted to classify the image. MLC, a supervised classification approach based on Bayesian theory, is used for classification (Matlhodi et al. 2021). Based on the probability function, the pixel is classified into a class using a discriminant function. The function's primary inputs, the class mean vector, and the covariance matrix can be calculated from the training pixels for a given class. MLC is a statistical decision-making tool that helps classify overlapping signatures by assigning pixels to the class with the highest probability. The MLC classifier uses the principle that an ellipsoid may frequently characterise the geometric shape of a set of pixels that belong to a class. As a result, pixels are organised into groups based on where they are located inside the class ellipsoid's effect zone. Visual analysis, classification accuracy, band correlation, and decision boundary are used to thoroughly evaluate the classification.

2.3.3 ACCURACY ASSESSMENT

Before a classification's correctness is evaluated, it is not complete. Error matrices are one of the most commonly used methods to convey categorisation (sometimes called a confusion matrix or a contingency table). Error matrices evaluate the correspondence between known reference data (ground truth) and the related outcomes of automated classification, comparing each category separately. A measure of agreement or accuracy between a classification map obtained from remote sensing and the reference data is the kappa coefficient. The kappa coefficient can be determined from the error matrix. The following equation can be used to determine the kappa statistics for each class:

$$Kappa\ statistics = \frac{Observed\ accuracy - change\ agreement}{1 - change\ agreement}$$

A kappa coefficient greater than 0.85 is considered to be the criterion of acceptance, which is also a minimum acceptable value of the United States Geological Survey (USGS) (Falah, Karimi, and Tavakoli 2020).

2.3.4 URBAN SPRAWL MODEL

A powerful technique for simulating the dynamic character of LULC is the CA-Markov model (Markov, Zulkarnain, and Kusratmoko 2016). The MARKOV and CA-MARKOV modules of Clarks Lab's IDRISI TerrSet software were used in the current study to anticipate and simulate the future growth and development of the city of Thoothukudi. According to Zhao et al. (2021), the model needs to include the spatial probability of the LULC class. To provide the model with a spatial component, CA is incorporated into the Markovian technique. The CA-Markov model for a pixel's transition probabilities depends on its surrounding pixels (Maithani 2010). Cross-tabulating two photos taken at various times yields a transition probability matrix that calculates the likelihood that a pixel in one land use class will switch to another during that time. On the other hand, a transition area matrix comprises the number of pixels that are anticipated to switch from one land use class to another during a given period. Suitability maps were created using the IDRISI TerrSet CELLATOM module. This is frequently applied in dynamic modelling, where a pixel's future state depends on its present state and the states of its immediate neighbours. The module uses these inputs to decide where the change will occur, how many pixels must go through each transition, and which pixels are most suited for which transition. In LULC modelling, validation is difficult since there is no consensus on how to evaluate the effectiveness of various land use models (Pontius and Malanson 2006). By contrasting the actual LULC change with the projected LULC change, the model was validated. To determine the kappa coefficient index of agreement: K_{no}, $K_{location}$, and $K_{quantity}$, IDRISI used a kappa coefficient study. The K_{no} represents the overall effectiveness of the simulation, $K_{location}$ is the precision of the location, and $K_{quantity}$ is the precision of the quantity (Pontius 2000). It serves as a gauge for how well the simulations can forecast quantity. A real prediction for the year 2021 was made using land use maps for the years 2007 and 2013, the transition areas matrix, a contiguity filter, and the transition suitability collection after calibrating the model and evaluating its validity.

2.4 RESULTS AND DISCUSSION

2.4.1 LAND USE/LAND COVER ANALYSIS

The MLC classification was adopted for the preparation of the LULC map. LULC maps for the years 2009, 2013 and 2017 are shown in Figures 2.3, 2.4, and 2.5, respectively. The LULC is divided into four categories: vegetation, aquatic bodies, urban, and others. The blue colour represents the water bodies like the sea, canals, etc. The green colour indicates vegetation like forests, shrubs, cropland, plantations, etc. The red colour shows the urban areas like roads, airports, residential areas, industrial

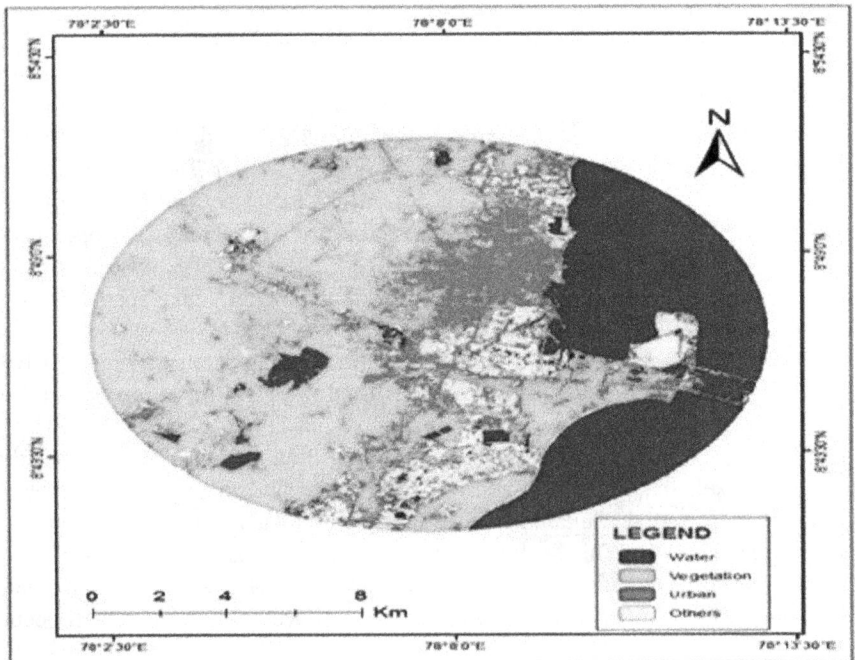

FIGURE 2.3   Land use and land cover map of 2009.

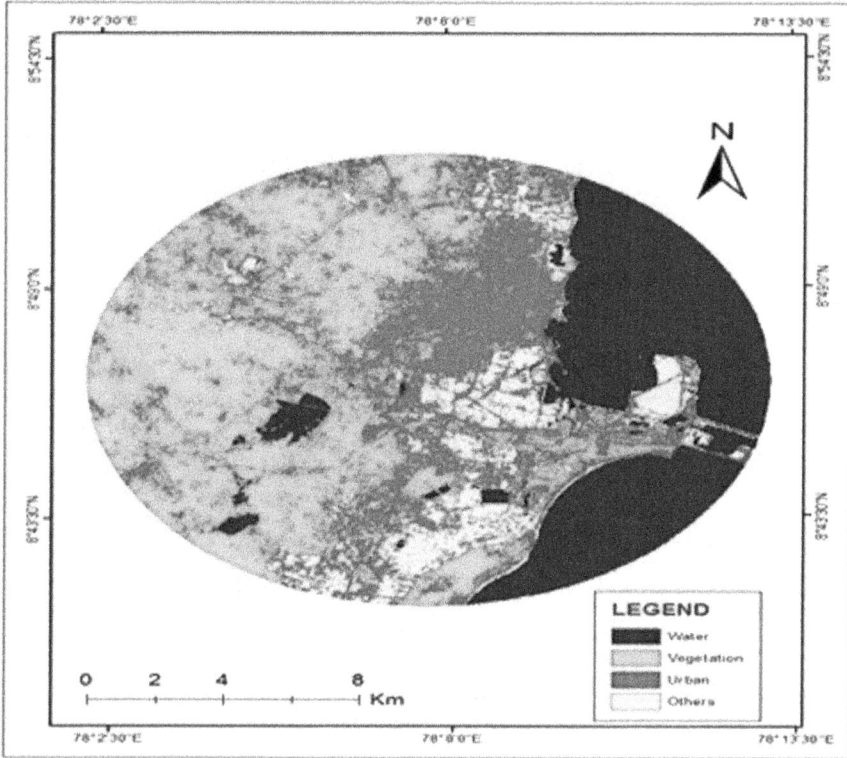

FIGURE 2.4   Land use and land cover map of 2013.

areas, paved surfaces, etc. Yellow colour indicates others, like salt pans, open ground, etc. Vegetation covers a greater area than the other classes of water, urban, and others in the LULC map of 2009. The visual interpretation makes it obvious that the urban area has grown considerably between 2009 and 2017. The urban areas increased along the roadways, like Palayamkottai road, Melavittan road, Ettayapuram road, and Thalamuthu Nagar main road, resulting in linear sprawl. Table 2.2 provides the quantitative elements of temporal land use dynamics. In 2017, 39% of the population lived in an urban region, up from 16% in 2009. Therefore, an overall urban expansion of 23% has been observed in the city. The vegetation has also declined from 48% in 2009 to 24% in 2017, and the water bodies decreased from 26% in 2009 to 24% in 2017. Land surface temperature (LST) rises with less vegetation and water bodies and more urban areas. Built-up land has the highest LST and serves as a source of heating, whereas aquatic bodies have the lowest LST and serve as a source of cooling (Li, Zhou, and Chen 2020). Using the error matrix and kappa statistics, the classification accuracy assessment of the image was completed. The overall accuracy and kappa statistics of the classified data are shown in Table 2.3. In the years 2009, 2013, and 2017, the overall accuracies for the classified photos were 94%, 90%, and 92%, respectively.

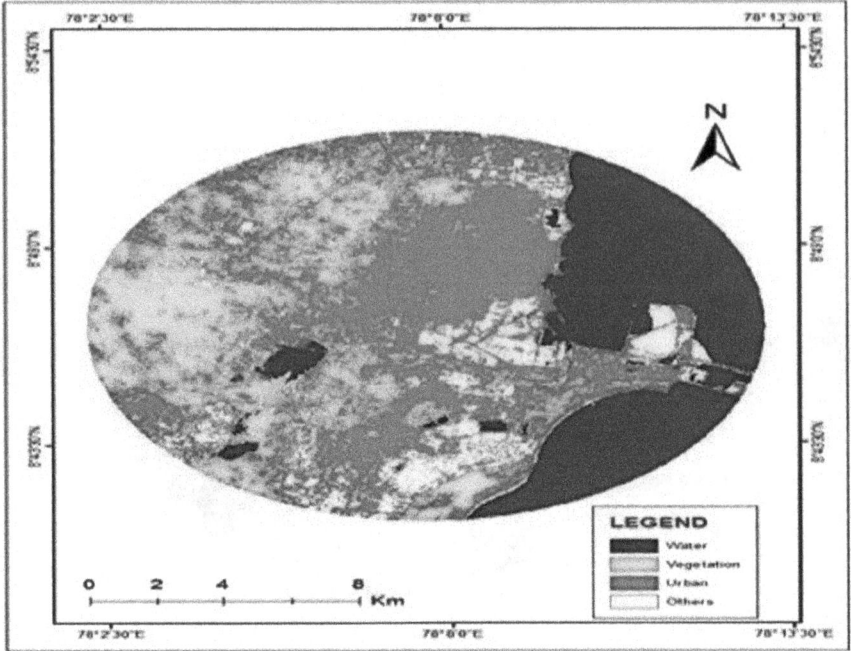

FIGURE 2.5   Land use and land cover map of 2017.

TABLE 2.2
The Temporal Land Use Dynamics from 2009 to 2017

Class	Water		Vegetation		Urban		Others	
Year	Ha	%	Ha	%	Ha	%	Ha	%
2009	8305.7	26.4	15014.1	47.7	5048.3	16	3150.5	10
2013	7800.8	24.8	11775	37.4	8868.51	28	3074.3	9.8
2017	7647.5	24.3	7631.37	24.2	12274.9	39	3964.9	12.6

TABLE 2.3
Kappa Values and Overall Accuracy

Year	Kappa coefficient	Overall accuracy %
2009	0.90	94
2013	0.86	90
2017	0.88	92

TABLE 2.4
Markov Transition Probability Matrix 2017

Land Use	Water	Vegetation	Urban	Others
Water	0.79	0.03	0.04	0.13
Vegetation	0	0.64	0.32	0.03
Urban	0.05	0.05	0.85	0.05
Others	0	0.18	0.17	0.65

2.4.2 MODELLING LULC DYNAMICS

As can be seen in Table 2.4, a Markov transition probability matrix was produced using LULC maps from 2009 and 2013. The LULC for the year 2017 was projected using this matrix. A raster group file called a transition suitability collection was made using the collection editor module, which also includes a suitability map. A raster group file collection is created using the module, or its members are defined using the module. According to the rule, a pixel that is close to one land use type (for instance, urban regions) is more likely to turn into an urban pixel and has a higher USp. The severity of USp will depend on the number of fragments, in the case of fragments increase the impact of USp will be very high in consideration of environmental aspects. This urban expansion may cause drastic environmental influences such as the formation of urban heat islands (Xinliang et al. 2017), poor water and air quality, energy inefficiencies, and a lack of multi-utility services, and also the ecosystem can be affected, leading to a decrease in ecosystem service values (Dey et al. 2021) because of this uneven development. The increase in built-up areas results in urban heat islands and as a result the temperature in the urban environment increases (Tariq and Shu 2020).

2.4.3 VALIDATION

Before using any model to make a prediction, validation is a crucial step. Figure 2.6 displays the forecasted map for 2017. There are some similarities between the spatial distributions of built-up areas in the two maps. However, class quantification can reveal the model's true performance. The areal distribution and percent coverage of each class for the expected and actual findings are shown in Table 2.5. The percentage of each class is provided for actual 2017, expected 2017, and predicted 2021. The four classes are represented in km², respectively. By 2021, 48% of the land area will be built up, up from 16% in 2009. Figure 2.7 depicts the projected land use and cover map for 2021. To compare the "prediction" with the "actual" LULC map of 2017, the validate module in IDRISI was used to compute the kappa index of agreement: K_{no}, $K_{location}$, and $K_{quantity}$. The findings are shown in Table 2.6. A good agreement between the "actual" and "predicted" map is indicated by values above 0.80. Table 2.7 shows the probability values for 2017 based on the transition matrix of 2009–2013. The reduction in vegetation and increase in built-up land could be a process of arable land

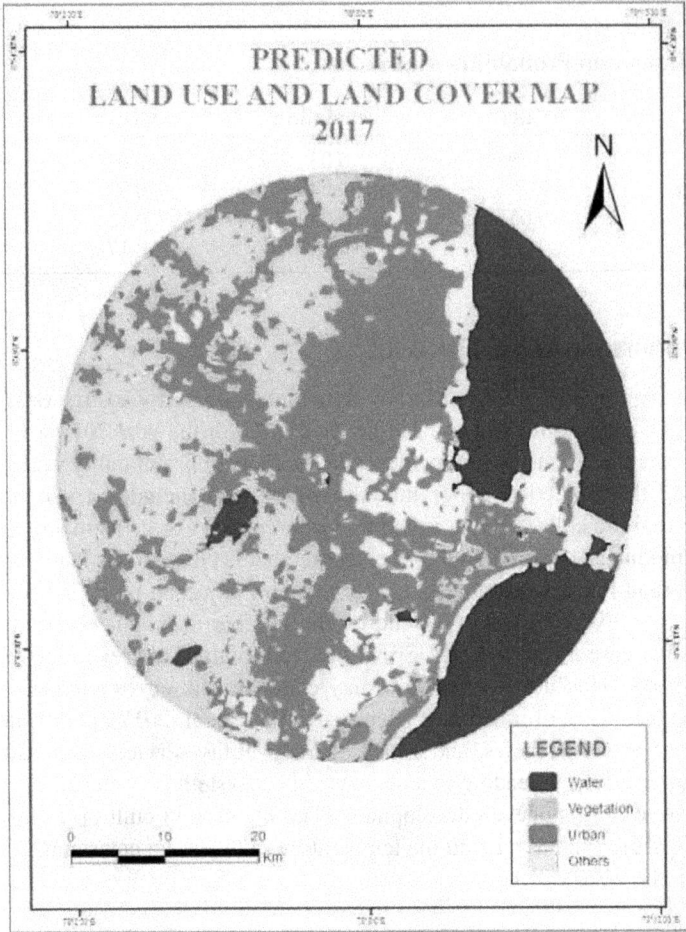

FIGURE 2.6   Land use map predicted for 2017.

TABLE 2.5
Land Use Statistics of the Predicted Results

Class	Water		Vegetation		Urban		Others	
Year	km²	%	km²	%	km²	%	km²	%
Actual 2017	76.47	24.3	76.31	24.2	122.75	39.1	39.65	12.61
Predicted 2017	66.71	21.2	87.83	27.9	122.08	38.8	38.56	12.2
Predicted 2021	76.14	24.2	43.59	13.8	149.58	47.6	45.85	14.5

TABLE 2.6
Accuracy Measures

K_{no}	0.8576
$K_{location}$	0.9096
$K_{standard}$	0.8551

conversion, indicating economic and population growth. Errors can be reduced by providing new types of factors, such as policy, environmental, and climatic factors, in addition to the existing factors.

2.4.4 Prediction

The prediction of urban sprawl for a future scenario was made through this model. Table 2.7 shows the probability transition index values for 2021. Based on Markov transition probability Cellular Automata rules, and neighbourhood, land use was predicted for the year 2021. The model is capable of capturing the dynamic nature of socio-economic factors, and local interaction rules to predict urban sprawl patterns (Deep and Saklani 2014). The proximity variables such as urban centres, bus terminuses, railway stations, and airports, and site attributes and constraints can be included while modelling. CA have the potential to support land use planning and policy analysis in developing countries and the promotion of sustainable development (Li, Zhou, and Chen 2020). The predicted sprawl map can help planners to devise a plan based on the forecast to achieve sustainable development (Feng et al. 2016) leading to preparation for sustainable cities through a proper LULC management plan (Rastogi and Sharma 2020). The study provided information about past land use and further illustrates the challenge due to LULC changes in the urban environment. Encroachment and land conversion due to urban expansion over agriculture and forest area are an environmental threat (Pour and Oja 2021). LULC prediction and regulatory measures should always be a part of sustainable land use planning and urban development.

2.5 CONCLUSION

Urbanisation is an inevitable part of a country's growth. The most critical issue related to urbanisation is inappropriate planning, which leads to multiple side effects on the environment and economy. Predictive techniques will be useful in predicting the country's growth because planning is difficult. Cellular Automata and the Markov model are both used in the hybrid CA-MARKOV model. The Cellular Automata model simulates the real-world process that is spatially distributed in nature. The Markov model simulates the stochastic processes in a future state that depends on the current state. This study shows how remote sensing and GIS may be used to track urban sprawl and changes to the urban land use system. According to a land use analysis, the urban area has grown by 23% overall, from 16% in 2009 to 39% in 2017. Using

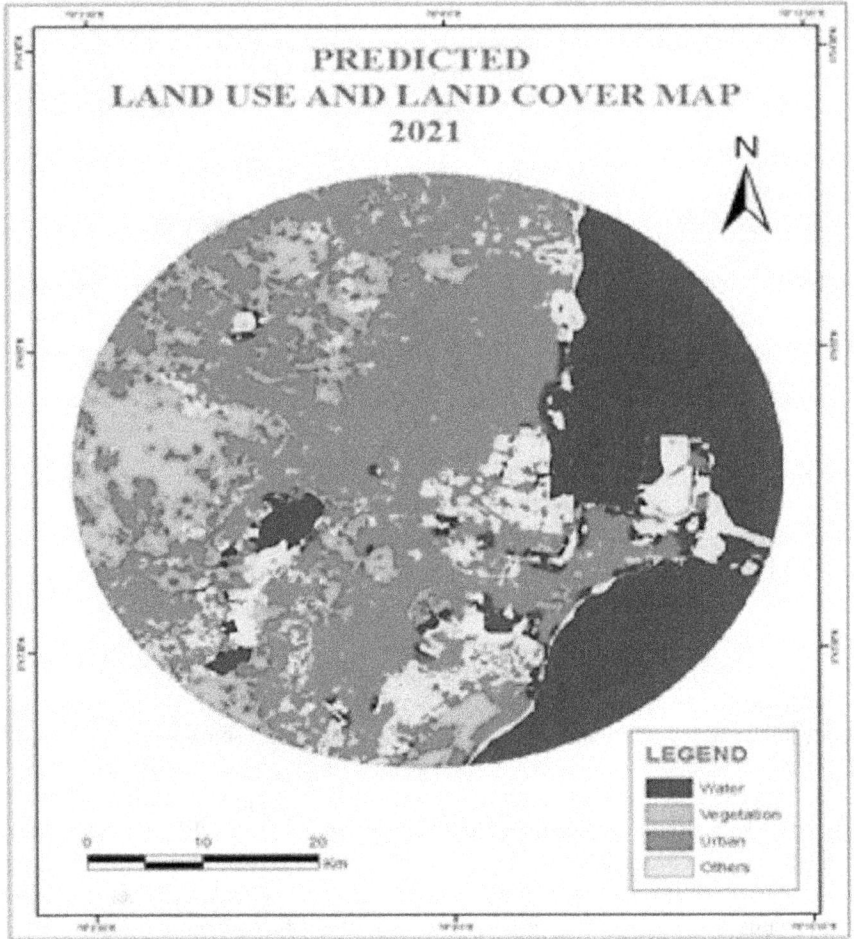

FIGURE 2.7   Predicted land use and land cover map for 2021.

TABLE 2.7
Markov Transition Probability Matrix for 2021

Land Use	Water	Vegetation	Urban	Others
Water	0.83	0.0004	0.09	0.08
Vegetation	0.0004	0.55	0.34	0.11
Urban	0.15	0.00	0.85	0.00
Others	0.0001	0.05	0.17	0.79

the CA-Markov model, land use dynamics have been modelled. The model was run in the year 2017 and the expected land use was for the year 2021. The predicted result shows that the urban area would be increased from 16% in 2009 to 46% in 2021. This was mainly due to expansion in the form of newly developed colonies such as Jothi Nagar, Muthammal colony, Eazhumalayan Nagar, Kumaragiri, and HWP colony in pre-existing wasteland and agricultural land. Declines in vegetation and water bodies were also observed in some regions. These temporal changes illustrate the need for the regulation of urban expansion over vegetated areas and water bodies. As a result, while developing a plan, development plans should take the effects into account to minimise or manage any detrimental effects on the urban environment. The model results help policymakers in framing policies, and help planners' avoid inefficient planning according to policies related to housing, infrastructure, and infrastructure.

REFERENCES

Alkaradaghi, Karwan, Salahalddin S Ali, Nadhir Al-ansari, and Jan Laue. 2018. "Evaluation of Land Use & Land Cover Change Using Multi-Temporal Landsat Imagery : A Case Study Sulaimaniyah Governorate , Iraq." *Journal of Geographical Information System* 10: 247–260. doi:10.4236/jgis.2018.103013

Deep, Shikhar, and Akansha Saklani. 2014. "Urban Sprawl Modeling Using Cellular Automata." *The Egyptian Journal of Remote Sensing and Space Sciences* 17 (2). Elsevier B.V.: 179–187. doi:10.1016/j.ejrs.2014.07.001

Dey, Nataraj Narayan, Abdullah Al Rakib, Abdulla Al Kafy, and Vinay Raikwar. 2021. "Geospatial Modelling of Changes in Land Use / Land Cover Dynamics Using Multi-Layer Perception Markov Chain Model in Rajshahi City , Bangladesh." *Environmental Challenges* 4. Elsevier B.V.: 100148. doi:10.1016/j.envc.2021.100148

Ebrahimipour, Ahmadreza, Mehdi Saadat, and Amirreza Farshchin. 2016. "Prediction of Urban Growth through Cellular Automata-Markov Chain." *Bulletin de La Societe Royale Des Sciences de Liege* 85 (January): 824–839. doi:10.25518/0037-9565.5677

Falah, Nahid, Alireza Karimi, and Ali Tavakoli. 2020. "Urban Growth Modeling Using Cellular Automata Model and AHP (Case Study : Qazvin City)." *Modeling Earth Systems and Environment* 6 (1). Springer International Publishing: 235–248. doi:10.1007/s40808-019-00674-z

Feng, Yongjiu, Zongbo Cai, Xiaohua Tong, Jiafeng Wang, Chen Gao, Shurui Chen, and Zhenkun Lei. 2016. "Urban Growth Modeling and Future Scenario Projection Using Cellular Automata (CA) Models and the R Package Optimx." *International Journal of Geo-Information* 6 (1). Elsevier B.V.: 1–23. doi:10.1016/j.envc.2021.100148

Hamad, Rahel, Heiko Balzter, and Kamal Kolo. 2018. "Predicting Land Use / Land Cover Changes Using a CA-Markov Model under Two Different Scenarios." *Sustainability* 10 (3421): 1–23. doi:10.3390/su10103421

Jadawala, Shrushti S, Shital H Shukla, and Poonam S Tiwari. 2021. "Cellular Automata and Markov Chain Based Urban Growth Prediction Cellular Automata and Markov Chain Based Urban Growth Prediction." *International Journal of Environment and Geoinformatics* 8 (3): 337–343.

Kumar, Mahesh, P K Garg, and Deepak Khare. 2008. "Monitoring and Modelling of Urban Sprawl Using Remote Sensing and GIS Techniques" 10: 26–43. doi:10.1016/j.jag.2007.04.002

Lambin, Eric F, B L Turner, Helmut J Geist, Samuel B Agbola, Arild Angelsen, Carl Folke, John W Bruce, et al. 2001. "The Causes of Land-Use and Land-Cover Change : Moving beyond the Myths." *Global Environmental Change* 11: 261–269.

Li, Xuecao, Yuyu Zhou, and Wei Chen. 2020. "An Improved Urban Cellular Automata Model by Using the Trend-Adjusted Neighborhood." *Ecological Processes* 9 (28). Ecological Processes: 1–13.

Liping, Chen, Sun Yujun, and Sajjad Saeed. 2018. "Monitoring and Predicting Land Use and Land Cover Changes Using Remote Sensing and GIS Techniques – A Case Study of a Hilly Area ,." *PLoS ONE* 13 (7): 1–23. doi:10.1371/journal.pone.0200493

Maithani, Sandeep. 2010. "Techniques in Urban Growth Modelling :" *Institute of Town Planners, India Journal* 7 (1): 36–49.

Markov, K., K. Zulkarnain, and E. Kusratmoko. 2016. "Coupling of Markov Chains and Cellular Automata Spatial Models to Predict Land Cover Changes (Case Study : Upper Ci Leungsi Catchment Area)." *IOP Conf. Series: Earth and Environmental Science* 47 (012032): 1–11. doi:10.1088/1755-1315/47/1/012032

Matlhodi, Botlhe, Piet K Kenabatho, Bhagabat P Parida, and Joyce G Maphanyane. 2021. "Analysis of the Future Land Use Land Cover Changes in the Gaborone Dam Catchment Using CA-Markov Model : Implications on Water Resources." *Remote Sensing* 13 (2427): 1–20.

Padmanaban, Rajchandar, Avit K Bhowmik, Pedro Cabral, Alexander Zamyatin, Oraib Almegdadi, and Shuangao Wang. 2017. "Modelling Urban Sprawl Using Remotely Sensed Data : A Case Study of Chennai City , Tamilnadu." *Entropy* 19 (163): 1–14. doi:10.3390/e19040163

Pontius, Gil R, and Jeffrey Malanson. 2006. "Comparison of the Structure and Accuracy of Two Land Change Models." *International Journal of Geographical Information Science* 19 (2): 243–265. doi:10.1080/13658810410001713434

Pontius, R Gll. 2000. "Quantification Error Versus Location Error in Comparison of Categorical Maps." *Photogrammetric Engineering & Remote Sensing* 66 (8): 1011–1016.

Pour, Najmeh Mozaffaree, and Tonu Oja. 2021. "Urban Expansion Simulated by Integrated Cellular Automata and Agent-Based Models ; An Example of Tallinn , Estonia." *Urban Science* 5 (85): 1–20.

Rastogi, Kriti, and Shashikant A Sharma. 2020. "Integration of Cellular Automata-Markov Chain and Artificial Neural Network Model for Urban Growth Simulation." *Journal of Geomatics* 14 (1): 55–60.

Stow, Douglas A, Allen Hope, David Mcguire, David Verbyla, John Gamon, Fred Huemmrich, Stan Houston, et al. 2004. "Remote Sensing of Vegetation and Land-Cover Change in Arctic Tundra Ecosystems." *Remote Sensing of Environment* 89: 281–308. doi:10.1016/j.rse.2003.10.018

Tariq, Aqil, and Hong Shu. 2020. "CA-Markov Chain Analysis of Seasonal Land Surface Temperature and Land Use Land Cover Change Using Optical Multi-Temporal Satellite Data Of." *Remote Sensing* 12 (3402): 1–22.

Xinliang, Xu, Hongyan Cai, Zhi Qiao, Wang Liang, Jin Cui, GE Yaning, Wang Luyao, and Xu Fengjiao. 2017. "Impacts of Park Landscape Structure on Thermal Environment Using QuickBird and Landsat Images." *Chin.Geogra.Sci.* 27 (5): 818–826. doi:10.1007/s11769-017-0910-x

Zhao, Yabo, Dixiang Xie, Xiwen Zhang, and Shifa Ma. 2021. "Integrating Spatial Markov Chains and Geographically Weighted Regression-Based Cellular Automata to Simulate Urban Agglomeration Growth : A Case Study of The." *Land* 10 (633): 1–19.

3 Urban Site Suitability Mapping Using GIS-Based Multi-Criteria Decision Analysis in Pathankot City, Punjab

Meghna Rout, Sumit Kumar, Reenu Sharma,
Sanjay Kumar Ghosh, and Brijendra Pateriya

CONTENTS

3.1 INTRODUCTION

In terms of population growth and urbanisation, emerging country cities have seen massive changes (Yeh, 1999). Uncontrolled and fast population growth poses enormous problems to government authorities in providing sufficient housing to the

DOI: 10.1201/9781003331001-3

thousands of homeless people living in metropolitan areas without suitable urban management practices. Slum and squatter colonies are rising in cities and metropolitan regions (FAO, 1976). As a result, there is a scarcity of utilities and an increase in the demand for urban land for residential reasons. This has also piqued the interest of city planners. Urbanisation has resulted in a rise in population density due to immigration. Land-use appropriateness studies are critical work for city planners and managers to determine the best spatial pattern for future land use (Hopkins, 1977; Collins et al., 2001). Rural people will continue to migrate to cities in search of better employment prospects, a greater standard of livelihood, and higher education. According to some reports, India is considered as one of the world's fastest-growing urbanisation countries. As a result, it is critical to govern the urbanisation process methodically and scientifically for future growth. Urbanisation is a dynamic phenomenon that continues to evolve throughout time. As a result, for good urban planning, precise and timely data are essential. To handle the challenges of metropolitan regions, urban planners must employ a range of data and approaches (Dutta, 2012). The availability of very high-resolution imagery provides a synoptic picture and vital planning tools.

For making dynamic judgements, decision-makers/urban planners require verified and accurate data and advanced computer tools. To help planners in this respect, remote sensing and GIS are important tools that correctly create and manage the data. The Earth observation data provide information regarding the land use category and elevation data. Remote sensing data sets are reliable, accurate, and periodic spatial data. The Geographic Information System is utilised as an analysis tool to create new derivative map layers to establish logical and mathematical connections among map layers. It is easy to update information in the GIS database. Through GIS, various spatial and non-spatial data can be managed and integrated using various decision-making tools to obtain a solution (Myers, 2010; Yadav and Ghosh, 2021). One of the most important methods for determining the optimum urban expansion places is conducting a suitability study in which various criteria and weights are assigned (Alexander et al., 2012). Land suitability evaluation entails determining proper development places based on mapping the area's suitability index (Joerin et al., 2021). GIS plays a significant role in managing and monitoring urban growth and its influence on the ecosystem (Rosli et al., 2014). Site suitability analysis using the GIS method provides optimal development areas while maintaining environmental sustainability (Park et al., 2011). Along with GIS, some other methods can be integrated to enhance the results of site suitability.

GIS technology combined with multi-criteria decision analysis (MCDA) is an important and helpful tool for determining land appropriate for urban expansion (Bagheri et al., 2013). Fechner proposed the pairwise comparison method in 1860, and Thurstone improved it in 1927 (Fechner, 1860). Saaty introduced the analytic hierarchy process (AHP) as a technique that comes under multi-criteria decision-making (MCDM) that incorporates qualitative data based on pairwise comparison (Saaty, 1980). AHP is a strategy that helps planners and decision-makers assess all evidence before deciding on future land-use changes. It has been employed in various decision-making scenarios over the last 25 years and in multiple applications in many different disciplines. Therefore, by combining the pairwise comparison matrix with the AHP method, valuable and more scientific results can be obtained. Based on this method, various weights have been assigned to each influencing factor. AHP is used to determine the importance of influencing elements on urban expansion. A pairwise comparison matrix is generally used to select the most appropriate option among various options. The approach uses a reciprocal decision matrix derived from paired comparisons to present the data linguistically. In this method, the influencing factors like physical, socioeconomic, commercial, and environmental data sets are considered to get the best result (Trung et al., 2007).

In this approach, a matrix is generated in which the weights of each criterion are assigned a value based on their relative significance to all other influencing factors. The weight of each criterion's adequate importance is computed (Beheshtifar, 2006). The biggest issue in implementing this technique is based on the correct professionals with their comprehensive view of knowledge and expertise in site suitability analysis and implementation to assess the elements' relevance according to their weights (Bagheri et al., 2013). A strategy that can accurately detect the weights is a difficult task. That is why AHP is one of the most critical tools for examining difficulties of spatial character. This research combines the GIS-based Analytical Hierarchy Process to create a site suitability map that considers influencing factors such as urban growth and the values assigned to each element.

3.2 STUDY AREA

The study area was situated in the Pathankot district of Punjab state (Figure 3.1). It lies at an elevation of 332 m and 32° 17′ 29″ N and 75° 37′ 20″ E. The total area covered by the study area was around 298 km² and there are around 190 villages within the study area. According to 2011 census data, the total population is 249,000. In the study area, the Ravi and Chakki rivers flow across both sides and there are drainage streams and canals present.

3.3 DATA SOURCES

High-resolution data sets of Planet satellite were used for land cover mapping. The Planet data consist of four spectral bands and a spatial resolution of approximate 3 metres, which makes the data suitable for LULC mapping. The elevation data from Alos Plasar have been used to generate a slope map. Population data were collected from the 2011 census data of India. The buffer zones were created around these thematic layers and processing was carried out using ArcGIS software. Some of the data, such as railway and road lines, were digitised from Google Earth data.

3.4 METHODS

In this case study, ten factors were used for the computation of appropriate areas for urban growth. The factors used for spatial interpretation are land use and land cover, slope, distance to stream, road, railway, residential, commercial, industrial, educational, and population density maps. These layers are used to create a criteria map. Identification of the appropriate regions for urban expansion and buffer zones was

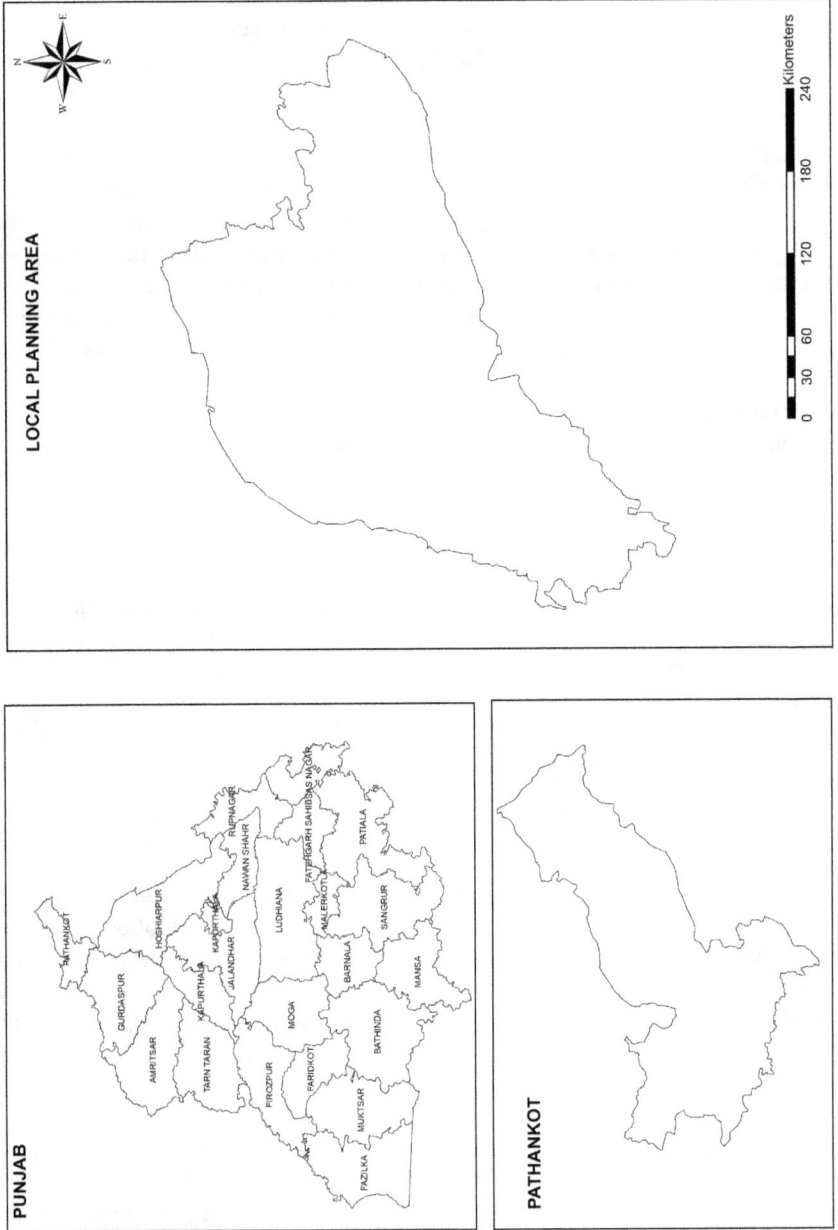

FIGURE 3.1 Overview of the study area.

TABLE 3.1
Scale to Compute a Pair-wise Comparison Matrix

Ranking of importance	Description	Suitability order
1	Equal importance	Least suitable
2	Equal to moderate importance	Very low suitable
3	Moderate importance	Low suitable
4	Moderate to high importance	Moderately low suitable
5	High importance	Moderately suitable
6	High to very high importance	Moderately high suitable
7	Very high importance	Highly suitable
8	Very to extremely high importance	Very highly suitable
9	Extremely high importance	Most suitable

carried out for each thematic layer. After this procedure, all the layers were reclassified into five classes based on their importance from high to low categories. Using Saaty's weighing scales of one to nine rank orders for calculating relative weights, a pairwise comparison matrix was generated (Table 3.1). To determine the consistency, the pairwise comparison matrix was computed. These weights were multiplied and then the weighted sum vector was calculated. All the components were divided by the equivalent component of their weights. To obtain a consistency vector, the weighted sum was divided by their weights. After this, the mean consistency vector was calculated to obtain lambda (λ). Lambda values should be equal to the number of criteria or greater. The consistency index can be expressed as

$$CI = (\lambda - n) / (n - 1) \qquad (3.1)$$

where CI = consistency index, λ = eigenvector, and n = size of matrix. Then CR was computed by using the following formula:

$$CR = CI / RI \qquad (3.2)$$

where CR = consistency ratio and RI = random index (Table 3.2).

According to the AHP methodology, a value of the consistency ratio (CR) less than 0.10 can be considered as a reliable level of consistency in the comparison matrix and is acceptable. However, in the case of CR higher than 0.10 it requires modification in the comparison matrix. The criterion layer was overlaid in ArcGIS software. Then, these layers were multiplied with weights to create the final map for urban suitability. The ranking was classified from the highest to the lowest suitable zone. The final map was generated by using the formula:

$$\text{Suitability map} = \Sigma \ [\text{criteria map} * \text{weight}] \qquad (3.3)$$

TABLE 3.2
Random Index

Order matrix	1	2	3	4	5	6	7	8	9	10
Ransom index	0.00	0.00	0.58	0.9	1.12	1.24	1.32	1.41	1.45	1.49

3.5   RESULTS AND DISCUSSION

The mapping of all the land suitability criteria was done by using remote sensing data. Then all the indicators were categorised according to their properties and importance for sustainable urban growth. The results obtained are described below.

3.5.1   DISTANCE TO STREAMS

Distance to stream for developing an urban region is an essential criterion (Figure 3.2(a)). Without water resources, the development of urban areas will not be possible. However, being too close to streams is also dangerous during rainy seasons, mainly due to sudden flash floods and other flood situations. The range from 200 m to 1 km was set in this study, with buffer zones at 200-m intervals being generated. Greater than 800 m has been given more priority than the 200 m buffer zone.

3.5.2   SLOPE

The slope is considered to be an essential factor in identifying suitable zones for urban area development (Figure 3.2(b)). Slopes with greater steepness were unsuitable for any construction work. Due to steeper slopes, the construction costs also increase and the floor area is reduced. Therefore, an area with a steeper slope is weighted as the lowest rank for urban expansion. Land erosion is a commonly evolved phenomenon in steeper slopes during construction works and environmental hazards. Less than 10° is considered a gentle slope for construction work and greater than 10° is viewed as a stepper slope that is not suitable for building construction (Rawat, 2010).

3.5.3   DISTANCE TO RESIDENTIAL AREAS

The closeness of built-up urban areas is critical in decreasing the costs of construction and services. This improves access to urban clustering, where facilities and services are already situated. The urban uniformity between current residential areas and potential growth regions is maintained by proximity to the urban bulk. Therefore, 250 m distance buffer zones are generated for better accuracy (Figure 3.2(c)).

3.5.4   DISTANCE TO ROADS

Roads are also considered to be an important criterion for urban suitability. For the development of urban areas, the presence of a road network is an important criterion.

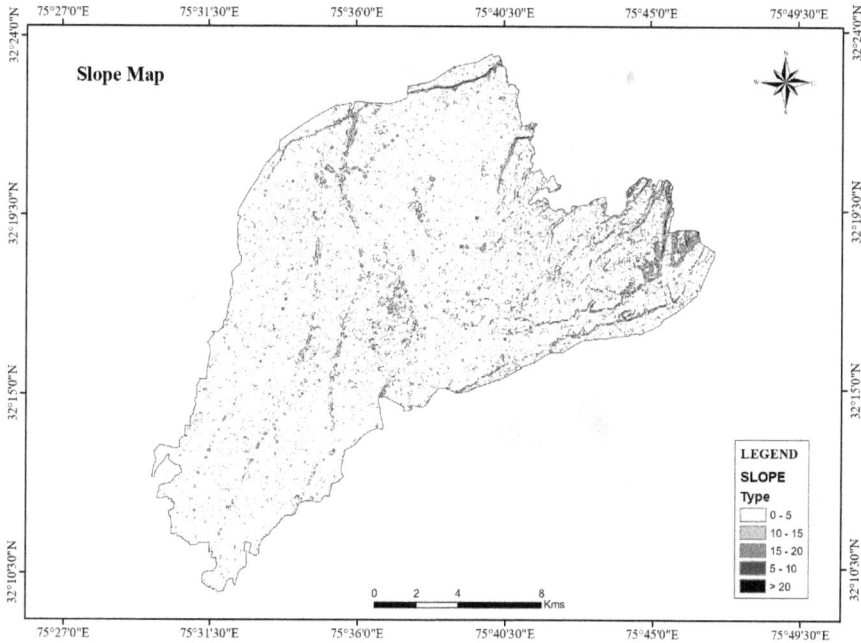

FIGURE 3.2 (a) Slope map.

Closeness to the road network gives greater importance to infrastructure development. To construct a building, road connectivity must be available. It is required for the transportation of raw products and finished materials as well as for citizens. In this case study, identification of better accessibility to existing roads can be generated through 200 m distance buffer zones from the existing road (Figure 3.2(d)).

3.5.5 LAND USE AND LAND COVER

The land use/land cover map (Figure 3.2(e)) of Pathankot district has been classified as barren land, built-up, agricultural land, plantation/forest, and water bodies. A supervised classification was carried out for the land use/land cover classification. In this study, the built-up area is already a developed region and heavily crowded zone. Therefore, it is unsuitable for future development. Thus, for future growth, barren land is highly suitable, and agricultural land is used as the medium-suitability category.

3.5.6 POPULATION DENSITY

In the Pathankot district, the main Pathankot municipal area is a highly populated city. Therefore, highly populated zones should be avoided for a better quality of life for residents. In highly populated cities, there is a lack of a healthy environment which is also a necessary factor. In the case study, areas with the highest population density have the lowest importance for urban growth (Figure 3.2(f)). The areas with the lowest population were considered as the suitable locations for urban expansion.

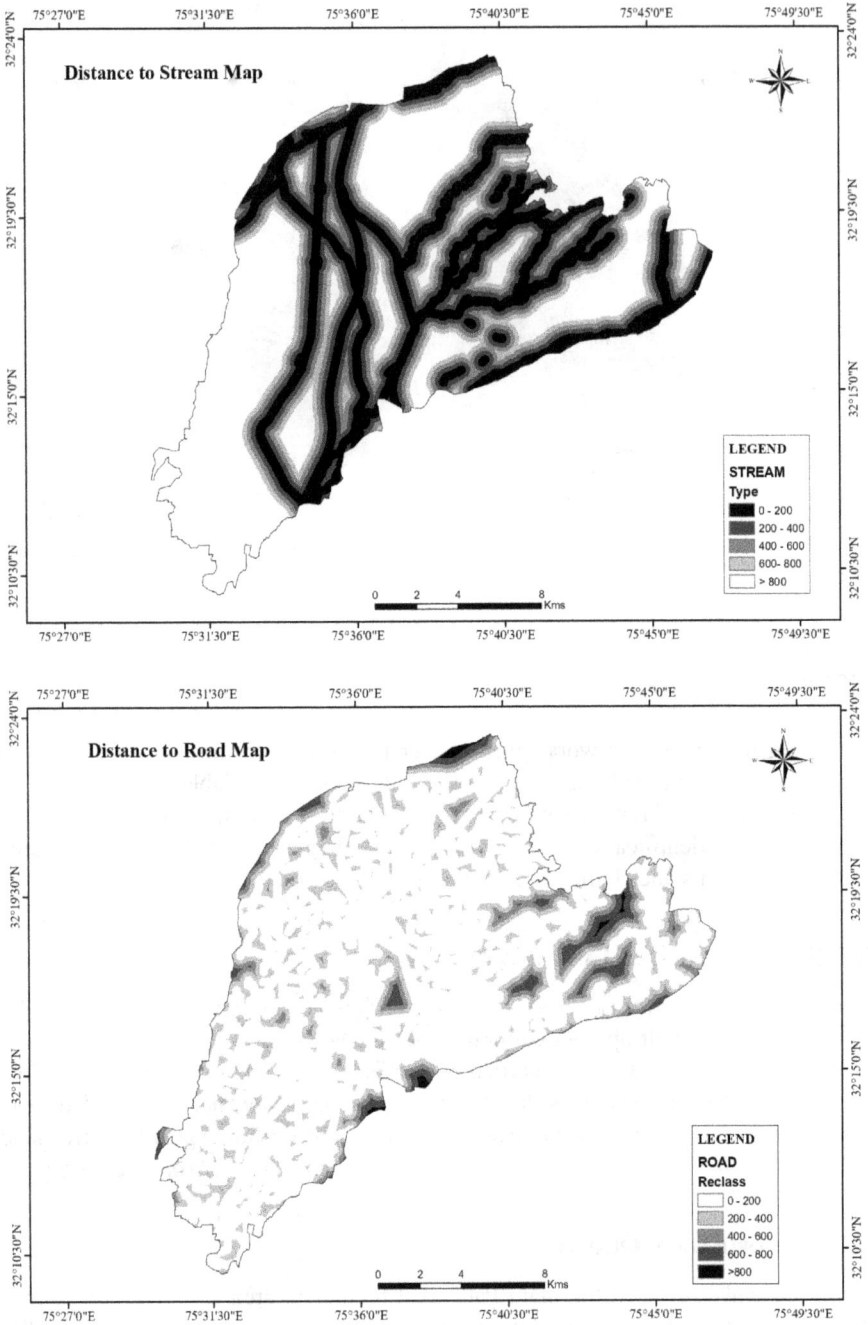

FIGURE 3.2   (b) Distance to stream map; (c) Distance to road map.

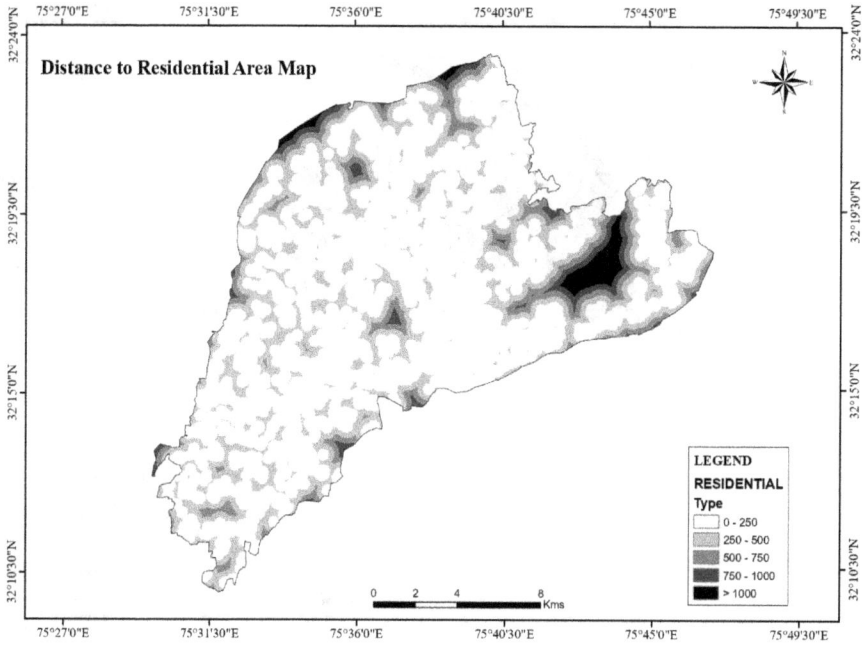

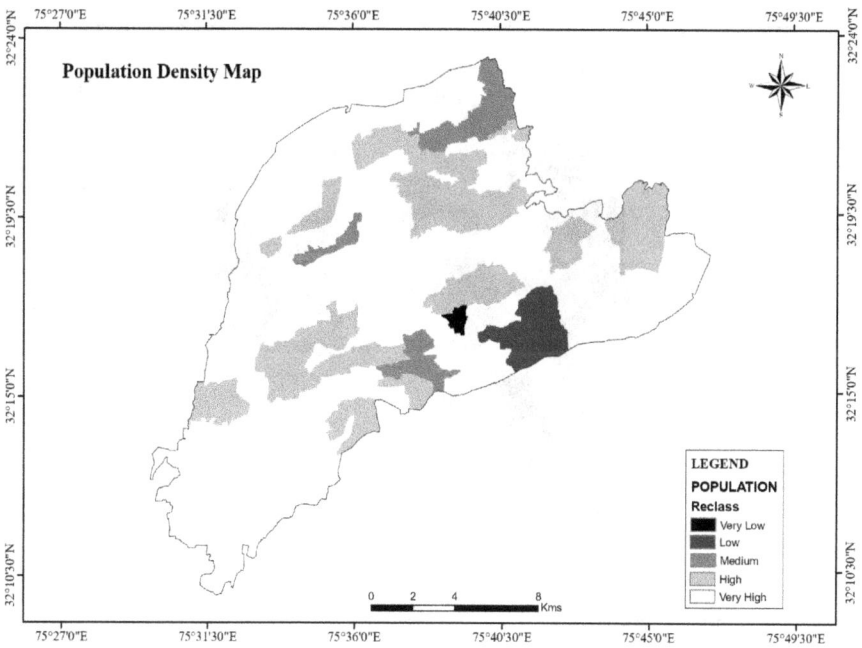

FIGURE 3.2   (d) Distance to the residential area map; (e) Population density map.

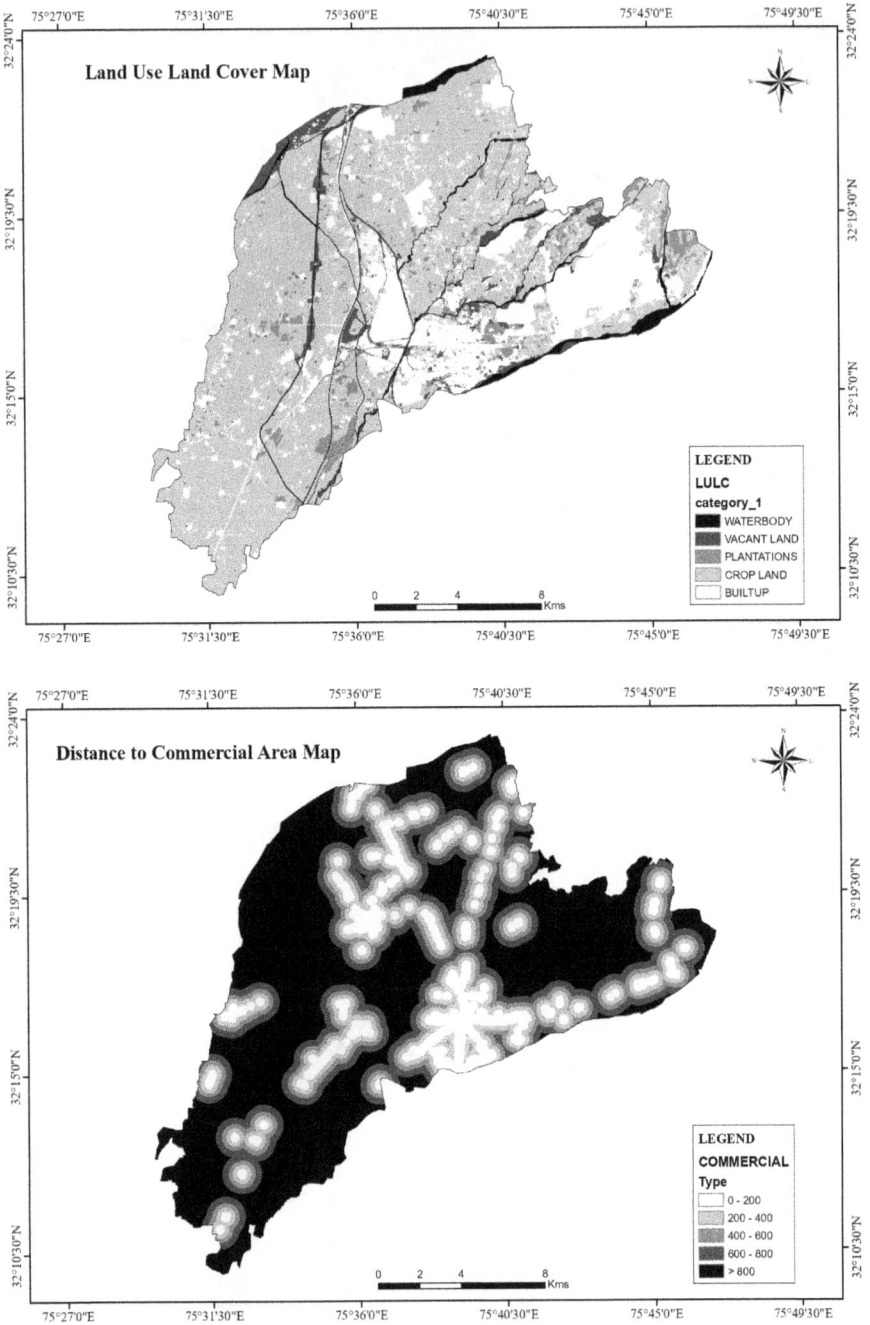

FIGURE 3.2   (f) Land use/land cover map; (g) Commercial area map.

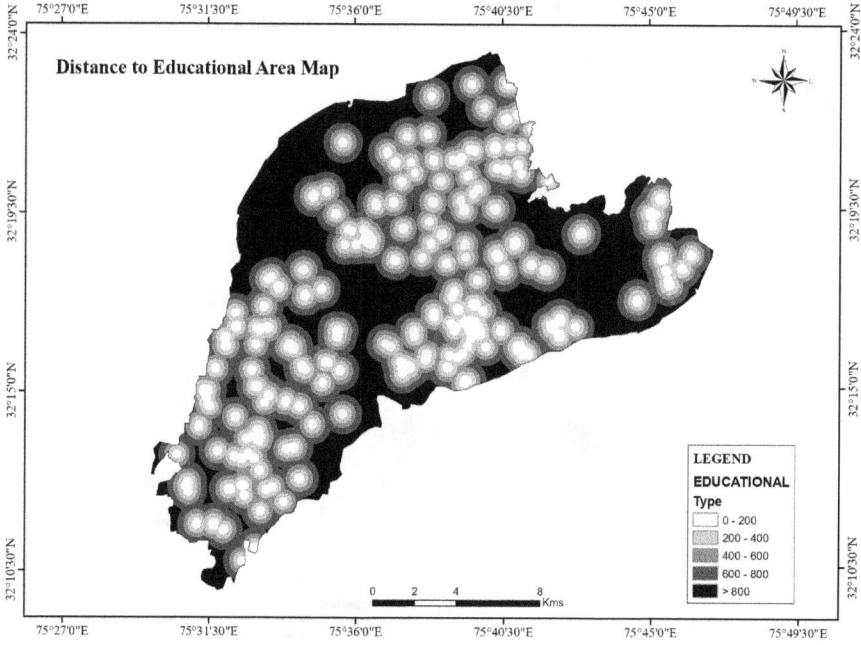

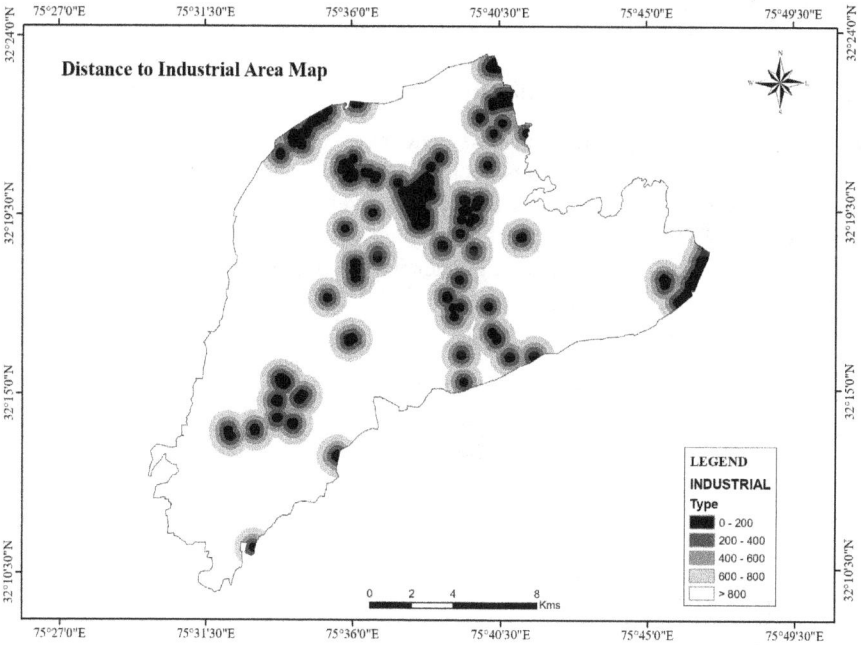

FIGURE 3.2 (h) Educational area map; (i) Industrial area map; (j) Railway map.

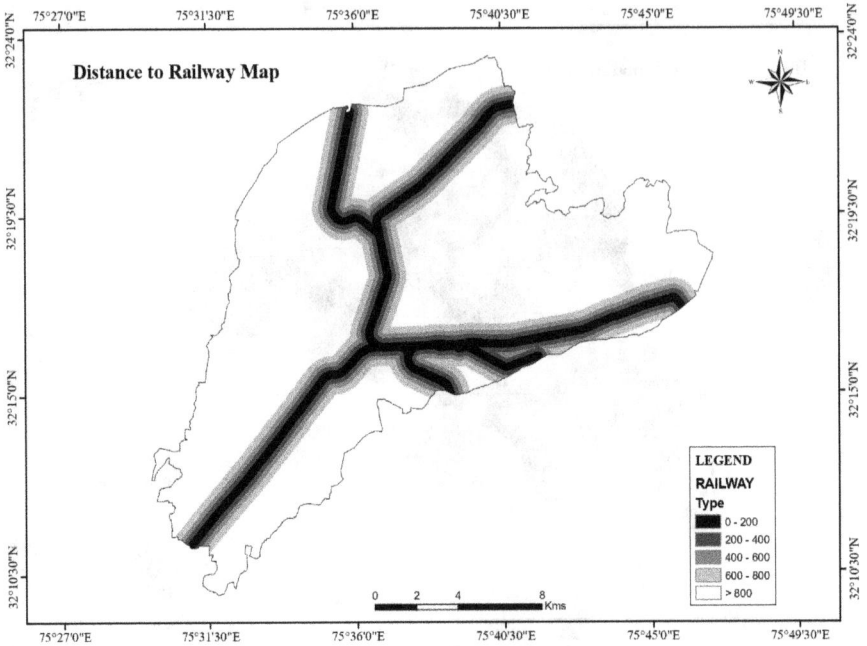

FIGURE 3.2   (Continued)

3.5.7 DISTANCE TO COMMERCIAL AREAS

The access distance for all services should not exceed 5 km. Population growth also explores close facility centres such as marketplaces in comparison to isolated locations where the availability of services is to distant. The multi-ring buffer zone of a distance of 200 m is created from commercial areas (Figure 3.2(g)). The areas near the commercial areas provided greater weight than the areas far from the commercial areas.

3.5.8 DISTANCE TO EDUCATIONAL LOCATIONS

As per the specified standards for educational institutes, the maximum distance to educational institutes should not be over 1.5 km (Menezes & Pizzolato, 2014). Therefore, buffer zone distances of 200 m from academic areas were given the highest priority (Figure 3.2(h)).

3.5.9 DISTANCE TO INDUSTRIAL AREAS

The establishment of industries can result in a significant number of job opportunities. It also enhances the quality of life. The need for social infrastructures such as health, education, and commerce also increase, resulting in further development. A buffer zone of greater than 800 m was used as a suitable area for urbanisation (Figure 3.2(i)).

3.5.10 Distance to Railways

Like roads, railways also play a vital role in transportation (Figure 3.2(j)). Industries, goods, and services are highly dependent on railway facilities. In addition, for inhabitants, railways can be a daily mode of transportation. This is a mass transportation facility that uses less vehicles per person, which helps in sustainable development.

3.5.11 Calculation of the Consistency Ratio (CR)

It is important to cross-check whether the pair-wise comparison is consistent or not. Table 3.5 shows the determination of the weighted sum overlay and consistency vector. After that, the average of the consistency vector (λ) should be equal to or more than the number of effective criteria. If the average of the consistency vector (λ) is less than the influencing criteria, then it is considered to be inconsistent

Calculation of consistency vector (λ)

$$= (10.30 + 10.24 + 10.18 + 10.16 + 10.15 + 10.15 + 10.14 +$$
$$10.12 + 10.08 / 10) = 10.17$$

The CI was computed using formula (3.1)

$$= (10.17 - 10)/(10 - 1) = \mathbf{0.02}$$

The CR was computed using formula (3.2)

$$= 0.02/1.49 \text{ (RI} = 1.49 \text{ for n} = 10) = \mathbf{0.01}$$

In this case, the study consistency ratio was 0.01<0.1. Therefore, this indicated a reasonable level of consistency in the pair-wise comparisons matrix (Table 3.3). Therefore, the values obtained from the output satisfy the above condition, which indicates that the weights obtained by the pair-wise comparison matrix are acceptable. The assigned weights show consistency and these weights were used for the computation of the final map (Table 3.4).

3.5.12 Generation of the Site Suitability Map

The variables for the mapping were firstly computed according to their importance and then the raster layers were generated. The classification of each pixel was carried out accordingly to determine the score. The classification of the raster layer ranges from one to five. The higher value was given to the variable with high importance for urban development and the lower value was used for the areas unsuitable for urban expansion. The criteria layer was combined and stacked for the site suitability mapping (Figure 3.3). The weight was computed from the AHP technique and the criteria layer was multiplied with the computed weights (Table 3.5). The following formula was used for the computation of the final maps:

TABLE 3.3
Pair-wise Comparison Matrix

Criteria	Distance to stream	Slope	Distance to residential	Distance to roads	LULC	Population density	Distance to commercial	Distance to educational	Distance to industrial	Distance to railways
Distance to stream	1	1	2	3	4	5	6	7	8	9
Slope		1	1	2	3	4	5	6	7	8
Distance to residential	0.5	1	1	1	2	3	4	5	6	7
Distance to roads	0.33	0.5	1	1	1	2	3	4	5	6
LULC	0.25	0.33	0.5	0.5	1	1	2	3	4	5
Population density	0.2	0.25	0.33	0.33	1	1	1	2	3	4
Distance to commercial	0.17	0.2	0.25	0.33	0.5	1	1	1	2	3
Distance to educated	0.14	0.17	0.2	0.25	0.3	0.5	1	1	1	2
Distance to industrial	0.12	0.14	0.17	0.2	0.3	0.3	0.5	1	1	1
Distance to railways	0.11	0.12	0.14	0.17	0.2	0.3	0.33	0.5	1	1

TABLE 3.4
Normalised Pair-wise Comparison Matrix and Computation of Criterion Weights

Criteria	Distance to stream	Slope	Distance to residential	Distance to roads	LULC	Population density	Distance to commercial	Distance to educational	Distance to industrial	Distance to railways	Criteria weights
Distance to stream	0.26	0.21	0.30	0.32	0.30	0.28	0.25	0.23	0.21	0.20	0.26
Slope	0.26	0.21	0.15	0.21	0.23	0.22	0.21	0.20	0.18	0.17	0.20
Distance to residential	0.13	0.21	0.15	0.11	0.15	0.17	0.17	0.16	0.16	0.15	0.16
Distance to roads	0.09	0.11	0.15	0.11	0.08	0.11	0.13	0.13	0.13	0.13	0.12
LULC	0.07	0.07	0.08	0.11	0.08	0.06	0.08	0.10	0.11	0.11	0.08
Population density	0.05	0.05	0.05	0.05	0.08	0.06	0.04	0.07	0.08	0.09	0.06
Distance to commercial	0.04	0.04	0.04	0.03	0.04	0.06	0.04	0.03	0.05	0.007	0.04
Distance to education	0.04	0.04	0.03	0.03	0.02	0.03	0.04	0.03	0.03	0.04	0.03
Distance to industrial	0.03	0.03	0.03	0.02	0.02	0.02	0.02	0.03	0.03	0.02	0.02
Distance to railways	0.03	0.03	0.02	0.02	0.02	0.01	0.01	0.02	0.03	0.02	0.02
	1	1	1	1	1	1	1	1	1	1	

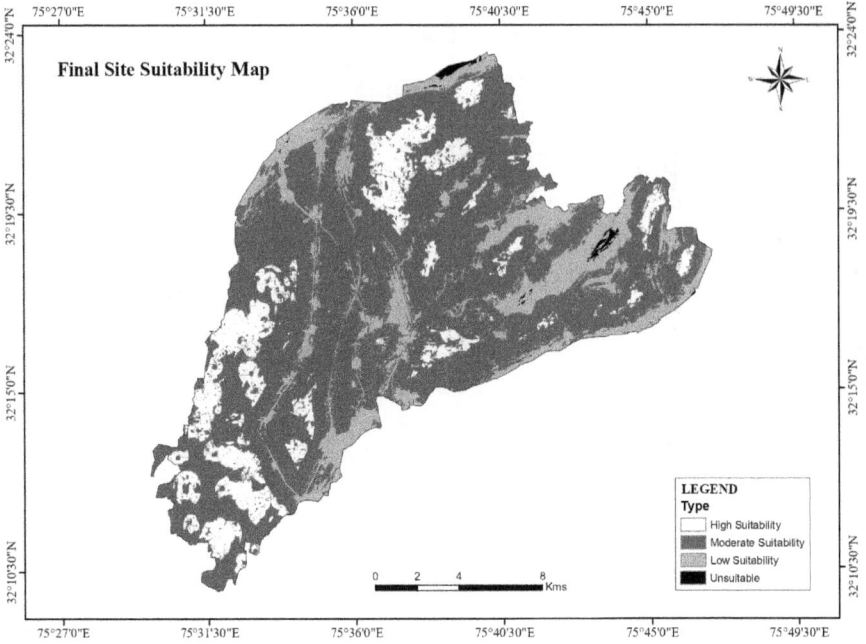

FIGURE 3.3   Final site suitability map.

TABLE 3.5
Computation of Consistency Vector

Criteria	Weighted sum overlay	Consistency vector
Distance to stream	2.64	10.30
Slope	2.10	10.24
Distance to residential	1.59	10.18
Distance to roads	1.17	10.16
LULC	0.86	10.15
Distance to commercial	0.45	10.15
Distance to educational	0.33	10.14
Distance to industrial	0.25	10.12
Distance to railways	0.20	10.08

Site suitability map = ([Distance to stream] *0.26) + ([Slope]*0.20) + ([Distance to residential area] *0.16) + ([Distance to road] * 0.12) + ([LULC] * 0.08) + ([Population density] * 0.06) +([Distance to commercial area] * 0.04) + ([Distance to educational area] * 0.03) + ([Distance to industrial area] * 0.02) + ([Distance to railway] * 0.02).

The output urban suitability map was divided into four suitable categories from the above site suitability criteria (Figure 3.3): high suitability, moderate suitability,

TABLE 3.6
Area Classified into Different Suitability Zones

Categories	Area (sq. km)	Area (%)
Unsuitable	1.36	0.46
Low suitability	48.05	16.45
Moderate suitability	208.50	69.90
High suitability	39.34	13.19
Total area	298.25	100

low suitability, and unsuitable regions. Only 39.25 km² (13.19%) area comes under the high suitability zone. Towards the southwest and north of the study area, there is a higher possibility of suitable zones. In comparison, 208.50 km² (69.90%) falls under the moderately suitable zone. The low and unsuitable locations came under a 50.41 km² area of 16.91%. The results show more than 80% of the area was suitable for urban growth. The region near the river or streamlines and high-density urban areas are under the less suitable category (Table 3.6).

3.6 CONCLUSIONS

This study provides an analysis concentrated on highly appropriate regions with the greatest potential for urban expansion. Suitable urban development in the adjacent areas of the existing Pathankot city has been analysed. With the availability of high-resolution satellite data and open-source digital elevation model data it is easy for analysts to generate several layers from these data to carry out the analysis. The study uses the AHP method for the computation of suitable areas for urban development. The integration of GIS and the AHP method offers decision-makers essential information and understanding about sustainable urban development utilising GIS technology. Because of the model's simplicity, which is easy to grasp and interpret, using an AHP-based GIS approach to discover ideal places for urban growth is beneficial. It is vital for planners to decide whether land should be developed immediately or be preserved for future development. Moreover, this method can be used with different types of criteria, whether these are qualitative factors like land-use and land-cover layers or quantitative factors like population density data. In this study, layers were identified using the GIS and remote sensing with the AHP model based on ten criteria. Using different influencing factors according to location, a suitable AHP model can be obtained for future development in urban planning. Expert suggestions are considered the most substantial factor in this type of assessment. This means that this model operates by amalgamating human and technical components. The combination of AHP with the help of the GIS technique is a suitable technique for site suitability area analysis. The AHP technique integrated with GIS increases the efficiency for site suitability. The study created a replica of spatial analysis with the help of ArcGIS and QGIS software, urban planning ideas and standards used for selecting potential areas for urban growth. These techniques and methods could efficiently help planners in the study sites choose the best locations for urban development, with the best knowledge to generate the proposed model that can be implemented in other areas with similar assets and situations. Based on its results, the research suggests that the GIS method, along with multi-criteria analysis, could be used for planning and decision-making purposes. This technique's most beneficial aspect relies mainly on urban planning and designing maps for potential zone identification and their development for urban growth. Therefore, this study can be recommended to guide individuals planning to using the multi-criteria decision-making process by combining GIS-based technology with the AHP as a robust urban development and planning approach.

REFERENCES

Alexander, K. W., Benjamin, M., & Grephas, O. P. (2012). Urban landuse suitability assessment using geoinformation techniques for Kisumu municipality in Kenya. Int. J. Research and Reviews in Applied Sciences, 13(2), 522–530.

Bagheri, M., Sulaiman, W. N. A., & Vaghefi, N. (2013). Application of geographic information system technique and analytical hierarchy process model for land-use suitability analysis on coastal area. Journal of coastal conservation, 17(1), 1–10.

Beheshtifar, M. (2006). Site selection the thermal power plants (Doctoral dissertation, Master thesis of GIS, KN Toosi University of Technology).

Collins, M. G., Steiner, F. R., & Rushman, M. J. (2001). Land-use suitability analysis in the United States: historical development and promising technological achievements. Environmental management, 28(5), 611–621.

Dutta, V. (2012, June). War on the dream–How land use dynamics and peri-urban growth characteristics of a sprawling city devour the master plan and urban suitability? In 13th annual global development conference, Budapest, Hungary.

FAO (1976). A framework for land evaluation. Food and Agriculture Organization of the United Nations, Soils Bulletin No. 32, FAO: Rome.

Fechner, G. (1860). Elements of psychophysics (Traduzidopor HE Adler) New York: Rinehart & Winston.

Hopkins, L. D. (1977). Methods for generating land suitability maps: a comparative evaluation. Journal of the American Institute of Planners, 43(4), 386–400.

Joerin, F., Thériault, M., & Musy, A. (2001). Using GIS and outranking multicriteria analysis for land-use suitability assessment. International Journal of Geographical information science, 15(2), 153–174.

Menezes, R. C., & Pizzolato, N. D. (2014). Locating public schools in fast expanding areas: application of the capacitated p-median and maximal covering location models. Pesquisa Operacional, 34(2), 301–317.

Myers, A. (2010). Camp Delta, Google Earth and the ethics of remote sensing in archaeology. World Archaeology, 42(3), 455–467.

Park, S., Jeon, S., Kim, S., & Choi, C. (2011). Prediction and comparison of urban growth by land suitability index mapping using GIS and RS in South Korea. Landscape and urban planning, 99(2), 104–114.

Rawat, J. S. (2010). Database management system for Khulgad Watershed, Kumaun Lesser Himalaya, Uttarakhand, India. Current Science, 98(10), 1340–1348.

Rosli, A. Z., Reba, M. N. M., Roslan, N., & Room, M. H. M. (2014, February). Sustainable urban forestry potential based quantitative and qualitative measurement using geospatial technique. In IOP Conference Series: Earth and Environmental Science (Vol. 18, No. 1, p. 012021). IOP Publishing.

Saaty, T. L. (1980). The analytic process: planning, priority setting, resources allocation. McGraw, New York.

Trung, N. H., van Mensvoort, M. E. F., & Bregt, A. K. (2007). Application of GIS in land-use planning: A case study in the coastal Mekong Delta of Vietnam. International Journal of Geoinformatics, 3(4), 1–8.

Yadav, V., & Ghosh, S. K. (2021). Assessment and prediction of urban growth for a mega–city using CA-Markov model. Geocarto International, 36(17), 1960–1992.

Yeh, A. G. (1999). Urban planning and GIS. Geographical information systems, 2(877–888), 1.

4 Strengthening the Cities of Maharashtra State to Become Climate Smart to Achieve Sustainable Development Goals

A Nature-Based Solutions (NbS) and Climate Adaptation Approach

Wasim Ayub Bagwan

CONTENTS

DOI: 10.1201/9781003331001-4

4.1 INTRODUCTION

Recent decades have seen rapid urbanisation of the planet. Only 30% of the world's
population lived in metropolises in 1950, but that figure had increased to 55% by
2018. The global urbanisation rate obscures important regional disparities in urbanisation levels [1]. Globally, cities currently house 3.5 billion people, with a predicted
population of 5 billion by 2030 [2]. Anthropogenic development aims to take into
account the potential offered by sustainable facilities and infrastructure, which will
increase adaptability even after a disaster or incident related to climate change. Cities
are erecting buildings, structures, and infrastructure made of artificial materials
since urbanisation is the preferred way of life around the world. In such a setting,
it is critical to prioritise the basic needs and to rethink the prototype of environmentally friendly place-making and urban design techniques. Nature-centric habitation
making, or beautifying places with particular eco-friendly characteristics to add to
the city's strong identity, can be a main shared goal for citizens and local officials
and can aid in the development of sustainable and resilient cities [3]. The goal of
NbS is to support society with help to achieve sustainability by safeguarding well-
being in a way that reflects cultural and societal values while balancing the ecosystem services and building resilience capacity [4]. Based on a prediction, India will
rank in third place till 2050 from an economic perspective, and the role of the urban
area contributes significantly to the same. Nearly 65% of India's GDP is created by
urban areas and this will increase to 70% by 2030; this collectively affects the socio-
economic and political landscape of the country [5].

Although urbanisation is viewed as a desirable transformation, unplanned
and unprecedented urbanisation can result in rapidly expanding slum regions
and increasing economic inequality. The main cause of rising urban poverty and
unemployment is considered to be rapid urbanisation without equal expansion in
work possibilities. The expansion of economic prospects in cities has lagged behind
the rate of rural-to-urban migration. Consistent mass migration from rural to urban
regions can place an added burden on urban services [6]. Sustainable urban development is an integrated way to bring people's participation, building, public spaces,
green mobility, and another transit system together to act against climate change. It

also ensures the rational use of resources, building resilience in mobility at lower financial and environmental costs. In cities, comprehensive development is critical for enduring sustainability, equity, shared prosperity, and civil culture [7]. Another way to handle the climate-based urban dynamic is a 'smart city'. All urban infrastructures are affected by a smart city project, including public and private buildings, factories, and transportation services. Knowledge management in the urban setting should be supported by a robust information and communication infrastructure, and the sustainability of a smarter city will have a positive impact on water, energy, and mobility [8]. NbS, Ecosystem-based Adaptation (EbA), Ecosystem Services (ESS), and Green Infrastructure (GI) are emerging concepts in the last two decades to involve and amplify the role of the environment in policy-making at various scales. They also encourage citizens to make a contribution towards environmental protection with a special focus on biodiversity in the anthropogenic world [9]. In the present study, the measurement of environmental activities is undertaken to build resilience across the cities of Maharashtra. Additionally, with the help of thematic maps, the study area and locations are depicted. The overall adopted technologies have been paving the way towards sustainable cities in Maharashtra state.

4.2   STUDY AREA

As portrayed in Figure 4.1, Maharashtra is located in western and central India, with a 720-kilometre coastline along the Arabian Sea and natural fortifications in the form of the Western Ghats, also known as Sahyadri (locally) and *Satpuda* mountain ranges. Maharashtra is bordered on the north by Madhya Pradesh, on the west by Gujarat, on the east by Chhattisgarh, and on the south by Telangana, Karnataka, and Goa. The state also shares a boundary with Dadra and Nagar Haveli (Union Territory). The state has 36 districts and six revenue divisions for administrative purposes. According to the 2011 Population Census, the state has a population of 11.24 crore people and spatial coverage of around 3.08 lakh km^2, ranking it second in terms of population and third in terms of geographic extent. Maharashtra is one of India's most urbanised states [10]. In Maharashtra, eight cities have been chosen for development as new-generation smart cities: Aurangabad, Kalyan-Dombivali, Nagpur, Nashik, Pune, Pimpri-Chinchwad, Thane, and Solapur [11].

4.3   LITERATURE SEARCH

In this study, urban activities-related data were collected by using structured keywords from online platforms including Scopus and Web of Science, searching for articles published on 'Urbanisation', 'Sustainable cities', 'Climate-resilient cities', 'future cities', and so on. Nature-based solutions and climate action for future climatic change-based scenario-related books, proceedings, comments, reports, review articles, newspaper articles, annual reports, and research papers from peer-reviewed journals from publishers including Nature, Springer, Elsevier, Taylor and Francis, etc. were considered in carrying out the literature survey.

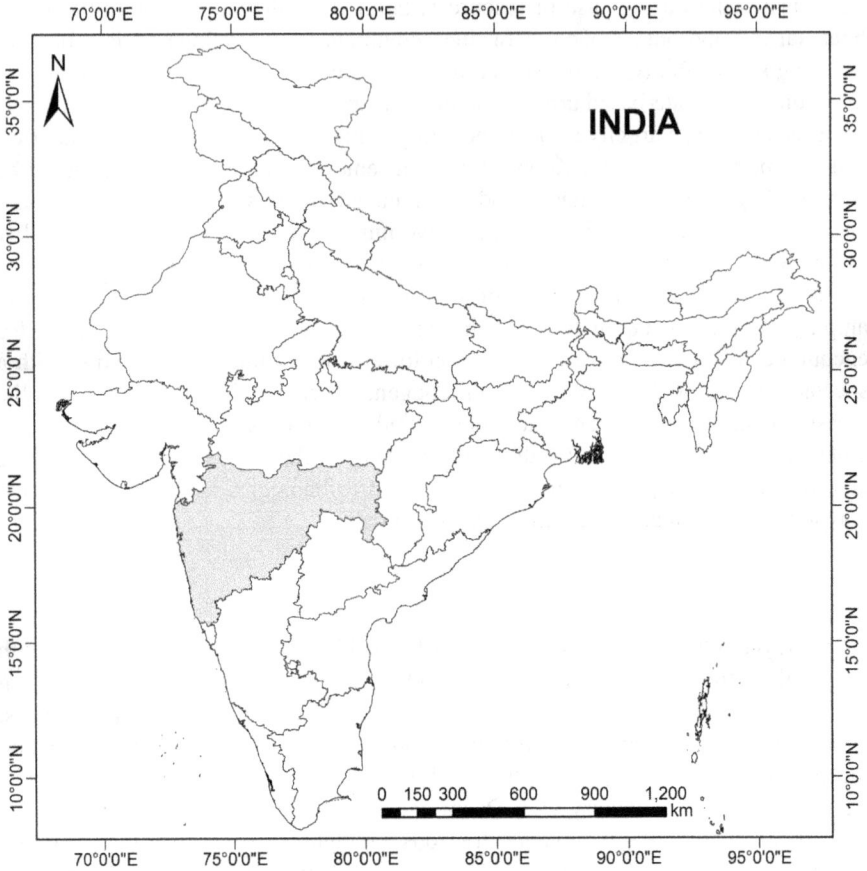

FIGURE 4.1   Location of Maharashtra state in India.

4.4   DATA

The study included some statistical data from Maharashtra state, including Economic Survey of India reports published annually by the Directorate of Economics and Statistics, Planning Department, Govt. of Maharashtra from 2019 to 2022 were taken into the consideration to gather the numerical data (source: https://mahades.maharash tra.gov.in/home.do?lang=en). Also, for the current analysis, urban population data were downloaded from the World Bank website (https://data.worldbank.org/indica tor/SP.URB.TOTL.IN.ZS).

4.5   URBAN POPULATION IN MAHARASHTRA

The constant relocation of rural people of Maharashtra to urban areas is a result of the relative attraction of cities. Cities offer a wide range of options and a huge number of facilities and services, including employment opportunities to earn money, find work, further education, receive health care, attend cultural and social events, and so on.

TABLE 4.1
Indicators of the Urban Population

Parameter	1960–61	1970–71	1980–81	1990–91	2000–01	2010–11	2020–21
Urban population (thousands)	11,163	15,711	21,993	30,542	41,101	50,818	50,818
Urban population (%)	28.2	31.2	35	38.7	42.4	45.2	45.2

Source: [10].

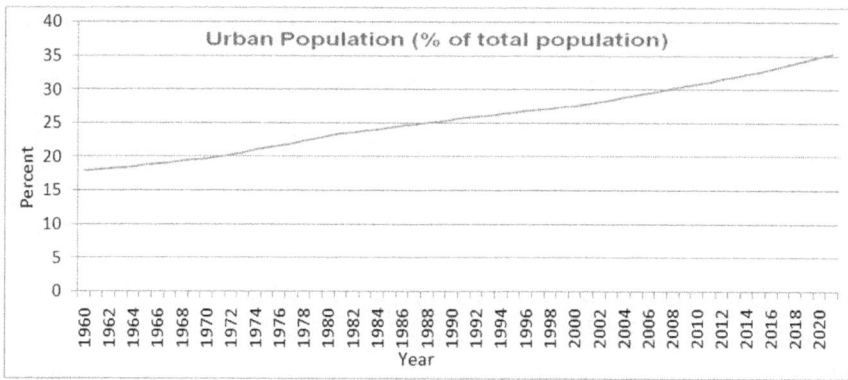

FIGURE 4.2 Urban population trend of India (%of the total population).

Source: World Bank [13].

Furthermore, urban infrastructures are more developed than countryside infrastructures, although some urban inhabitants are unable to access or benefit from these infrastructures [12]. The urban population trend is illustrated in Figure 4.2 and Table 4.1.

4.6 RURAL–URBAN POPULATION GROWTH RATE

Between 1991 and 2001, the population growth rate in Maharashtra was 15.25% in rural areas and 34.57% in urban areas, with a total growth rate of 22.73% for the Maharashtra state. The same pattern can be seen in the 2011 provisional totalities, with rural growth of 10.34% and urban growth of 23.67%, for a cumulative rate of 15.99% between 2001 and 2011 [14].

4.7 CLIMATE CHANGE RISK TO CITIES

Climate extremes are projected to have a greater impact on human settlements in the future. Worldwide, there has been a substantial increase in the population residing in cities and the infrastructure that supports them. This expansion is taking place in the face of rising risks of high heat, rainfall, and flooding. If landscape planning and

TABLE 4.2
Smart Cities in Maharashtra with Their Area

Sr. no.	Name of the city	Area (km²)	Reference
1	Aurangabad City	141.1	https://aurangabad.gov.in/en/district-profile/
2	Kalyan-Dombivali	67	www.kdmc.gov.in/RtsPortal/smart-city/ about-kalyan-dombivli.jsp
3	Nagpur	217.65	www.nmcnagpur.gov.in/demographic-profile
4	Nashik	259.13	https://nashik.gov.in/about-district/
5	Pimpri-Chinchwad	181	www.pcmcindia.gov.in/location_info.php
6	Pune	331.26	https://en.wikipedia.org/wiki/Pune_ Municipal_Corporation#cite_note-:11-9
7	Solapur	178.57	https://smartnet.niua.org/sites/default/files/ tdtransferstation.pdf
8	Thane	128	https://thanesmartcity.in/category/projects/ area-based-development/

urban design continue to operate in the same 'business as usual' manner, without considering the rising hotspots of susceptibility, the consequences will almost certainly be far worse [15]. The resources that are accessible will be impacted by the changing climate and affect wildlife that inhabits urban areas, such as birds, insects, and other larger animals, by altering the quality, quantity, and time of feeding of the fauna. This can affect the rate at which new diseases and other infections, as well as alien/invasive species, penetrate the expanded territories surrounding cities. Changes in forage and species mix will affect species competition and pest management strategies [16].

4.8   SMART CITIES IN MAHARASHTRA

On June 25, 2015, the Indian government announced the Smart Cities Mission. This mission aimed to support communities that deploy 'smart solutions' to offer fundamental structure, healthy environmental conditions that are suitable, as well as adequate living standards for the local population. The programme aims to promote financial growth and improve quality of life through holistic efforts on social, physical, economic, and institutional pillars of the city [17]. The list of smart cities is shown in Table 4.2 and its GIS-based map is displayed in Figure 4.3. The implications of this study are not only limited to the above-mentioned cities but also other semi-urban areas which are probable to become cities in near future.

4.9   LAWS AND LEGISLATION ON URBAN DEVELOPMENT

The Government of India's Ministry of Urban Development has established a guiding principle on Urban Development Plans Formulation and Implementation (UDPFI), which aims to preserve eco-sensitive zones from urban development while also

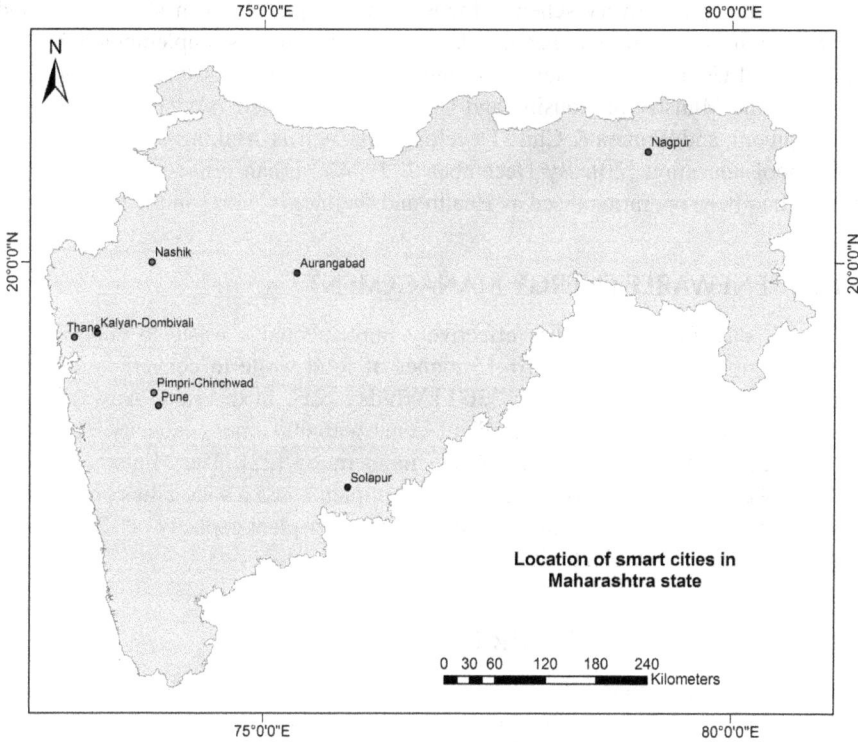

FIGURE 4.3 Location of smart cities across Maharashtra state.

providing an adequate open space network. Several metropolitan cities' urban development specialists have also produced standards for preserving open spaces during urban growth [18].

4.10 THE INTERNATIONAL VIEW OF NBS AND ECOSYSTEM SERVICES

The Paris Agreement intends to boost the international response to the challenge of climate change by aligning monetary flows with a low-carbon footprint, climate-resilient growth path. The collaboration of the government, corporate sector, and non-governmental organisations (NGOs) can improve the flow of funds and aid in the development of the capacities of local communities [19].

4.11 PUBLIC HEALTH SYSTEM

The programme of the National Urban Health Mission (NUHM) is mainly intended for the healthcare needs of the urban population, with a particular focus on the urban poor, by building critical primary healthcare centres available to them and lowering their out-of-budget treatment costs. This will be accomplished by facilitating the present healthcare delivery scheme, focusing on people living in slum areas, and combining it with different health-related welfare schemes implemented by the Ministries of Urban Development, Housing and Urban Poverty Alleviation (currently known as the Ministry of Housing and Urban Affairs [MoHUA]), Human Resource Development, and Women & Child Development, such as drinking water, sanitation, and school education [20]. By December 2021, 460 urban primary health centres (UPCs) had been operationalised as Health and Wellness Centres in the state [10].

4.12   RENEWABLE ENERGY MANAGEMENT

In Nashik city, GIZ GmbH has effectively implemented a waste to energy plant (WTE), which daily processes 10–15 tonnes of food waste to convert into energy since December 2017, generating 3300 kWh/day [21]. In the same way, Mumbai also has a proposal for a WTE located at Deonar with 600 tonnes capacity. This plant will generate 4-megawatt energy through incineration [22]. The Ministry of New and Renewable Energy (MNRE) successfully implemented a solar cities programme under this; Nagpur city has a solar photovoltaic power plant capacity of 125 kW and Thane city has 136 kW electricity generation [23].

4.13   IMPROVEMENT IN AIR QUALITY

It has been reported that 78% of air pollution, including GHG emissions, is from cities in India. The transportation sector in India contributes roughly 13.2% of overall CO_2 emissions, with road transport contributing up to 90% of total GHGs emitted by transportation industry in India [24]. In Pune city, the Municipal Corporation operates 150 electric buses. e-Buses can reduce transportation-related carbon dioxide (CO_2) and particulate matter (PM 2.5) emissions. An electric bus is a cost-effective transportation method from an economic point of view. As per the environment report of 2020, it has been found that the emission of CO_2 is from 7% to 4% [25].

4.14   SOLID WASTE MANAGEMENT

Maharashtra state has proper implementation and execution of the management of solid waste. The Municipal Solid Waste Rules (SWR) 2000 and Solid Waste Management Rules (SWM) 2016 are followed in the state. A total of 19,882.08 tonnes per day of solid waste is generated in the 12 municipal corporations/urban local bodies of the state [26]. Further processing on the same dry garbage received from the city is separated and categorised as paper, plastic, glass, metal, and other materials by a secondary isolation process at transfer stations or the city's selected Material Recovery Facility (MRF) or solid waste disposal and treatment plant. Dry waste is recycled through local merchants in the city, to recyclers (including formal and informal) in the area, or to waste processing factories in surrounding metro-cities after secondary segregation into paper, plastic, glass, metal, and so on. In the last few years due to technological advancements, the rate of obsolescence of this electronic equipment is growing, resulting in increased electronic waste (E-waste) output. The disposal of E-waste scientifically and safely is regarded as a most difficult task. According to research done by the Maharashtra Pollution Control Board (MPCB) in 2007 for the Mumbai and Pune regions, Mumbai Metropolitan Region is India's greatest source of E-waste [27].

4.15   WASTEWATER MANAGEMENT

Urban wastewater or sewage management entails the planning, construction, and operation of infrastructure to meet standards for potable water and sanitary usable, surface runoff, infiltration and stormwater control, recreational parks, and urban ecosystems in maintained condition [28]. In the periphery of Aurangabad city, there are three sewage treatment plants in which 205 million litres per day (MLD) gets treated which the municipal corporation uses for agriculture activities [29]. Reuse-quality wastewater from sewage treatment plants meets the requirements of reclaimed water quality with the bacterial and chemical standards under reuse guidelines. Such reuse-quality water helps in the achievement of sustainable development goals and can be utilised for sustainable agriculture to help meet the targets of SDGs 2, 3, 6, 12, 14, and 16 [30]. A total of 25 common effluent treatment plants (CETPs)are operating in the state, with a total capacity of 203 MLD. CETPs help enterprises control pollutants more easily, and they also serve as a step towards a cleaner environment and societal service [31]. Proper wastewater treatment saves the urban water bodies/wetlands from eutrophication.

4.16   URBAN BLUE AND GREEN SPACES

As a part of 'green infrastructure', green spaces in urban land add to the city's aesthetics. The biodiversity of the city can be safeguarded in this area, and the healthy well-being of residents can be maintained. Urban green spaces help psychologically by providing a touch of nature. The initiative in expanding urban green space reminds people about the values of trees, animals, and peaceful ambience needed in their busy lifestyle. From a policy-making view, future cities will be designed with re-thinking by considering the role of each species by adopting the 'live and let live' approach.

Investment in metropolitan areas, through the creation of scientifically managed biodiversity hotspots, is seen as a kind of uninterrupted natural area protection for expected positive and healthy environmental change. Blue green infrastructure (BGI), such as playgrounds, gardens, lakes, ponds, mangroves, and wetlands, are some of the essential markers of quality of life in the face of increasing urban development. The urban BGI is threatened by infrastructural development, land-use changes, and challenging land use. It has been repeatedly stated and reinforced that natural resource management (NRM) techniques should be given top priority when developing development plans for urban expansion and that they should constitute the heart of mandatory binding rules. Due to several tightly packed and severely concretised metropolitan centres lacking the open areas required for large-scale plantation, the NRM approach involves, but is not limited to, tree planting. It is critical to analyse the sustainability of a project from a long-term perspective [32].

4.17   ADVANCED TOOLS FOR URBAN ADAPTATION

Climate Action for Urban Sustainability (CURB) is intended to help guide cities through the process of planning and implementing a range of adaption strategies through the action of minimising energy use, reducing GHG emissions, and economic feasibility. The methods and policy actions included by CURB can deliver important local quality of life and socioeconomic benefits such as job opportunities [33]. A research analysis conducted by Egerer et al. [14] showed that the Urban Socio-Ecological-Technological System (SETS) has been explicitly helpful in building strategies in the case of a cause–effect relationship of the climate change adaptation with the ability to handle the urban complexity more dynamically. This tool has a crucial part in managing urban resources, infrastructure management, and design.

The ecological mappings in the numerous action plans either cited existing research on climate change, adaptation, flexibility, and susceptibilities, or presented a status report on ecological circumstances in the city, focusing on topics like noisy environments, solid waste management and treatment, and drinking water quality standards. Despite the frequent inclusion of these concepts in the texts, most of the plans reviewed had inadequate definitional transparency of terms such as sustainability, resilience, sustainable development, and climatic change [34].

4.18   USE OF GEOSPATIAL TECHNOLOGY, REMOTE SENSING, AND EARTH OBSERVATION

Low spatial resolution data collected through remote sensing can also be valuable in conceptualising the knowledge about a variety of assets such as land cover, type of ecosystem, and recognising the patterns of agriculture and forests. Moreover, it has the potential to analyse climate-related variables at the global scale [35]. The ArcGIS software suite from ESRI is widely used by planners in the business and public sectors. With advancements in technology and user demands, many GIS programmes have been developed over time. The shift towards web GIS, such as ESRI's ArcGIS Online and their current desktop application, ArcGIS Pro, which connects with the online platform, may be the most significant recent change [36]. In the case of visualisation of urban settlements in Maharashtra, Sentinel 2 imagery-based Land Use Land Cover for the year 2020 was employed to map the urban expansion. Figure 4.4 shows the respective map. From this map, there is a clear indication of the expansion that is dominated in the western part of the state. Nagpur district in the eastern part also is covered as built-up.

4.19   SUSTAINABLE URBAN MOBILITY

Maharashtra state is pioneering and has laid the ideal example for cities that are citizen-caring by providing streets for walking and cycling. The Ministry of Housing

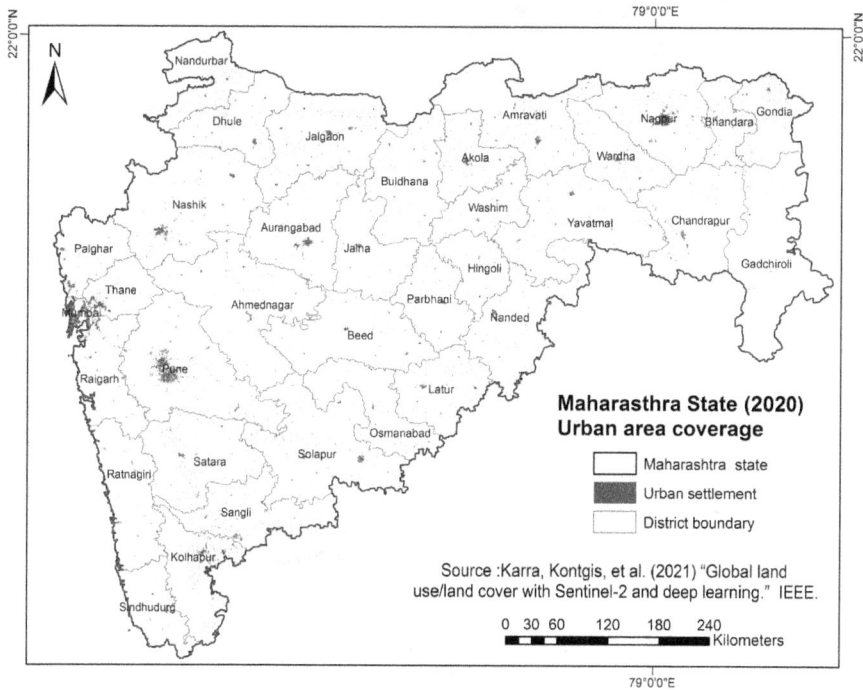

FIGURE 4.4 Sentinel-2 imagery for the year 2020-based settlement map of Maharashtra state.
Source: Extracted data from Ref. [37].

and Urban Affairs (MoHUA) initiated the #Streets4People campaign intending to provide a people-friendly environment in urban cities. The eight cities chosen for scaling up their experimental initiatives and constructing stable public infrastructure include Pune, Pimpri-Chinchwad, Nagpur, and Aurangabad. The Streets4People competition required communities to rethink their streets as public spaces using affordable, innovative techniques that made them safer for citizens, particularly seniors and children [38].

4.20 GREEN BUILDING

An ecology-oriented building certification programme, GBCI-LEED and TERI-GRIHA, that fulfils the sustainable approach to the built environment has provided incentives to the developer as well as property buyer to opt for green building [39]. Maharashtra state appears among the ten states of the LEED program. As per the data availability, it was calculated that 106,057,234 gross square feet (GSF) have been created with green certification under 334 projects in urban areas of the state [40]. Recent infrastructure systems have significant and unequalled consequences on diverse communities through mechanisms such as dislocation, ecological risk, and access to critical services such as water and health-related facilities, and are thus crucial to the issue of socioenvironmental/ecological justice [41].

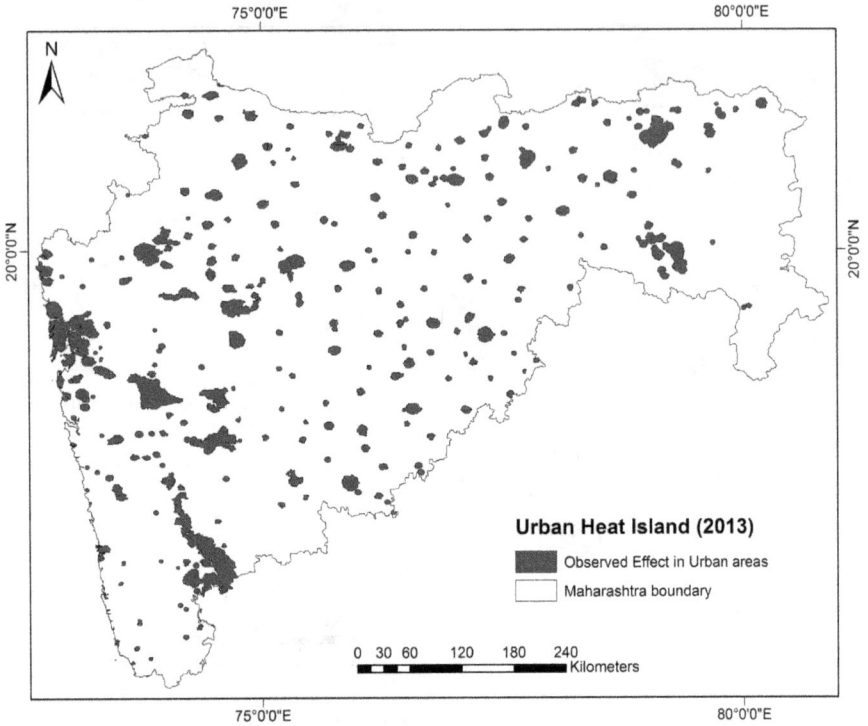

FIGURE 4.5   Urban heat island (UHI) observed in the year 2013.

Source: [42].

4.21   INDUSTRIALISATION

As part of the Delhi–Mumbai Industrial Corridor (DMIC), Aurangabad Industrial City (AURIC), a well-planned and greenfield smart industrial city that follows the notion of 'walk to work', is being developed across an area of 4039 acres in the state. AURIC had allotted 126 plots with a total area of around 337 acres to investors by November 2021. AURIC has a total investment of about 5500 crores and employs over 5900 people [10].

4.22   MITIGATION OF URBAN HEAT ISLANDS (UHIS)

To map the urban heat island area of Maharashtra state, the Global Urban Heat Island map from the 2013 data was downloaded from https://sedac.ciesin.columbia. edu/downloads/data/sdei/sdei-global-uhi-2013/sdei-global-uhi-2013-shp.zip [42]. Figure 4.5 shows the UHI of Maharashtra state. From the map, it can be computed that a total of 22,894 km^2 has a UHI effect.

4.23 CONSERVATION OF URBAN/METROPOLITAN WETLANDS

Urban wetlands are located nearby populous areas of a city and provide ecological, economic, and social benefits. They can be natural or manmade, and benefit the environment by functioning with many ecosystem services such as reducing the UHI effect, groundwater recharge, reducing flood risk, etc. Wetland restoration helps to revive the aquatic ecosystem by filtering stormwater and wastewater to provide valuable quality-related restoration [43]. Urban wetland restoration can be done by many types such as riverine, lacustrine, palustrine, green roofing, ecological bridging, replacement of wetland, and by maintaining natural drainage systems [44]. The safeguarding of the existing natural ecosystem is also followed in the state. In the Thane district of Maharashtra, Thane Creek is a 26-kilometre-long creek that begins at the southwest approach to Mumbai Harbour and links to the Ulhas River in the north. Despite high levels of domestic and industrial pollution, Thane Creek is home to a diverse range of flora and fauna. On August 6, 2015, it was designated as an Important Bird and Biodiversity Area and Thane Creek Flamingo Sanctuary [45].

4.24 NOISE REDUCTION

Noise pollution can be buffered by soothing water sounds or absorption in blue spaces and adjacent vegetation or green space. As a result, good blue space design addresses distance and noise shielding from sources such as traffic or recreation. Noise pollution is a serious environmental health hazard, particularly in metropolitan areas, and it has negative health consequences. Noise created by vehicles and recreational activities is the most common source of environmental noise [46].

4.25 WATER, SANITATION, AND HYGIENE (WASH) IN URBAN AREAS

UNICEF's mission focuses on increasing access to water, sanitation, and hygiene (WASH) services for the poorest people in cities. Our Global Support for Urban Water, Sanitation, and Hygiene lays out our long-term strategy at the global, national, and regional levels [47]. Individuals' health and hygiene are heavily reliant on the accessibility to safe drinking water, improved sanitation, and improved sanitary behaviours. Despite being preventable, water- and sanitation-related infections continue to be one of the most serious child well-being issues and causes of malnourishment around the world. To provide clean drinking water and sanitation to everybody, the Indian government maintains two leading programmes: the National Rural Drinking Water Programme (NRDWP) and the Total Sanitation Campaign (TSC). The NRDWP is also concerned with the long-term supply and quality of water. Moreover, under the *Swachh Bharat* mission, the state government has established the *Swachh Maharashtra (Urban)* mission to make cities open defecation free (ODF). Maharashtra is the state that is leading the way in terms of establishing state-level policies for faecal sludge treatment. The government has approved 311 faecal sludge treatment plants (FSTPs), with 189 cities having finished the project and begun their desired operations, with the remaining FSTPs under construction [10].

4.26 THE EMERGING CONCEPT OF ECOSYSTEM-BASED DISASTER RISK REDUCTION (DRR)

Resilience is related to the sustainability of an objective over the long term that is capable of coping with capacity from a disaster point of view. Meanwhile, adaptation strategies have the flexibility to change with the elements of the system which could have certain implications for resilience [48]. For disaster and climate challenges, Eco DRR and EbA integrate ecosystem-based executive methodologies, tools, and principles. These techniques and practices help to preserve and improve ecosystems and their services, while also increasing ecosystem and community resilience to catastrophes and climate change [49]. Ecosystem-based DRR encourages the preservation and improvement of ecosystems and their services, with an emphasis on minimising vulnerability and building sustainable modes of livelihood to improve human catastrophe adaptability. This viewpoint considers the interconnectedness of social and natural systems, putting humans at the centre of policy-making [50]. From an ecosystem restoration point of view, the plantation has a key role connected to the existing resources. Mumbai and suburban Thane are districts with mangrove cover. It has been observed that the sparse mangrove cover has been converted into dense cover, and Mumbai has converted its portion into dense cover formation [51].

4.27 CONCLUSIONS

For sound and sustainable urban development, a combination of NBS and smart climate-resilient techniques has a potential role in minimising urbanisation pressure. Such sustainable cities reduce the stress on overexploited resources and make progress towards climate-resilient cities. Greening cities not only provides solutions to the current urban issues, in some cases, such as waste-to-energy and E-waste recycling it has a role in the generation of employment. Further, the city development plan and climate action plans for the cities are people-centric policies with a futuristic vision to adapt to climate change. The overall sustainable development of urban environmental conditions can have positive impacts on public health and create secure and healthy environments for current and future generations.

REFERENCES

[1] UN DESA, "UN Department of Economic and Social Affairs. Population Facts 2018/1," 2018.

[2] U. N. org, "Goal 11: Make cities inclusive, safe, resilient and sustainable. www.un.org/sustainabledevelopment/cities/." 2022.

[3] M. Mukherjee and R. Shaw, "Forward-Looking Lens to Mainstream Blue-Green Infrastructure," in *Ecosystem-Based Disaster and Climate Resilience*, M. Mukherjee and R. Shaw, Eds. Singapore: Springer Singapore, 2021, pp. 501–512. doi: 10.1007/978-981-16-4815-1_23.

[4] E. Cohen-Shacham, G. Walters, C. Janzen, and S. Maginnis, Eds., *Nature-based solutions to address global societal challenges*. IUCN International Union for Conservation of Nature, 2016. doi: 10.2305/IUCN.CH.2016.13.en.

[5] MoHUA, "Annual Report 2021–22. Ministry of Housing and Urban Affairs. http:// mohua.gov.in." 2022.

[6] IIPS and UNICEF, "Assessing Urban Vulnerabilities in Maharashtra. www.iipsindia. ac.in/sites/default/files/Urban_Vulnerability_Analysis_Report_121021.pdf." 2022.

[7] itdp.org, "Sustainable Urban Development. www.itdp.org/our-work/sustainable-urban-development/." 2022.

[8] R. P. Dameri, "Smart City Definition, Goals and Performance," in *Smart City Implementation*, Cham: Springer International Publishing, 2017, pp. 1–22. doi: 10.1007/978-3-319-45766-6_1

[9] S. Pauleit, T. Zölch, R. Hansen, T. B. Randrup, and C. Konijnendijk van den Bosch, "Nature-Based Solutions and Climate Change–Four Shades of Green," in *Nature-Based Solutions to Climate Change Adaptation in Urban Areas*, N. Kabisch, H. Korn, J. Stadler, and A. Bonn, Eds. Cham: Springer International Publishing, 2017, pp. 29–49. doi: 10.1007/978-3-319-56091-5_3

[10] MAHADES, "Economic Survey of Maharashtra 2021–22.Directorate of Economics and Statistics, Planning Department, Government of Maharashtra, Mumbai.https:// Mahades.Maharashtra.Gov.in/ESM1920/Chapter/English/Esm2122_e.Pdf." 2022.

[11] PIB, "PIB. https://pib.gov.in/PressReleasePage.aspx?PRID=1744761." 2021.

[12] M. van Maarseveen, J. Martinez, and J. Flacke, Eds., *GIS in Sustainable Urban Planning and Management: A Global Perspective*. Boston London New York: CRC Press, Taylor & Francis, 2019.

[13] World Bank, "Population, total–India. https://data.worldbank.org/indicator/SP.URB. TOTL.IN.ZS." 2022.

[14] M. Egerer *et al.*, "Urban change as an untapped opportunity for climate adaptation," *npj Urban Sustainability*, vol. 1, no. 1, p. 22, Dec. 2021, doi: 10.1038/s42949-021-00024-y

[15] X. Ye and D. Niyogi, "Resilience of human settlements to climate change needs the convergence of urban planning and urban climate science," *Comput.Urban Sci.*, vol. 2, no. 1, p. 6, Dec. 2022, doi: 10.1007/s43762-022-00035-0

[16] W. Solecki and P. J. Marcotullio, "Climate Change and Urban Biodiversity Vulnerability," in *Urbanization, Biodiversity and Ecosystem Services: Challenges and Opportunities*, T. Elmqvist, M. Fragkias, J. Goodness, B. Güneralp, P. J. Marcotullio, R. I. McDonald, S. Parnell, M. Schewenius, M. Sendstad, K. C. Seto, and C. Wilkinson, Eds. Dordrecht: Springer Netherlands, 2013, pp. 485–504. doi: 10.1007/978-94-007-7088-1_25

[17] smartcities.gov.in, "About smart cities mission. https://smartcities.gov.in/about-the-mission." 2022.

[18] D. Govindarajulu, "Urban green space planning for climate adaptation in Indian cities," *Urban Climate*, vol. 10, pp. 35–41, Dec. 2014, doi: 10.1016/j.uclim.2014.09.006

[19] H. Gupta and L. C. Dube, "Benefits of Evaluating Ecosystem Services for Implementation of Nature-based Solutions Under the Paris Agreement," in *Social-Ecological Systems (SES)*, M. Behnassi, H. Gupta, M. El Haiba, and G. Ramachandran, Eds. Cham: Springer International Publishing, 2021, pp. 39–56. doi: 10.1007/978-3-030-76247-6_2

[20] nhm.gov.in, "National Urban Health Mission. https://nhm.gov.in/index1.php?lang= 1&level=1&sublinkid=970&lid=137." 2022.

[21] cseindia, "Waste to Energy Plant, Nashik, Maharashtra. www.cseindia.org/waste-to-energy-plant-nashik-maharashtra-8412." n.d.

[22] L. Singh, "Mumbai's first waste to energy plant gets Centre's green nod. https://indian express.com/article/cities/mumbai/mumbais-first-waste-to-energy-plant-gets-centres-green-nod-7704745/," Mumbai, 2022.

[23] mahaurja, "Renewable Energy Projects implemented so far under Solar Cities Programme. www.mahaurja.com/meda/data/off_grid_solar/solar%20city%20list-2.pdf." 2022.

[24] GIZ, "Training needs assessment for electric buses in India. Volume I–Identification of training needs. Deutsche Gesellschaft für Internationale Zusammenarbeit (GIZ) GmbH." 2021.

[25] PMC, "Pune Municipal Corporation-Electric buses. www.pmc.gov.in/en/electric-buses." 2022.

[26] MPCB, "A summary statement on progress made by local body in respect of solid waste management. www.mpcb.gov.in/sites/default/files/solid-waste/Abstract_of_SWM_Region_wise%202018.pdf." 2018.

[27] MMRDA, "e-Waste Processing & Disposal Facility. https://mmrda.maharashtra.gov.in/e-waste-processing-disposal-facility." 2022.

[28] D. P. Loucks and E. van Beek, "Urban Water Systems," in *Water Resource Systems Planning and Management*, Cham: Springer International Publishing, 2017, pp. 527–565. doi: 10.1007/978-3-319-44234-1_12

[29] https://maharashtratimes.com, "agriculture will get stp water. https://maharashtratimes.com/maharashtra/aurangabad/agriculture-will-get-stp-water/articleshow/90440279.cms." 2022.

[30] H. M. K. Delanka-Pedige, S. P. Munasinghe-Arachchige, I. S. A. Abeysiriwardana-Arachchige, and N. Nirmalakhandan, "Wastewater infrastructure for sustainable cities: assessment based on UN sustainable development goals (SDGs)," *International Journal of Sustainable Development & World Ecology*, vol. 28, no. 3, pp. 203–209, Apr. 2021, doi: 10.1080/13504509.2020.1795006

[31] MPCB, "Concept of CETP. https://mpcb.gov.in/sites/default/files/common-effluent-treatment-plant/general/Concept_of_CETP.pdf." 2022.

[32] S. Dhyani, R. Majumdar, and H. Santhanam, "Scaling-up Nature-Based Solutions for Mainstreaming Resilience in Indian Cities," in *Ecosystem-Based Disaster and Climate Resilience*, M. Mukherjee and R. Shaw, Eds. Singapore: Springer Singapore, 2021, pp. 279–306. doi: 10.1007/978-981-16-4815-1_12

[33] World Bank, "CURB, Climate action for urban sustainability." 2017.

[34] H. Unnikrishnan and H. Nagendra, "(2021) Building climate resilient cities in the global South: assessing city adaptation plans in India, The Round Table, 110:5, 575–586, DOI: 10.1080/00358533.2021.1985268," *The Round Table*, vol. 110, no. 5, pp. 575–586, 2021, doi: DOI: 10.1080/00358533.2021.1985268

[35] European Commission. Directorate General for Research and Innovation., *Evaluating the impact of nature-based solutions: a handbook for practitioners*. LU: Publications Office, 2021. Accessed: Apr. 20, 2022. [Online]. Available: https://data.europa.eu/doi/10.2777/244577

[36] E. Soward and J. Li, "ArcGIS Urban: an application for plan assessment," *Comput. Urban Sci.*, vol. 1, no. 1, p. 15, Dec. 2021, doi: 10.1007/s43762-021-00016-9

[37] K. Karra, C. Kontgis, Z. Statman-Weil, J. C. Mazzariello, M. Mathis, and S. P. Brumby, "Global land use / land cover with Sentinel 2 and deep learning," in *2021 IEEE International Geoscience and Remote Sensing Symposium IGARSS*, Brussels, Belgium, Jul. 2021, pp. 4704–4707. doi: 10.1109/IGARSS47720.2021.9553499

[38] M. Bose, "Citizen-friendly initiatives by four Maharashtra cities win accolades from the Centre. Read more at: www.deccanherald.com/national/west/citizen-friendly-initiatives-by-four-maharashtra-cities-win-accolades-from-the-centre-1072605.html." 2022.

[39] maharashtra.mygov.in, "Maharashtra Green Building Policy . https://maharashtra.mygov.in/en/task/maharashtra-green-building-policy/." 2022.

[40] constructionworld.in, "Maharashtra: 1st rank among 10 Indian LEED states. www.constructionworld.in/latest-construction-news/real-estate-news/Maharashtra-1st-rank-among-10-Indian-LEED-states/21692." 2019.

[41] H. Pearsall *et al.*, "Advancing equitable health and well-being across urban–rural sustainable infrastructure systems–npj Urban Sustainability," *npj Urban Sustainability*, 2021, doi: https://doi.org/10.1038/s42949-021-00028-8

[42] CIESIN, "Center for International Earth Science Information Network (CIESIN), Columbia University. Documentation for the Global Urban Heat Island (UHI) Data Set, 2013. Palisades NY: NASA Socioeconomic Data and Applications Center (SEDAC). http://doi.org/10.7927/H44M92HC." 2013.

[43] W. Dooley and M. Stelk, "Urban Wetlands Protection and Restoration Guide. Association of State Wetland Managers. Windham, Maine." 2021.

[44] K.-G. Kim, H. Lee, and D.-H. Lee, "Wetland restoration to enhance biodiversity in urban areas: a comparative analysis," *Landscape Ecol Eng*, vol. 7, no. 1, pp. 27–32, Jan. 2011, doi: 10.1007/s11355-010-0144-x

[45] N. Vasudevan, "Conservation and Sustainable Management of Coastal and Marine Areas in Maharashtra, India." 2017.

[46] S. Bell, L. E. Fleming, J. Grellier, F. Kuhlmann, M. J. Nieuwenhuijsen, and M. P. White, *Urban Blue Spaces: Planning and Design for Water, Health and Well-Being*, 1st ed. London: Routledge, 2021. doi: 10.4324/9780429056161

[47] UNICEF, "Water, Sanitation and Hygiene (WASH) Safe water, toilets and good hygiene keep children alive and healthy. www.unicef.org/wash#." 2022.

[48] I. Pal, R. Shaw, S. Shrestha, R. Djalante, and R. A. W. C. Cavuilati, "Toward sustainable development: Risk-informed and disaster-resilient development in Asia," in *Disaster Resilience and Sustainability*, Elsevier, 2021, pp. 1–20. doi: 10.1016/B978-0-323-85195-4.00001-9

[49] S. Bhardwaj and A. K. Gupta, "Ecosystem-Based Approaches and Policy Perspective from India," in *Ecosystem-Based Disaster and Climate Resilience*, M. Mukherjee and R. Shaw, Eds. Singapore: Springer Singapore, 2021, pp. 101–125. doi: 10.1007/978-981-16-4815-1_5

[50] A. Gupta and S. Nair, "Ecosystem Approach to Disaster Risk Reduction, National Institute of Disaster Management, New Delhi, Pages 202." 2012.

[51] S. Bhalerao, "Study shows increase in overall mangrove forest cover by 48.79 sq km: Forest dept. https://indianexpress.com/article/cities/mumbai/study-shows-increase-in-overall-mangrove-forest-cover-by-48-79-sq-km-forest-dept-7213077/." 2021.

5 Urban Heat Island

A Review of Effects and Predictions with Deep and Machine Learning Technologies

R. Pushpa Lakshmi

CONTENTS

5.1   INTRODUCTION

The global population is rapidly rising; new cities are being developed, and existing ones are becoming overcrowded. Currently, 35.4% of India's population lives in cities (World Data Atlas, 2021). In recent years, the population in urban areas has increased. According to a UN population assessment, the number of citizens in urban areas in India will surpass the number of citizens in rural areas by the year 2050 (Population

DOI: 10.1201/9781003331001-5

71

Division, 2014). Increased urban growth has led to changes in urban surfaces and form, along with a rise in the quantity of anthropogenic heat discharged into the atmosphere. As a result of the combined effects of these changes, metropolitan areas experience higher air temperatures than rural or suburban areas, which results in the development of heat islands.

An urban heat island (UHI) occurs when the heat in a highly populated metropolis is up to 2 degrees higher than in suburban or exurban areas. This is due to opaque materials used for roads, roofs, pavements, and roads, such as tar, bricks, and concrete, which do not transmit light and have a greater heat capacity and thermal conductivity than in rural areas, which have more grass, trees, and open space.

A study conducted by IIT Kharagpur examined the weather differences in metropolitan and neighbouring exurban areas across 44 main metropolises throughout all seasons from 2001 to 2017 (Raj et al., 2019). Cities with more vegetation cover in their surrounding rural areas, such as Chennai, Thiruvananthapuram, and Kolkata, were found to have a greater heat island effect during the day. Cities with little green cover, such as Delhi, experienced a significant heat island impact at night. The heat island effect will have an even greater impact on pollution levels in a metropolis. UHIs exist in Bengaluru neighbourhoods such as Jayanagar and Koramangala as a result of the growth of high-rise flats, industrial parks, and suburban structures.

Cities are more likely to experience heat waves as a result of UHI, which can cause respiratory discomfort, heat cramps, heat strokes, sleep deprivation, and increased mortality rates in humans and animals. UHIs also influence neighbouring aquatic bodies, as warmer water is moved from the city to sewer drains and then discharged into neighbouring ponds and streams, lowering the quality of their water. According to a study, excessive temperatures cause over 7 lakh deaths in India each year. From 2009 to 2019, global temperatures grew by 0.9 degrees Celsius every decade (Qi Zhao et al., 2021), with hot temperatures increasing in all locations. From 2009 to 2019, cold-associated mortality was reduced by 0.51%, whereas heat-associated deaths increased by 0.21%.

This study provides a thorough outline of the literature on survival in UHIs, and deep learning and machine learning methodologies associated with the prediction of UHIs. It explores surveys conducted related to UHIs in India, published papers on the application of various algorithms in UHIs, and erstwhile correlated resources. This literature review's primary goals are to identify significant trends, the latest threats, and research needs.

5.2 OUTLINE OF MACHINE LEARNING AND DEEP LEARNING

The need for air conditioning to keep buildings cool rises as a result of heat islands. As a result of the increased demand, electricity costs have increased. Heat islands raise total electricity demand as well as peak electricity demand. Peak demand commonly increases on hot summer workday afternoons, when air conditioners, lights, and appliances are turned on in businesses and houses. As a result, air pollution and greenhouse gas discharges are increased. Elevated temperatures have the potential to accelerate the development of ground-level ozone. They also have an impact on water quality and aquatic life in general, including affecting many aquatic species' metabolism and reproduction.

With technological advancements, urban planners may now employ data models, deep learning, and artificial intelligence to not only assess the demand but also foresee the repercussions of rapid expansion. Although urban planning has numerous areas of attention, there has recently been a surge in scholarly studies into UHIs and urban cold islands.

Machine learning (ML), which combines computer science and statistics and is a key component of data science and artificial intelligence, has its scope of use expanding by the day. In recent years, cities have become more active in these applications, largely to address demands for intelligence, sustainability, pollution control, environmental protection, and so on. In general, past research has shown that machine learning methods offer a lot of potential for modelling modern and intelligent urban formations (Ma et al., 2020; Choung & Kim, 2019). Convolution neural networks (CNNs) have been demonstrated to be efficient in removing characteristics from geographical data, whereas recurrent neural networks (RNN) have been found to be effective in extracting features from temporal data (Geng et al., 2019; Gómez et al., 2020). In numerous studies of smart urban forms, ensemble approaches have been shown to be very helpful (Jochem et al., 2018; Geiß et al., 2020; Ma et al., 2020). Other types of ML practices, such as basic supervised and unsupervised approaches, are also frequently employed for a variety of municipal planning functions (Li et al., 2020; Gao et al., 2020).

The replication of human intelligence into robots is what AI is all about, as the name suggests. ML is a part of AI that entails mathematical approaches, and functions that enable learning from prior data and correspondingly generate output. The artificial neural network is used in DL models to produce predictions that are completely independent of humans. DL models have already been used successfully in a variety of applications. Because of its processing power, rapid understanding, accuracy, and robustness in model generation, DL has become increasingly important in today's environment. Consequently, the purpose of this work is to present a critical overview of the literature on contemporary ML applications and DL to predict UHIs, along with the corresponding difficulties, possibilities, and future research areas.

An overview of UHIs, metropolitan information resources, and a summary of the used DL and ML methods are provided in this review. It also highlights their operational concepts, positive and negative aspects, and prospective applications. The focal offerings of this study are as follows: a thorough analysis of UHIs, a detailed analysis of the ML/DL algorithms applied in UHI prediction, and several resources of data for ML applications.

5.2.1 UHI ELEMENTS

Several urban characteristics are linked and utilised to forecast UHIs. The selection of these parameters is critical since it allows for the optimum calibration and, as a result, the model's efficiency. This area of expertise covers all the variables that are directly connected to the architectural and structural shape of the city. The size of the urban area, population density, weather and climate, urban design, ratio of green coverage, heat due to environmental pollution, buildings' cooling set points, wall construction properties, water bodies, land use/land cover (LULC) descriptions, and substances used in infrastructure are all factors that influence UHI intensity (Abulibdeh, 2021; Salvati & Cecere, 2017).

5.2.2 DATA SOURCES FOR UHI ANALYSIS

Data gathering and structuring in cities are in high demand, particularly for machine learning applications. ML models cannot be trained or calibrated without the relevant data. Remote sensing techniques such as thermal sensors, onboard satellites such as NASA Landsat, drones, aircraft, or mobile devices (IoT) can be used to determine the land surface temperature. Field surveys and internet surveys are also used to obtain data on urban forms.

5.3   OVERVIEW OF ML METHODS FOR UHI ANALYSIS

5.3.1 SUPERVISED ML MODELS

The supervised learning models produce the desired output based on the training carried out with the training dataset. The performance of the model is improved through training with different combinations of data. The model weights are changed during cross-validation to better fit the model. Based on prior knowledge, the outcomes are projected. Classification, which classifies the dataset and regression, a simple prediction methodology, is a basic facet of supervised education. For categorical output, classification algorithms are used to appropriately allocate test results to certain categories. Random forests, decision trees, and support vector machines are typically employed classifiers. Regression algorithms are used to figure out the association between the conditional and detached variables. Linear regression and polynomial regression are two extensively used regression algorithms.

5.3.1.1   Random Forest (RF) Model

RF is a powerful ML technique which assesses noisy data or partial data while also being anti-interference. It is a feature selection method for high dimensional data to determine a variable's relevance, anticipate and categorise correlated facets, and so on. Yao et al. (2020) proposed a machine-learning algorithm for monitoring UHI in their study. It uses the Landsat dataset (LST) to figure out LST estimates by combining an ML with an oscillatory model. LST data were initially downscaled using the RF technique. The RF algorithm performs well and produces precise outcomes in both exurban and metropolitan locations (Li et al., 2019).

The following are the four groups of predictor variables: (1) surface reflectance bands; (2) biophysical indicators (vegetation-related indices, water indices, built-up related indices, soil-related indices, albedo); (3) terrain factors (aspect, DEM, slope, hill shade); and (4) the land categorisation map was chosen based on the relationships between LST and various biophysical characteristics. SVM was used to extract land cover data from a cloud-free Landsat image. The high spatial resolution predictors were aggregated with low spatial resolution predictors during LST downscaling. To anticipate LST with the uplifted resolution, an RF model was used, which was then re-aggregated and downscaled to LST with high resolution.

Chen et al. (2022) proposed a UHI monitoring technique in which an RF model was constructed to predict air temperature based on independent environmental variables such as land cover and surface albedo. Based on the RF model's output, they calculated the canopy UHI intensity (CUHII). Anthropogenic, geometric, and physical factors were among the 18 independent variables included in the model's training.

The performance was evaluated using the fivefold cross-validation (CV) method, where four subsets of the dataset are used for training and the other is used for validation. They looked at the relationship between wind speed and CUHI and identified the relationship between offset and wind direction. To detect the pattern between land space and surface UHI, Zhang Yao et al. (2021) employed numerous decision trees, each of which was created by bootstrap random sampling. Monte Carlo was used before bootstrapping in RFR to improve bootstrap. RMSE and R2 values were used to assess the model's accuracy. To simulate the UHI intensity, the RF model proposed by Thomas Gardes et al. (2020) includes six predictors: French climatic zone, total population, elevation variations, distance to the shore, core of the city, and local climate zone (UHII).

Each classification and regression tree (CART) is constructed using subsampled predictors. It uses 42 agglomerations to avoid overfitting, and each urban agglomeration was given its own RF model. With an error rate of 0.85, the RF model predicts UHII, but the quality of the prediction varies greatly depending on the urban agglomeration under consideration. Duncan et al. (2019) investigated the complicated and non-linear interactions amid diverse municipal layouts and heat using the RF model. It was discovered that trees and shrubs are major predictors of LSTs.

Yanwei Sun et al. (2019) used ordinary least squares and RF regression to separate the features of metropolitan characteristics: nighttime light intensity, building morphology, transportation, public infrastructure, and ecological infrastructure on LST in their study. Utilising RF, metropolitans will experience greater than 90% variance in LST. The ecological infrastructure is recognised as an important driver of chilling effects among the five characteristics of urban form metrics.

5.3.1.2 Support Vector Machine (SVM)

The problems related to classification and regression are solvable using a supervised model (SVM). It is mostly applied for classification challenges. The vector support generates a hyperplane in linear classification (Smola & Schölkopf, 2004) which uses a maximum margin to divide the dataset into two subgroups (Tekouabou et al., 2021). Here, data are mapped into high-dimensional space for categorisation (Figure 5.1).

Sherafati et al. (2013) employed an artificial neural network (ANN) with regression based on a support vector to model UHI growth in their research. This considers urban sprawl criteria such as neighbourhood development level, distance to nearby roads, elevation, angle, and features to establish the relationship between LST in 1984 and 2007.

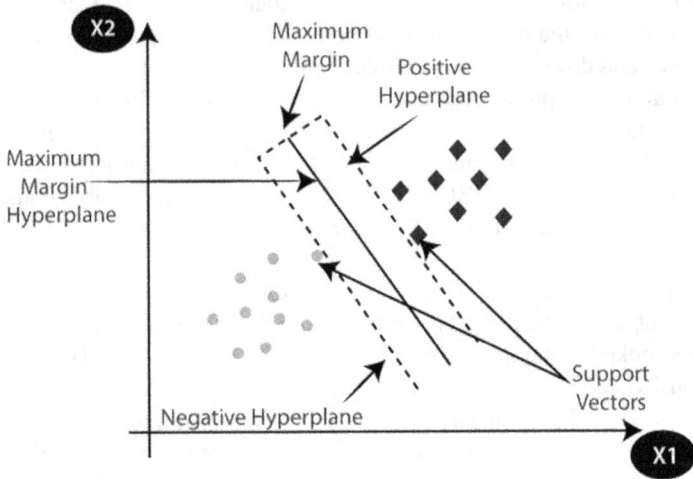

FIGURE 5.1 Hyperplane representation of SVM.

Source: javatpoint.

5.3.1.3   Naïve Bayes (NB)

The naïve-Bayes method based on the Bayes theorem is another supervised classi-fication algorithm. The classifier predicts the outcome based on the probability of objects. Garzon et al. (2021) proposed a remote sensing-based solution for SUHI that uses the weighted NBML algorithm for the identification of places that are likely to have excessive heat. It uses NBML decision rules to categorise the urban environment based on constraints and allocate precise action to temperature groups.

5.3.1.4   Decision Trees (DT)

The decision tree is a supervised classification method that discourses issues related to classification and regression. It uses a tree illustration, where the leaf node represents the class label and the interior nodes represent properties. The information gain and Gini index metrics are used to choose the split attribute for categorisation. Using decision trees, Dai et al. (2010) proposed a method for mining the UHI effect. The measurable correlations between the municipal heat environment and the affecting factors are environmental state, built-up mass and concentration, and heat established using a decision tree. It uses 14 classification rules to show the association amid LST and the parameters that influence it.

5.3.1.5   Regression Analysis

Regression, a supervised procedure identifies the output using one or more predictor variables and discovers the level of correlation that exists between variables. Linear and logistic regression algorithms are the most regularly used algorithms for cor-relation study. Logistic regression is generally applied to solve classification issues, whereas linear regression is meant for solving general regression problems. A linear relationship between the independent and dependent variables is represented by linear regression. The output of linear regression is continuous and can take any value. The projected outcome in logistic regression, on the other hand, is discrete and limited to a small number of values.

A regression model based on spatial analysis proposed by Yin et al. (2018) studied UHI impacts of terrestrial usage and municipal form on LST. For geographical analysis, the spatial lag and error model (SLM and SEM) are the broadly used spatial-regression models (Anselin & Rey, 1991; Anselin, 1990). This study found that SEM outperforms SLM in predicting the association between LST and the parameters affecting it. Terrestrial usage is calculated based on the proportions of impervious surface area, water, and loafed area. Three urban form metrics were also measured: factor of sky view (SVF), the density of the building, and the ratio of floor area (FAR). The results suggest that when compared to OLS, SEM is most suitable for evaluating spatial correlation. It concludes that efforts to mitigate UHI effects should focus on urban form metrics.

Chun et al. (2011) used both geographical information systems (GIS) and statistical methodologies to investigate the impact of urban factors on surface temperatures. It investigates the association between UHI and urban factors at various hierarchical levels using various regression models. Four independent factors are considered: building information, 2-D surface information, 3-D space information, and energy usage, with averaged surface temperatures (ASTs) as the dependent variable. Spatial autoregressive model (SAR), ordinary least squares (OLS), and an amalgam of SEM and SAR models are used. The spatial model for the 480 m grid and the GSM models for the 120 m and 60 m grids are utilised for sensitivity analysis, according to the results. Lu et al. (2012) used regression analysis to find a link between six shelter features: density of the building (BD), percentage of water sources (WBP), impermeable rate (IR), rate of planting (PR), an average of ambient air temperature (AAT), and UHI effect. The order of impact of these factors on wind heat was determined using the MLR approach. In tested areas, the order of relevance of shelter features was BD<PR<WBP<IR on day-to-day average wind heat.

5.3.2 Unsupervised ML Models

5.3.2.1 Cluster Analysis

Clustering is an unsupervised machine-learning problem that requires natural data grouping. A set of related data points in closed proximity are grouped into a cluster. K-means, DBSCAN, BIRCH, and OPTICS are some of the most commonly used clustering methods. Based on the computed UHI intensity, Luo et al. (2010) used the k-means clustering technique to analyse the UHI effect (UHII). Li et al. (2021) used k-means clustering to find changes and patterns in the intensity and frequency of surface urban heat islands. The k-means clustering method was used to classify SUHI patterns based on quantity and time, and the results were classified into three classes high, medium, and low patterns. According to the results of the comparison, SUHIF is possibly consistent with SUHII in the same city.

5.3.2.2 Association Rules

The learning of the association rule is a form of unsupervised learning technique that examines whether one data item is dependent on another. It analyses datasets using ML algorithms to derive patterns or associated information. It represents association rules in an if–then format. Apriori, F-P Growth, and Eclat are some of the most extensively utilised association rule learning algorithms. Using the association rule mining technique, Rajasekar et al. (2009) investigated the relationship between remote detection measures of LST and biophysical/socioeconomic variables. Land use land cover (LULC), LST, and scaled normalised difference vegetation index are all included in the dataset (SNDVI). According to the findings, there was no significant relationship between the LULC variable and the transportation buffer zones. Forest and impervious surfaces were found to have a substantial relationship with temperature and SNDV. Furthermore, certain zones, such as hospitals and universities, have a negative relationship with water.

5.4 OVERVIEW OF DL METHODS FOR UHI ANALYSIS

Deep learning uses several layers of neural networks for processing data and is a part of ML methods. DL algorithms work similarly to the functions of the human brain. DL algorithms rely on neural networks to function. A neural network is made up of artificial neurons, also known as nodes, and is organised like the human brain. The input, hidden, and output layers are the three layers of these nodes placed on top of each other. The input layer receives features gathered from observations and is processed with random weights. At the end of the process, to decide which neuron to fire, activation functions are used. The number of hidden layers varies depending on the problem's complexity. As a result, the input values are processed via all hidden layers, and the output is generated. Long short-term memory networks (LSTMs), convolutional neural networks (CNNs), and recurrent neural networks (RNNs) are some of the most common deep learning techniques.

5.4.1 SUPERVISED DL METHODS

5.4.1.1 Artificial Neural Network (ANN)

The basic component of an ANN, as well as its operating principle, is represented in Figure 5.2. Artificial neurons are at the heart of ANNs. The neuron receives information from other neurons, processes with weights assigned, summarises, and outputs the net result to other neurons. Before transferring the output to the next variable, certain artificial neurons may apply an activation function to it. In the activation functions, a set of transfer functions are applied to obtain the expected output. The binary, linear, non-linear, and sigmoidal activation functions are generally used to obtain the output.

For the prediction of UHII, Gobakis et al. (2011) proposed a model that used ANN. The neural network supplied data, time, temperature, and solar radiation as input parameters. Three NNs were used: cascade NN, Elman NN, and feed-forward NN. Each NN is made up of one to three hidden layers, each containing 20–40

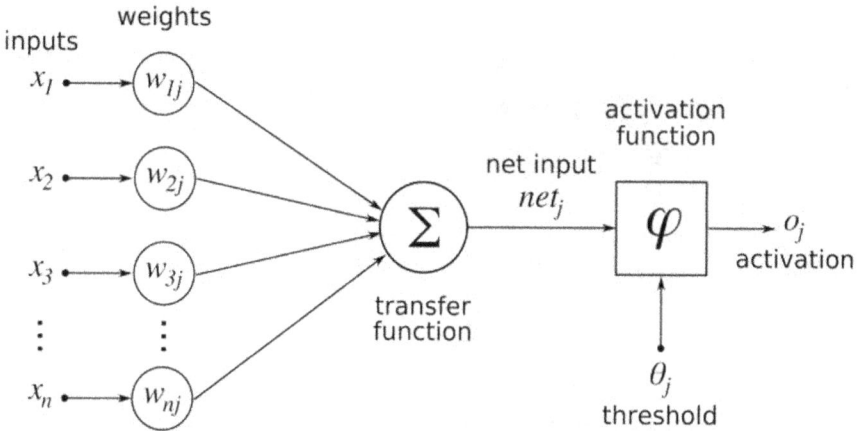

FIGURE 5.2   Operating principle of ANN.

Source: Wikipedia.

neurons. Gradient descent, Levenberg-Marquardt, BFGS quasi-Newton, and sigmoid function were utilised as the training and transfer functions, respectively. The findings revealed that the chosen NN architecture and methods are highly accurate over a 24-hour prediction horizon. The ANN model was trained with the following input parameters: wind velocity, maximum temperature, relative humidity, minimum temperature, and solar radiations as output in the research work of Kaur et al. (2021). It employs three hidden layers, each of which has a rectifier function that serves as an activation function. The sigmoid function with the Adam optimiser is employed in the final output for optimisation. The goal of this research was to find the best weather station that accurately connects solar radiation with input parameters.

It also states that direct sun radiation has a beneficial impact on UHI, whereas wind velocity has a negative impact. In Sherafati et al. (2013), the input and output layers of the ANN model comprised six and one neurons, respectively. It accepts as inputs neighbourhood development level, length to the proximate road, slope, elevation model, aspect, and LST in 1984, and as output, LST in 2007. Temperature, green cover, open space, water cover, built-up space, and paved surface are all input elements in the Simultaneous Extent Analysis (SEA) model (Gopinath 7 Thippesh, 2018), which predicts UHII using an ANN model with UHII as an output parameter.

5.4.1.2   Convolutional Neural Network (CNN)

Data with a grid-like architecture, such as an image, are processable using a convolutional neural network (CNN). It includes convolutional, pooling, and fully connected layers. The architecture of CNNs used in animal classification is shown in Figure 5.3. The input image is convoluted by applying filters in CNNs, which results in a feature map. The rectified linear unit (ReLU) is the most often applied for CNN, which allows accounting for non-linear interactions. It is used after the convolution. The pooling layer conducts downsampling by lowering the height and width of feature

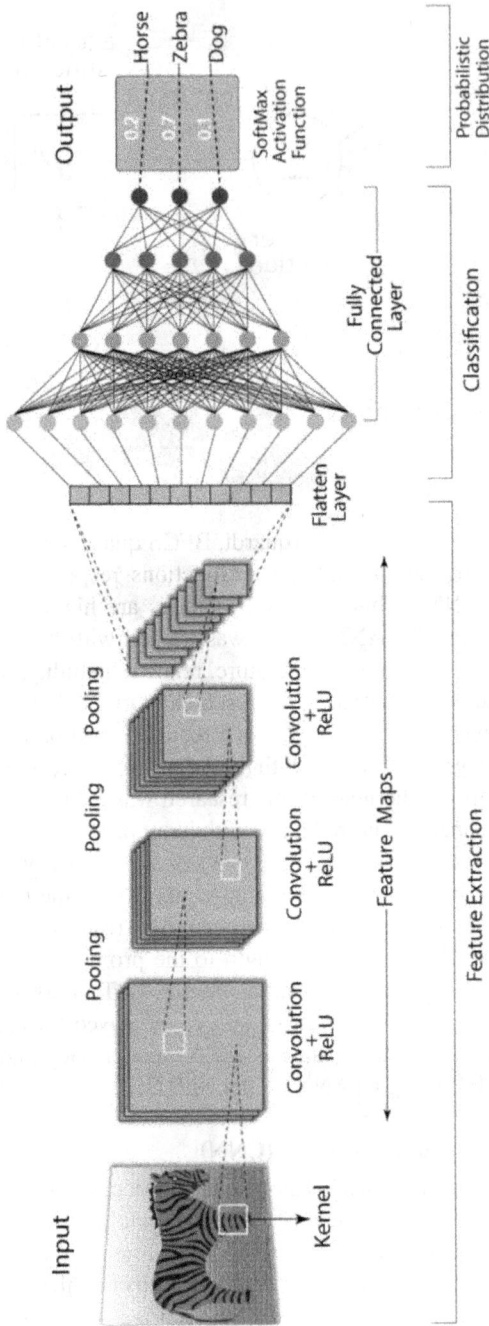

FIGURE 5.3 The architecture of CNN.

Source: developersbreach.com.

maps while retaining the map's classification information. The output feature map (matrix) will be turned into a vector after the final convolution layer, ReLU, and pooling layer (one-dimensional array). This is referred to as a flattened layer.

This fully connected layer receives the output from the flattened layer. The images are classified based on the feature vector obtained from the fully connected layer. The softmax layer used in the last level acts as a classifier. For binary classification problems, the logistic function is typically utilised, while soft-max is commonly used for multi-classification problems.

Yoo et al. (2020) classified local weather regions using three CNN-based systems. Each scheme's input variables include 10 m resolution RS pictures with 10 bands, as well as the building's size and height. In scheme 2, CNN encounters a mix of satellite photos and auxiliary building data, making it challenging to identify LCZ classes. As a result, they devised scheme fusion, which combines the outputs of schemes 1 and 2. By giving the building's structural details, this study found that integrating structural data increased urban-type LCZ classification accuracy. Zhang et al. (2021) used cellular automata (CA) and artificial neural networks (ANNs) to estimate forthcoming variations in LULC and LST in their research. Using the previous years' LST, LULC, NDBI, BDBSI pictures, latitude, and longitude, this model predicts LST. This model is insufficient to accurately forecast the evolution of LULC and LST.

5.4.1.3 Recurrent Neural Network (RNN)

The recurrent neural network (RNN) is a type of ANN that employs time series or sequential data. The preceding phase's outputs are supplied into the current stage as input. The sequential details are maintained in the hidden state. RNN keeps track of all information along with time. As it remembers all past inputs, it is referred to as long short-term memory (LSTM). Bidirectional RNN (BRNN), gated recurrent units, and LSTM are examples of RNN architecture modifications (GRUs).

The LSTM remembers only the prior relevant information, which is experienced by adding several activation function levels identified as 'gates'. Maddu et al. (2021) presented an LSTM and bidirectional LSTM (BiLSTM), a hybrid DL model for predicting surface temperature (ST) that captures multi-dimension characteristics from inputs. It predicts ST using input data such as wind speed and direction, humidity, the dew point temperature, and pressure. It has one BiLSTM layer and three LSTM layers. The main affecting criteria for predicting ST over Indian coastal towns are the dew point temperature, humidity, and pressure, according to this study.

5.5 CONCLUSIONS

The urban heat island is a significant environmental issue that should be taken into account while designing cities. To meet the demands of modern urban forms, intelligent predictive modelling is an important part of current urban planning. The use of more efficient ML and DL approaches allows for improved data processing in both urban and rural settings. It can aid in the modelling, monitoring, and, more importantly, forecasting of the finest indicators to aid the decision-making process in intelligent planning based on a city's desired future.

As a result, we have presented a comprehensive overview of the common factors which permit the construction of different learning models for UHI prediction in this chapter. We began with an overview of UHI, examining the parameters of city form, and then moved on to an outline of urban data resources. Then, we looked at the most common machine learning methods for UHI analysis, focusing on supervised and unsupervised models, as well as their possible applicability to the problem of urban heat island prediction. We also looked at the most common DL approaches for UHI analysis, focusing on supervised models and their use in UHI prediction.

REFERENCES

Ammar Abulibdeh, 2021. Analysis of urban heat island characteristics and mitigation strategies for eight arid and semiarid gulf region cities, Environmental Earth Sciences, 80, 259.

Anselin, L., & Rey, S., 1991. Properties of tests for spatial dependence in linear regression models. Geogr. Anal. 23 (2), 112–131.

Anselin, L., 1990. Spatial econometric: methods and models. J. Am. Stat. Assoc. 85 (411), 160.

Chen, S., Yang, Y., Deng, F., Zhang, Y., Liu, D., Liu, C. & Gao, Z., 2022. A high-resolution monitoring approach of canopy urban heat island using a random forest model and multi-platform observations. Atmospheric Measurement Techniques, 15(3), 735–756.

Choung, Y.-J., & Kim, J.-M., 2019. Study of the relationship between urban expansion and pm10 concentration using multi-temporal spatial datasets and the machine learning technique: Case study for Daegu, South Korea. Appl. Sci. 9 (6), 1098.

Chun, B., & Guldmann, J. M., 2011. Urban Core Determinants Of The Urban Heat Island: Spatial Regression Models, 52nd ACSP Annual Conference, Salt Lake City, Utah, October 13.

Dai, X., Guo, Z., Liu, X., Li, X., & Zhu, Y. 2010. An approach for mining the causation of heat island effect based on decision tree. In 2010 Seventh International Conference on Fuzzy Systems and Knowledge Discovery (Vol. 6, pp. 2746–2750). IEEE.

Developersbreach. n.d. www.developersbreach.com

Duncan, J.M.A., Boruffa, B., Saunders, A., Sun, Q., Hurley, J., & Amati, M. 2019. Turning down the heat: An enhanced understanding of the relationship between urban vegetation and surface temperature at the city scale, Science of the Total Environment 656,118–128.

Gao, S, Zhan, Q., Yang, C., & Liu, H., 2020. The diversified impacts of urban morphology on land surface temperature among urban functional zones. Int. J. Environ. Res. Public Health 17 (24), 9578.

Garzón, J., Molina, I., Velasco, J., & Calabia, A 2021. A Remote Sensing Approach for Surface Urban Heat Island Modeling in a Tropical Colombian City Using Regression Analysis and Machine Learning Algorithms. Remote Sens. 13, 4256. https://doi.org/ 10.3390/ rs13214256.

Geiß, C., Schrade, H., Pelizari, P.A., & Taubenböck, H., 2020. Multistrategy ensemble regression for mapping of built-up density and height with sentinel-2 data. ISPRS J. Photogrammetry Remote Sensing 170, 57–71.

Geng, X., Li, Y., Wang, L., Zhang, L., Yang, Q., Ye, J., & Liu, Y., 2019. Spatiotemporal multi-graph convolution network for ride-hailing demand forecasting. Proceedings of the AAAI conference on artificial intelligence (Vol. 33, pp. 3656–3663). AAAI Press.

Gobakis, K., Kolokotsa, D., Synnefa, A., Saliari, M., Giannopoulou, K., & Santamouris, M. 2011.Development of a model for urban heat island prediction using neural network techniques. Sustainable Cities and Society, 1(2), 104–115.

Gómez, J.A., Patiño, J.E., Duque, J.C., & Passos, S., 2020. Spatiotemporal modeling of urban growth using machine learning. Remote Sensing 12(1), 109.

Gopinath, R., & Thippesh, V. 2018. Artificial neural network trained 'simultaneous extent analysis' as a logical tool in computation of urban heat island intensity. Science & Technology Asia, 23(4), 18–22.

Javatpoint. n.d. www.javatpoint.com

Jochem, W.C., Bird, T.J., & Tatem, A.J., 2018. Identifying residential neighbourhood types from settlement points in a machine learning approach. Computers, Environ. Urban Syst. 69, 104–113.

Kaur, G., Sharma, G., Vijarania, M., & Gupta, S, 2021. Urban Heat Island Prediction Using ANN, International Journal of Innovative Science and Research Technology, 6(6), 252–258.

Li, K., Chen, Y., & Gao, S. 2021. Comparative analysis of variations and patterns between surface urban heat island intensity and frequency across 305 Chinese cities. Remote Sensing, 13(17), 3505.

Li, W., Ni, L., Li, Z.-L., Duan, S.-B., & Wu, H. 2019. "Evaluation of machine learning algorithms in spatial downscaling of MODIS Land Surface Temperature," IEEE J. Sel. Topics Appl. Earth Observ. Remote Sens. (Vol. 12, No. 7, pp. 2299–2307). IEEE.

Li, X., Cheng, S., Lv, Z., Song, H., Jia, T., & Lu, N., 2020. Data analytics of urban fabric metrics for smart cities. Future Generation Computer Syst. 107, 871–882

Lu, J., Li, C., Yu, C., Jin, M., & Dong, S. 2012. Regression analysis of the relationship between urban heat island effect and urban canopy characteristics in a mountainous city, Chongqing. Indoor and Built Environment, 21(6), 821–836.

Luo, Y., Jiang, Y., Khan, S., Peng, S., Feng, Y., & Han, B. 2010. Analysis of urban heat island effect using k-means clustering. In The 2nd International Conference on Information Science and Engineering (pp. 3543–3546). IEEE.

Ma, J., Cheng, J.C., Jiang, F., Chen, W., & Zhang, J., 2020. Analyzing driving factors of land values in urban scale based on big data and non-linear machine learning techniques. Land Use Policy 94, 104537.

Maddu, R., Vanga, A.R., Sajja, J.K., Basha, G. & Shaik, R., 2021. Prediction of land surface temperature of major coastal cities of India using bidirectional LSTM neural networks. Journal of Water and Climate Change, 12(8), pp.3801–3819.

Qi Zhao et al. 2021. Global, regional, and national burden of mortality associated with non-optimal ambient temperatures from 2000 to 2019: a three-stage modelling study, The Lancet Planetary Health Journal, 5(7), 415–425.

Raj, S., Paul, S., Chakraborty, A., & Kuttippurath, J. 2019. Anthropogenic forcing exacerbating the urban heat islands in India, Journal of Environmental Management, https://doi.org/10.1016/j.jenvman.2019.110006

Rajasekar, U., & Weng, Q. 2009. Application of association rule mining for exploring the relationship between urban land surface temperature and biophysical/social parameters. Photogrammetric Engineering & Remote Sensing, 75(4), 385–396.

Salvati, A. Cecere, C.2017. Assessing the urban heat island and its energy impact on residential buildings in Mediterranean climate: Barcelona case study. Energy Build., 146, 38–54.

Sherafati, S.H., M. R. Saradjian, S. Niazmardi, 2013. Urban Heat Island Growth Modeling Using Artificial Neural Networks and Support Vector Regression: A case study of Tehran, Iran, International Archives of the Photogrammetry, Remote Sensing and Spatial Information Sciences, Volume XL-1/W3, pp. 399–403.

Smola, A. J., & Schölkopf, B. 2004. A tutorial on support vector regression. Statistics and computing, 14(3), 199–222.

Tekouabou, S. C. K. 2021. Intelligent management of bike sharing in smart cities using machine learning and Internet of Things. Sustainable Cities and Society, 67, 102702.

Thomas Gardes, Robert Schoetter, Julia Hidalgo, Nathalie Long, & Eva Marquès, Valéry Masson. 2020. Statistical prediction of the nocturnal urban heat island intensity based on urban morphology and geographical factors–An investigation based on numerical model results for a large ensemble of French cities, Science of the Total Environment 737, 139253.

United Nations, Department of Economic and Social Affairs, Population Division. 2014. World Urbanization Prospects: The 2014 Revision, Highlights.

Wikipedia. n.d. www.wikipedia.com

World Data Atlas. 2021. India–Urban population as a share of total population.

Yanwei Sun, Chao Gao, Jialin Li, Run Wang, & Jian Liu, 2019. Quantifying the Effects of Urban Form on Land Surface Temperature in Subtropical High-Density Urban Areas Using Machine Learning, Remote Sensing, 11, 959; doi:10.3390/rs11080959

Yao, Y., Chang, C., Ndayisaba, F., & Wang, S. 2020. A New Approach for Surface Urban Heat Island Monitoring Based on Machine Learning Algorithm and Spatiotemporal Fusion Model, in *IEEE Access*, 8, 164268–164281; doi: 10.1109/ACCESS.2020.3022047

Yin, C., Yuan, M., Lu, Y., Huang, Y., & Liu, Y. 2018.Effects of urban form on the urban heat island effect based on spatial regression model. Science of the Total Environment, 634, 696–704.

Yoo, C., Lee, Y., Cho, D., Im, J. & Han, D., 2020. Improving local climate zone classification using incomplete building data and Sentinel 2 images based on convolutional neural networks. Remote Sensing, 12(21), p.3552.

Zhang Yao, Liu Jiafu, & Wen Zhuyun, 2021. Predicting Surface Urban Heat Island in Meihekou City, China: A Combination Method of Monte Carlo and Random Forest, Chin. Geogra. Sci. 2021, 31(4), 659–670.

Zhang, M., Zhang, C., Kafy, A.A. & Tan, S., 2021. Simulating the Relationship between Land Use/Cover Change and Urban Thermal Environment Using Machine Learning Algorithms in Wuhan City, China. Land, 11(1), 14.

6 Anatomisation of Land Use/Land Cover (LULC) Dynamics with a Focus on Land Surface Temperature in Lucknow City Using Geospatial Techniques

Akanksha, Pranjal Pandey, and Ambrina Sardar Khan

CONTENTS

6.1   INTRODUCTION

Urbanisation is an adjustable, global cycle that transforms a rural territory into a metropolitan zone. The UN research into the worldwide population forecasts that by 2050 the majority the population of India will reside in urban areas (UN 2014). Urban development can be defined as the structural improvement of paved surfaces brought about by the migration or convergence of rural areas into urban areas or settlements, as well as the addition of financial pressures (Mumford 1961). Population growth involves the intentional construction of new housing, commercial, administrative, and transportation infrastructure. These commonly permit vegetation habitats and open fields to be changed, creating an increment in land pattern transformation and urban growth. When large systematic and non-systematic modifications are made to the land, a city tends to progress even if the city itself remains constant. Excessive and unexpected urban sprawl beyond its most significant level causes a reduction in the quality of the environment. A well-planned and composed urban development is required for the progress of a statistically, financially, and environment-friendly society (Cabral et al. 2013, Somvanshi et al. 2018). Fast development in peri-urban areas prompts an expansion of the metropolitan region and a decrease in horticultural land availability (Census 2011). This, thus, prompts huge changes in the trends of rural land development, the advancement of farming, and employment opportunities. The greater part of Indian cities is currently confronting this transition situation, which shifts the economic balance between rural and urban areas in individual provinces (Arif and Gupta 2018, Bharath et al. 2019, Munda 2006). Rapid urbanisation in India may also be linked to communities' migration from rural areas to larger settlement areas for upgradation and better socioeconomic opportunities, etc. The government of India has begun the hi-tech city mission for housing for everyone by 2022, the 'Atal mission' for urban transformation and rejuvenation, the 'Jawaharlal Nehru national urban renewal mission (JNNURAM)', and the 'national heritage city development and augmentation yojana (HRIDAY)' (Rojas 2013). This highlighted the need of recognising the recent patterns of increasing Indian urban settlements and visualising these trends in altering the pattern of land use as it relates to sustainable urban planning. An important factor contributing to the rapid growth of metropolitan communities' industries and businesses. Population expansion is resulting in a gradual increase in the demand for natural resources, which looks to be exerting increasing strain on natural land (FAO 1997). Augmentations of the land use pattern should be regarded as the most essential information on land development and effective management (Census 1991). Such variations are the result of the fast expansion of vast territories (Rembold et al. 2000). This facilitates strategy and decision-makers to project the trends of anthropogenic climatic changes (Taubenböck et al. 2014). These landscape changes and impacts reflect the risks created by current conditions, such as climatic issues, for example, environmental contamination and land surface temperature changes (Chen et al. 2002, Deng et al. 2009). This impacts the way of life of rural and urban people.

Scene diversity stands in contrast to the growth of cities and the resources they bring with them. This makes it important to study, analyse, and evaluate the growth of cities for specific exploration areas (Taubenböck et al. 2014, Census 1991). Changing climate was initially a determined need for scholarly information toward the start of the 19th century, yet it has become a widely debated topic all over the world in the last couple of decades (Eastman 2009). Today, it is viewed as the most critical factor for all living beings (UN 2018). Like many other nations, India is a fast growing and developing nation and it is vulnerable to climate change because of its enormously growing population and its reliance on climatic-sensitive features, e.g. agriculture, farming, and natural resources (Lawrence & Fatima 2013). This has created the need to measure Indian cities' spatio-temporal land use pattern to visualise the change in use. Remote sensing and GIS techniques have been used in several studies across the world to understand for the challenges related to changes in land use, its consequences on the environments of urban cities, and its solution (Census 2001, Liu et al. 2016). Indian equivalents are stored by Tian et al. (2014) and Mumford & Copeland (1961). Different studies around the world have used this technique to discover the solution to challenges occurring in land use planning and its impact on land surface temperature (LST) (Liu et al. 2016, Singh et al. 2017). This study was carried out using geospatial tools, spatial indices, and statistical analysis. For this 'normalised difference vegetation index (NDVI)', 'normalised difference built-up index (NDBI)', and 'modified normalised difference water index (MNDWI)' were generated to highlight the spatial variation of multiple types of LULC patterns and their correlation with LST. This study, therefore, aims to illustrate the LULC transition trends and their impact on LST with historical satellite images.

6.2   AREA USED FOR THE STUDY

Lucknow is the capital city of Uttar Pradesh, which is the fifth largest state in India and the second largest metropolitan city in northern and central India (Grimm et al. 2008). In northern India, Lucknow lies at 26.84°N latitude and 80.94°E and 123 meters above mean sea level. The district covers an area of 2528 square kilometres but only the major urban area (city area) approximately 404 square kilometres is considered for the study area (shown in Figure 6.1). The region receives approximately 1001 mm of rainfall annually. The most extreme temperature rises above 45°C throughout the summer season (April–June) and goes down below 5°C throughout the winter season (Dec–Jan) (Shukla and Jain 2019). By 2016, Lucknow had turned into an enormous, urbanised city from a relatively small low-population place in 1972. It is now India's eleventh most crowded city (Rojas et al. 2013). The district is recognised by assorted social and socioeconomic attributes and it is one of the most rapidly growing and developing urban communities in central India (Rosenzweig et al. 2018). Over these

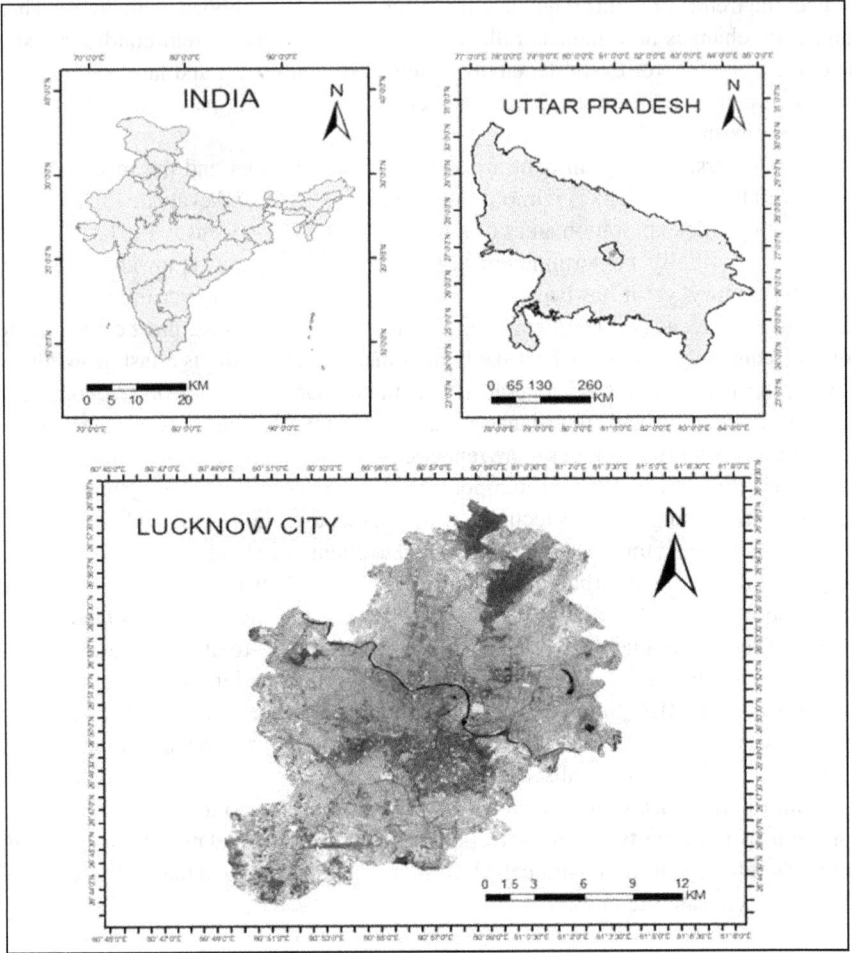

FIGURE 6.1   Study area.

19 years, the city has experienced several foreseen and unforeseen changes due to population growth, creating many social challenges.

6.3 METHODOLOGIES ADOPTED FOR ANATOMISATION OF LAND USE/LAND COVER DYNAMICS (FIGURE 6.2)

6.3.1 USE OF SATELLITE DATA

This study has been carried out using multi-temporal and remotely sensed LANDSAT data which were procured from the USGS earth explorer site (https://earthexplorer.usgs.gov/) in GeoTIFF format. It is particularly helpful in observing LULC change as it provides significant help for large-scale landscape research and this information is continually altered and generally accessible. The data have been taken for the years

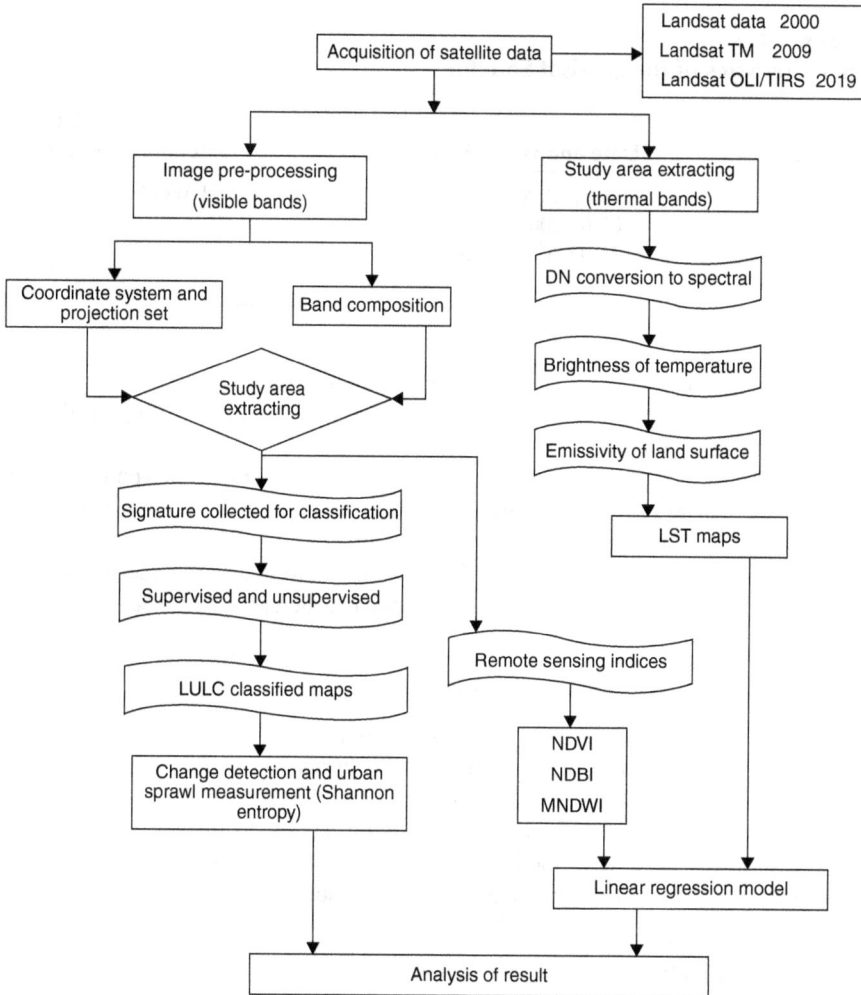

FIGURE 6.2 Methodology flowchart.

2000, 2009, and 2019 (Table 6.1). The maximum and minimum surface temperatures were analysed using a thermal band in a Landsat satellite.

6.3.2 USE OF IMAGE PROCESSING

Utilising ArcGIS 10.5, radiometric and geometric rectification was carried out to reduce atmospheric consequences then composite (FCC) of RGB bands was prepared as upgradation of satellite images for further classification. The images are sub-settled to the city boundary shapefile and both were communicated as ArcMap layers. World Geodetic System (WGS) 1984 was used for the projection of satellite images (Bharath et al. 2019). Supervised and unsupervised classification methods were used for image

TABLE 6.1
Characteristics of the Landsat Satellite

Satellite sensor	Date and year	Source	Projection	Spatial resolution
Landsat-7 (ETM)	29.02.2000	USGS Earth	UTM-WGS	30 METERS
Landsat-5 (TM)	03.03.2009	Explorer	1984	
Landsat-8 (OLI/TIRS)	25.02.2019	https://earthexplorer. usgs.gov/		

classification. This method is utilised as an essential spatial tool to acquire genuine and authentic data from images (Proptiger 2018). Four classes were recognised in our city: waterbodies, built up, vegetation, and barren land.

Without accuracy assessment, the classification of images cannot be considered final (Kalnay & Cai 2003). To ensure the accuracy of images, the Kappa khat method was used to create an error matrix using Erdas Imagine 15 software. More than 100 reference points were generated over land use maps to assess the user's accuracy, producer's accuracy, overall accuracy, and kappa coefficient.

6.3.3 USE OF LULC PATTERN CHANGE

The LULC change from class to class was detected using the reclassification method on Arcgis10.5. Changing trends of land use patterns were acquired for 19 years from 2000–2009 and 2009–2019, which were placed in a statistically classification structure to distinguish the changes for different classes. Comparative analyses of classified images were performed, and pixel-based changing data were observed to understand the changes over an undefined time frame.

6.3.4 USE OF URBAN SPRAWL MEASUREMENT

Urban sprawl estimation is required to manage an urbanisation development study. It is estimated by the spatial movements in developed zones per inhabitant or restrained development in the population (Sankhala & Singh 2014). Shannon entropy is a wonderful tool for finding the most suitable expansion of a developed urban area. The estimation of entropy is needed for region estimation with the assistance of the RS and GIS strategy (Chen et al. 2002). We implemented this framework for evaluating the urban sprawl for the duration of 2000–2019. With this method, the depth of differences and the creation of fresh land use categories become more transparent.

$$Hn = -\sum PiLogn(Pi) \qquad (6.1)$$

where, Pi = built up area proportion in buffer-zone
n = number of buffer zone
$Log\ n$ = entropy zone maximum threshold

The Shannon entropy values are in the range of 0 to *Log n*, where 0 denotes a densely populated area, and *Log n* signifies the built-up area.

6.3.5 USING LST COMPUTATIONS

The climatic impact is a significant concern when we think about urbanisation. Quickly developed urban zones and diminishing vegetation cover result in a warmer climate. Examining the LST is an exceptionally basic approach to obtaining data about the effect of urban sprawl over a period of time. The availability of thermal bands from satellites (B6 in TM/ETM and B10 and 11 in OLI/TIRS), which are extremely helpful to enable the use of proper information about LST. The equations below were followed for the estimation of LST.

The DN value of band conversion to spectral radiance (NASA 2011) is:

$$L\lambda = \left[\left(L\max - L\min\right)/\left(Qcal\max - Qcal\min\right)\right]^{*}$$
$$\left(Qcal - Qcal\min\right) + L\min\lambda$$

where

L_{λ} = spectral radiance at the sensor's aperture in W/(m^2 sr um)
L_{max} = spectral radiance that is scaled to Q_{calmax} in W/(m^2 sr um)
L_{min} = spectral radiance that is scaled to Q_{calmin} in W/(m^2 sr um)
Q_{cal} = quantised calibrated pixel value in digital number
$Q_{cal\,max}$ = maximum quantised calibrated pixel value in digital number
$Q_{cal\,min}$ = minimum quantised calibrated pixel value in digital number

It has been established that radiation correction improves the accuracy of LST (Song 2000).

6.3.6 USE OF BRIGHTNESS TEMPERATURE

The spectral radiance converted from the pixel DN value is also used to calculate brightness temperature under the supposition of unit emissivity and using pre-launch estimation constant:

$$TBK = K_{2}/\ln(1 + (K1/L\lambda)$$

where,
TBK = effective satellite temperature in kelvin
K_1 and K_2 = calibration constants
For TM, K_1 = 607.76 and K_2 = 1260.56
For OLI/TIRS, K_1 = 774.88 and K_2 = 1321.07

6.3.7 Use of Emissivity of the Land Surface (LSE)

Land surface emissivity shows the average emissivity of an element of the land surface which is calculated from the formalised variance foliage index (NDVI):

$$\varepsilon = 0.004 * Pv + 0.986$$

P_v is the foliage percentage taken according to the set equation

$$Pv = \frac{NDVI - NDV \operatorname{Im} in}{NDV \operatorname{Im} ax - NDV \operatorname{Im} in}$$

6.3.8 Use of Land Surface Temperature

$$LST = (BT/1) + W * (BT/14380) * \ln(LSE)$$

where BT = brightness
W = wavelength
LSE = land surface emissivity

6.3.9 Use of Spatial Indices Calculation (Spectral Image Enhancement)

There are various spatial indices that have been created with certain combinations of bands that are useful to observe the temporal variation of objects. These indices have the advantage of reducing the impact of external factors, i.e., solar irradiance and atmospheric effect. One vegetation-based index 'normalized difference vegetation index (NDVI)', and also 'built up based indices', 'normalised difference built-up index (NDBI)', and water-based indices including the 'modified normalised difference water index (MNDWI)' have been generated using ArcGIS 10.5 software in this study (Table 6.2). These enhanced images were correlated with the land surface temperature.

6.4 STATISTICAL ANALYSES AND CORRELATION BETWEEN LST AND INDICES

Shannon entropy was evaluated by creating buffers in the study area using equation (5.1). The Shannon coefficient for all years was measured out. More than 70 location points were generated in the study area to get the variation of LST. The mean LST for each point was extracted for all years using the zonal statistics tool in ArcGIS 10.5 software. Further, the LST profile was graphically projected for the mean LST and location points using MS Excel. To establish a correlation between LST and spatial indices, the regression model was used with the help of measured mean values of LST and indices from location points over the study area. The correlation was assessed from the slope of regression by incorporating the linear trend lines using MS Excel.

TABLE 6.2
Spectral Indices

S. no.	Indices	Formula
1.	NDVI	$(NIR_{Band} - R_{Band})/(NIR_{Band} + R_{Band})$
2.	NDBI	$(NIR_{Band} - SWIR_{Band})/(NIR_{Band} + SWIR_{Band})$
3.	MNDWI	$GREEN_{Band} - NIR_{Band})/(GREEN_{Band} - NIR_{Band})$

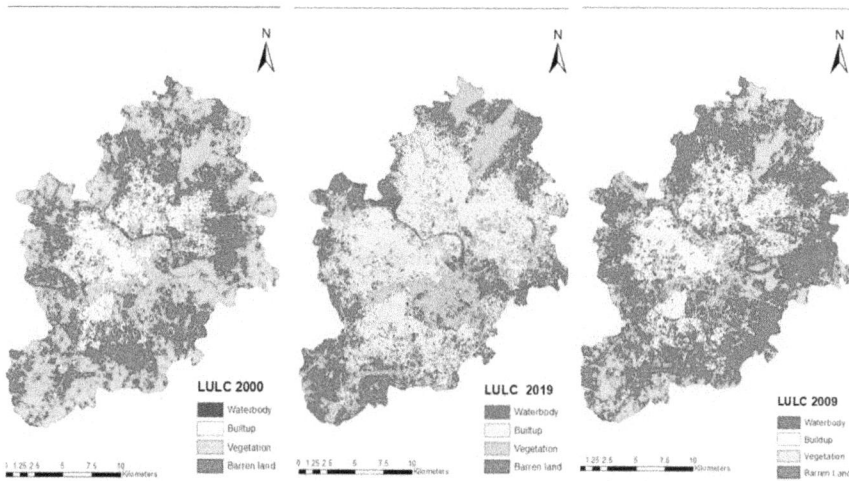

FIGURE 6.3 LULC maps for 2000 (a), 2009 (b), and 2019 (c).

6.5 LULC ANALYSES AND OUTCOME

For classification, two timeframe periods were considered, which displayed the decline of vegetation cover during the development of the urban zone. The developed region increased from 75.4 to 171.9 sq. km, which is tremendous growth, with a decline of the vegetative zone from 188.7 to 124.2 sq. km in 19 years. The first period of study, from 2000–2009 (Figure 6.3b), demonstrated that vegetation cover was changed to barren and urban areas. Expansion of barren areas was from 138 to 190.9 sq. km. In the decade from 2009–2019 (Figure 6.3c) growth of the urban area was shown with a decline of some of the barren areas. The urban area showed exceptional growth from 109.5 to 171.9 sq. km and the barren area declined from 190.3 to 103.3 sq. km. These LULC analyses for 19 years are shown in Table 6.3. The accuracy of the classified image was determined at 89.2% and the kappa coefficient is 0.84 as per the confusion matrix.

Land change implies changing trends of land cover with one class and then onto the next class. It gives harsh implications regarding the changing patterns of land cover and utilisation over a period of time so that we can without much of a stretch comprehend class-to-class change. Typically, the man-made zone is developed over

TABLE 6.3
LULC Statistics for Lucknow City

		2000		2009		2019	
S. no.	LULC classes	Area (km²)	Area (%)	Area (km²)	Area (%)	Area (km²)	Area (%)
1.	Water	2.12	0.52	2.35	0.58	4.68	1.19
2.	Built up	75.4	18.65	109.5	27.08	171.9	42.5
3.	Vegetation	188.7	48.68	101.4	25.10	124.2	30.7
4.	Barren land	138.0	34.13	190.0	47.23	103.3	25.5
5.	Overall	404.32	100	404.32	100	404.32	100

TABLE 6.4
LULC Change Detection

S. no.	LULC change	From the year 2000 to 2009 – Area (km²)	From the year 2009 to 2019 – Area (km²)
1.	Vegetation–water	0.59	1.49
2.	Barren–water	0.05	1.11
3.	Vegetation–built up	10.47	28.72
4.	Barren–vegetation	10.15	33.20
5.	Vegetation–barren	26.89	45.89
6.	Barren to built up	31.41	66.30
	Overall	80.08	176.71

the decrement of vegetation cover, which is a simple method to manufacture urban areas. Vegetation cover has been converted to a barren area from 26.89 sq. km to 72.78 sq. km, and barren areas have been converted from 31.41 sq. km to an urban area 97.72 sq. km over the 19 years (2000–2019). The change effect is shown in Table 6.4.

6.5.1 SHANNON ENTROPY

Shannon entropy is an approach to measuring randomness, i.e., it illustrates whether urban growth is dispersed or dense and concentrated. The values were derived for all years and are mentioned below in Table 6.5. Log (18) measured as 1.25 is the highest value (Table 6.5). The entropy results obtained for 3 years are 0.97, 1.02, and 1.05, respectively, which revealed that there was an exponential urban expansion in the area since 2000. The entropy values show the overall frequency of massive transformation of an area that will have a devastating socioeconomic, ecological, and environmental impact (Bhatta, B 2009). Figure 6.4 shows the different entropy zones after applying Shannon entropy over the study area.

TABLE 6.5
Shannon Entropy

Years	2000	2009	2019
Area of built-up (km²)	75.45	109.50	171.90
Entropy values	0.97	1.02	1.05

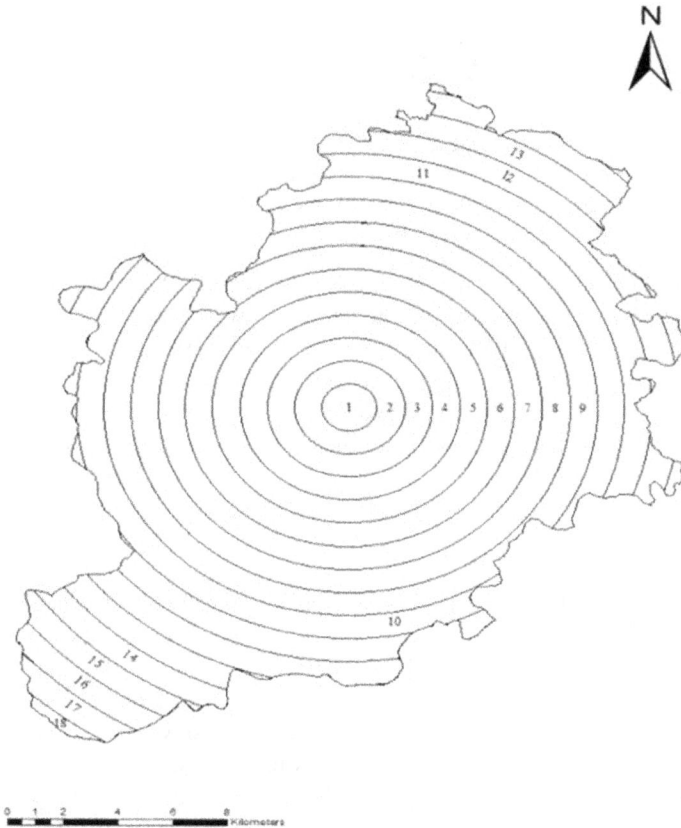

FIGURE 6.4   Entropy map.

6.5.2   LST

The innovation of remote sensing and GIS illustrates how the heat island impact is spatially dispersed. Heat-trapping in building material is somewhat discharged as warmth prompts an expansion in the surface temperature (Kumari et al. 2017). LST could be understood as the coming heat that falls on the surface of the earth which changes according to LULC types (Kayet et al. 2016). This surface heat

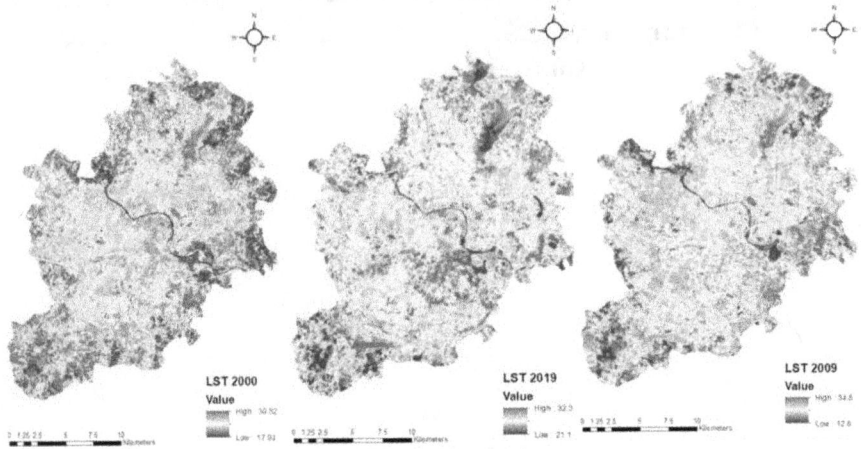

FIGURE 6.5   LST maps for 2000 (a), 2009 (b), and 2019 (c).

pattern was determined for the study area in the years 2000, 2009, and 2019 individually. During this time the base temperature was discovered to be 17.92°C
in 2000 and 21.1°C° in 2019, which demonstrates an increase of approximately
3.26°C. These progressions happen because of the critical development of the
urban territory and different classes. Vegetation cover and waterbodies reflect the
lower temperature which declined over the 19 years. Figure 6.5 shows the spatial
dissemination of LST.

6.5.3   VARIATION OF LST WITH LULC

Whatever action is taken to modify the land pattern, greater effects can appear in the
environment. The positive link between LST and various land use types was observed
from the complete analysis. The LST shift can be related to LULC transitions over
the specified timeframe (Bharath et al. 2019). The barren to urban zone reflects a
medium to higher temperature, while waterbodies and vegetation cover reflects
the lower temperature. There are some places in the city, as graphically projected
(Figure 6.6), which observed increases in temperature, such as Janki Puram,
Nishatganj, Hazratganj, Aishbagh, Husianabad, and Mahanagar, etc. There are some
parts of the low-temperature zone shown in the urban area (Figure 6.5) which are
known as urban green space. This urban green space, including recreational parks
with lots of plantations, is the major contributor to lowering the LST due to high
evapotranspiration, as seen in earlier studies, and this is known as the oasis effect in
cities. The dominant reason for decreased evaporative cooling in cities is the preva-
lence of barren land in the context of the following land types (Fan et al. 2017).
While the built-up region contains high-rise towers, residential houses, transportation
routes, commercial areas, and others, urban development has also been observed with
high LST due to the trapping of heat.

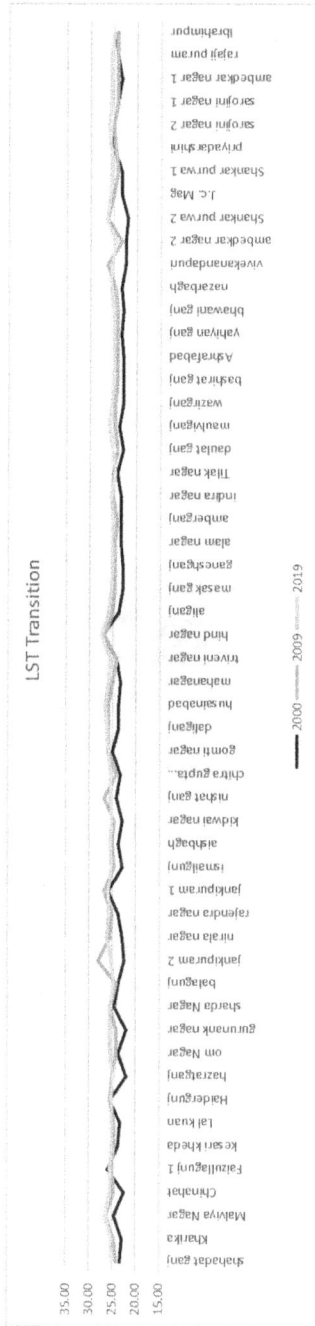

FIGURE 6.6 Mean LST transition for study area.

6.6   ANALYSES OF SPATIAL INDICES NDVI, NDBI, AND MNDWI

Temporal maps of all spectral indices have been prepared in ArcGIS 10.5 software. The regression analyses between mean LST and spectral indices (NDVI, NDBI, and MNDWI) for 2000, 2009, and 2019 were carried out considering the entire study area. The graphically represented figure shows the Y axis as mean LST values and the X axis as the spectral indices value (Figure 6.7).

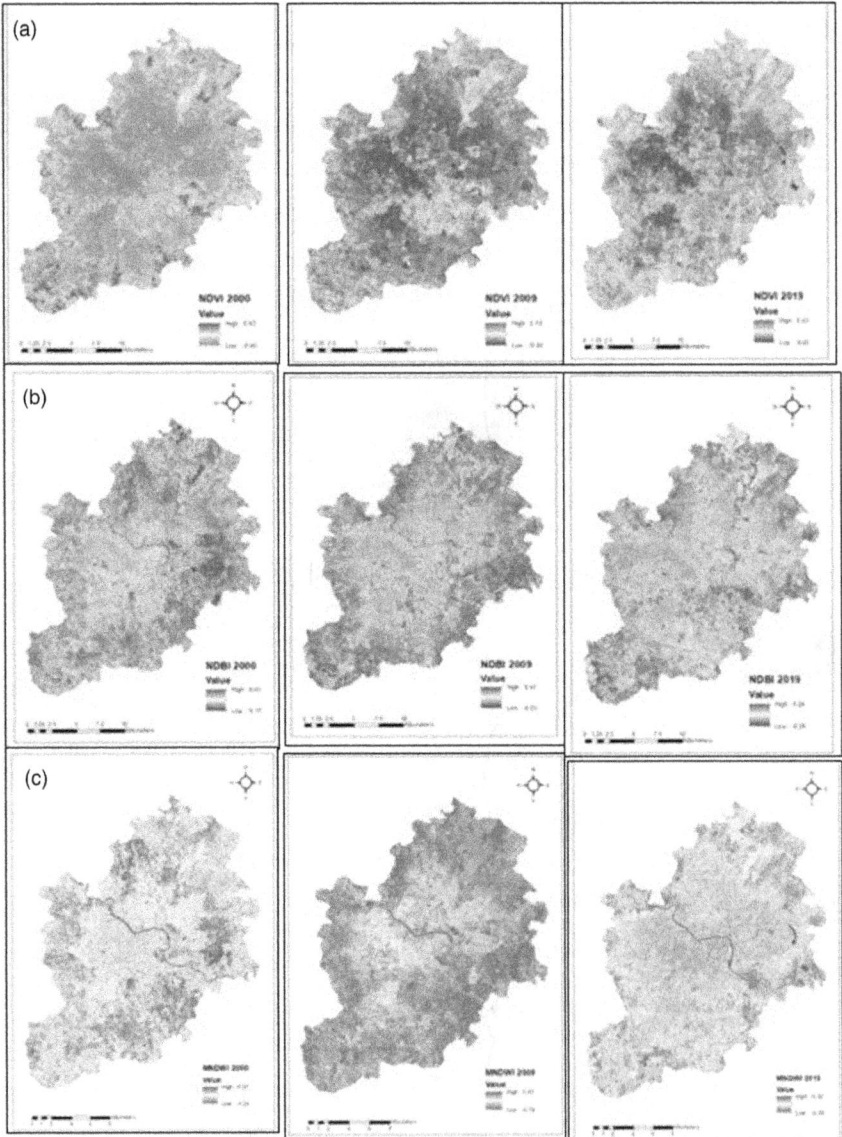

FIGURE 6.7   Spatial distribution of NDVI (a), NDBI (b), and MNDWI (c) for years 2000, 2009, and 2019.

6.6.1   LST CHANGE RELATION WITH NDVI

The spatial pattern of NDVI is generated using multi-temporal satellite images illustrated in Figure 6.7(a). The squeezing of vegetation cover within the city can be seen clearly. Lower NDVI values were also shown in water bodies. Higher NDVI values show dark vegetation in the study area. In 2000, the NDVI value ranged between −0.45 and 0.42, which gradually increased in 2009 from −0.40 to 0.71. Therefore, NDVI decreased in urban area land. NDVI demonstrates an inverse effect on LST, so both are opposite each other, thus as the NDVI value goes higher, the LST value decreases. The relationship between NDVI and LST is negative as the R^2 value decreases from 0.36 to 0.21 and 0.16 as observed from 2000 to 2019 in Figure 6.8(a).

6.6.2   LST CHANGE RELATION WITH NDBI

High-temperature variations are strongly correlated with built-up or barren lands, and heavily urbanised and industrialised areas (Xiong et al. 2012). Lower vegetation cover is the main reason for an increase in the urban heat effect (Weng and Yang 2004). The spatial variation of NDBI is performed from the years 2000 to 2019 to establish the relation with LST. NDBI highlights the built-up and barren land in the study area. Figure 6.7(b) shows that the NDBI value increased from 0.45 to 0.47 from 2000 to 2009 due to an increase in barren land and it drops to 0.24 in 2019 due to the increase in built-up zones. In Figure 6.8(b) the regression model observed an R^2 value between NDBI and LST of 0.25 to 0.22 from 2000 to 2009 and highest correlated in 2019 with an R^2 value of 0.42 which shows a positive relationship. This indicated that an impervious surface has a significant impact on increasing the surface temperature.

6.6.3   LST CHANGE RELATION WITH MNDWI

The temperature in the water is generally below that of other types of land use (Hathway and Sharples 2012). The water body proportion has a negative correlation with mean LST due to low temperature. Figure 6.7(c) shows the MNDWI condition in 2000, 2009, and 2019. The maximum MNDWI value was observed as 0.37 in 2000 and it decreased to 0.32 in 2019, implying that a water body contributes to reducing the LST. In all the years' regression analyses it was observed that MNDWI negatively controls LST (Figure 6.8(c)). The model observed the R^2 value as 0.46 in 2000, 0.25 in 2009, and 0.18 in 2019. This shows that the depth of water bodies decreased with time (Pal & Ziaul 2017). The existence of water sources at any point on land tends to reduce its temperature as well as that of its surroundings.

6.7   CONCLUSIONS

Increasing populations consistently require additional space. These requirements become an essential evil by restricting asset accessibility in any urbanising city. This analysis has reflected the present versus the past situation of the area with the valuable utilisation of satellite data to discover the change in LULC pattern in various class study areas. The incorporated methodology has been used to evaluate the urban turn of

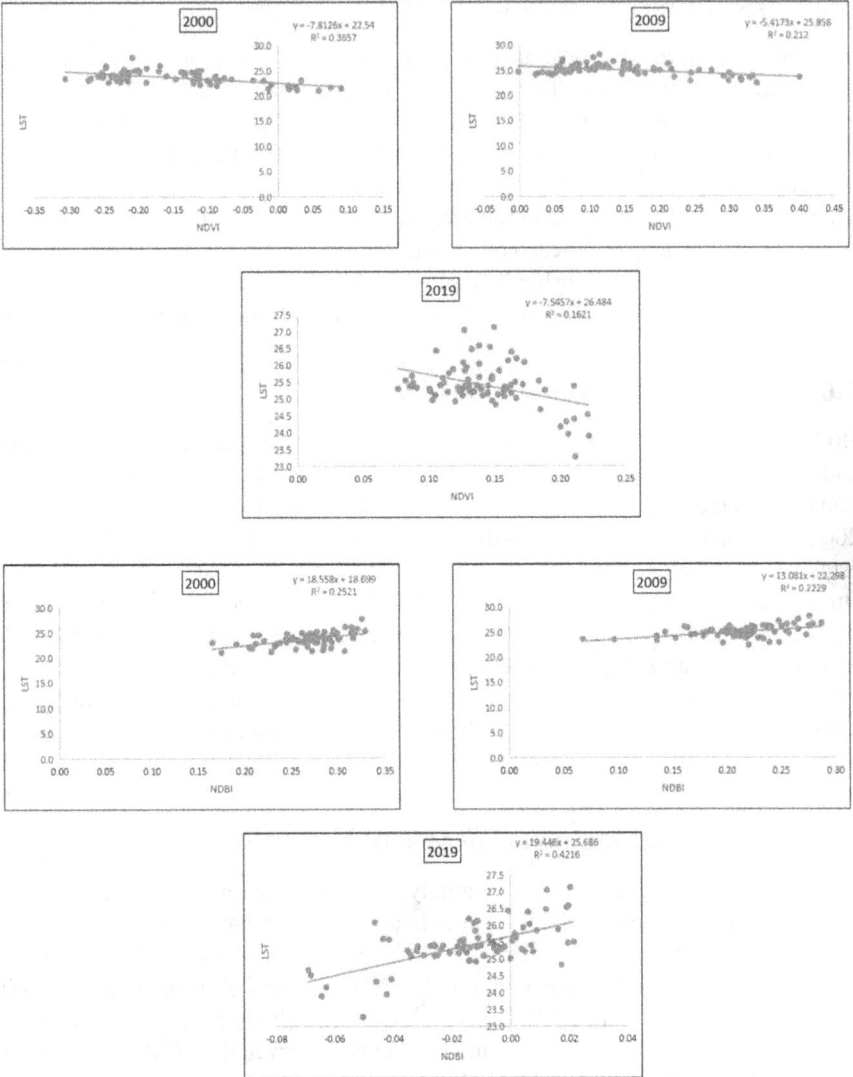

FIGURE 6.8 Regression correlation between LST and NDVI (a), NDBI (b), and MNDWI (c).

events and its effect on neighbourhood conditions. This study reveals that the sudden urban land use pattern shifted from 75.4 km² (2000) to 171.1 km² (2019), an almost 33.7% land pattern change in 19 years. Change in the class-to-class LULC pattern is also found. The combined study of satellite information and Shannon entropy is used to measure the spatial land pattern changes at the local and provincial levels in the city. The components of Lucknow city have changed because of the unplanned urban sprawl and the consumption of vegetation areas. Shivjaipuram, Jankipuram Gomti Nagar, Cantonment Zone, and Krishna Nagar are known as the city's major hubs and

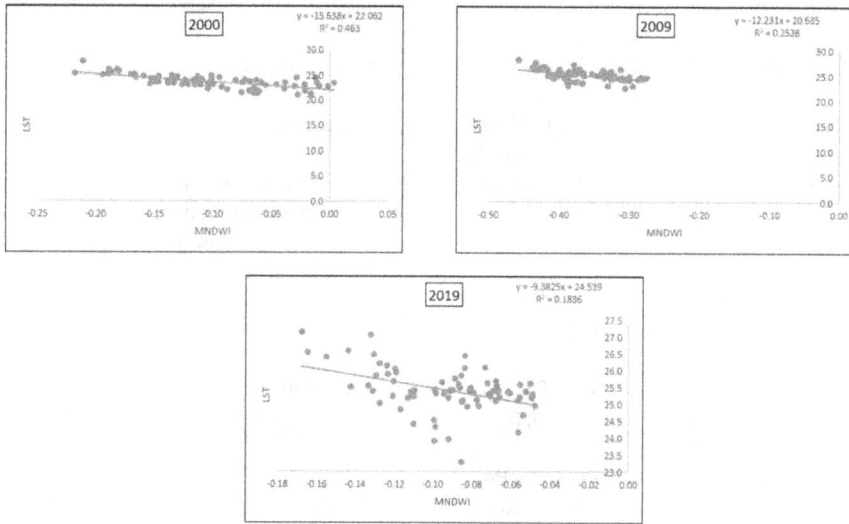

FIGURE 6.8   (Continued)

the nearby areas have been drastically converted. Built-up areas tend to increase over the years because local government bodies make fundamental plans for future development and population flow from rural to urban areas close to the city. The surface temperature of land (LST) variability for specific land uses showed that the built-up region experienced higher temperatures owing to the trapping of heat because of the materials used in the construction of buildings, houses, roads, and other developments relevant to commerce and infrastructure, etc. Some of the barren land also showed high temperatures due to low vegetation coverage. Recreational parks for playing and walking with lots of vegetation in residential open spaces contribute to decreasing the heat in the city, which is a great solution for managing the LST. The established regression correlation among LST and NDVI, NDBI and MNDWI show a significant result in understanding the impact of urbanisation in a better way (Ghosh et al. 2019). The correlation that emerges between NDBI and LST is positive, meanwhile the correlations between NDVI, MNDWI, and LST are negative. Thus, a detailed study of shifting land use trends and their implications induced by urbanisation may provide clarity about proper land use planning to mitigate the higher LST impact. Based on the obtained results, it is observed that this research would be beneficial for proper urban/environmental management plans for sustainable development.

REFERENCES

Aithal, Bharath H., M. C. Chandan, and G. Nimish. "Assessing land surface temperature and land use change through spatio-temporal analysis: a case study of select major cities of India." *Arabian Journal of Geosciences* 12, no. 11 (2019): 1–16.

Arif, Mohammad., and K. Gupta. "Mapping Peri-Urbanization in a Non-Primate City: A Case Study of Burdwan, India." *European Academic Research* 5, no. 11 (2018): 6065–6081.

Bhatta, Basu. "Analysis of urban growth pattern using remote sensing and GIS: a case study of Kolkata, India." *International Journal of Remote Sensing* 30, no. 18 (2009): 4733–4746.

Brown, Sandra. *Estimating biomass and biomass change of tropical forests: a primer.* Vol. 134. Food & Agriculture Org., 1997.

Cabral, Pedro, Gabriela Augusto, Mussie Tewolde, and Yikalo Araya. "Entropy in urban systems." *Entropy* 15, no. 12 (2013): 5223–5236.

Cao, Liqin, Pingxiang Li, Liangpei Zhang, and Tao Chen. "Remote sensing image-based analysis of the relationship between urban heat island and vegetation fraction." *The international Archives of Photogrammetry, remote sensing and spatial information sciences* 37 (2008).

Census of India, 1991 Census of India Provisional population totals Registrar General & Census Commissioner, Ministry of Home Affairs, Government of India, India, New Delhi (1991).

Census of India, 2001 Census of India Provisional population totals Registrar General & Census Commissioner, Ministry of Home Affairs, Government of India, India, New Delhi (2001).

Census of India, 2011 Census of India Provisional population total Registrar General & Census Commissioner, Ministry of Home Affairs, Government of India, India, New Delhi (2011).

Chaudhuri, Gargi, and Niti B. Mishra. "Spatio-temporal dynamics of land cover and land surface temperature in Ganges-Brahmaputra delta: A comparative analysis between India and Bangladesh." *Applied Geography* 68 (2016): 68–83.

Chen, Yonghua, Jennifer A. Francis, and James R. Miller. "Surface temperature of the Arctic: Comparison of TOVS satellite retrievals with surface observations." *Journal of Climate* 15, no. 24 (2002): 3698–3708.

Deng, Jin S., Ke Wang, Yang Hong, and Jia G. Qi. "Spatio-temporal dynamics and evolution of land use change and landscape pattern in response to rapid urbanization." *Landscape and urban planning* 92, no. 3–4 (2009): 187–198.

Eastman, J. R. "IDRISI Taiga, guide to GIS and remote processing. Guide to GIS and remote processing." *Worcester: Clark University* (2009).

Fan, Chao, Soe W. Myint, Shai Kaplan, Ariane Middel, Baojuan Zheng, Atiqur Rahman, Hue-Ping Huang, Anthony Brazel, and Dan G. Blumberg. "Understanding the impact of urbanization on surface urban heat islands – A longitudinal analysis of the oasis effect in subtropical desert cities." *Remote Sensing* 9, no. 7 (2017): 672.

Geerken, R., B. Zaitchik, and J. P. Evans. "Classifying rangeland vegetation type and coverage from NDVI time series using Fourier Filtered Cycle Similarity." *International Journal of Remote Sensing* 26, no. 24 (2005): 5535–5554.

Ghosh, Subrata, Nilanjana Das Chatterjee, and SantanuDinda. "Relation between urban biophysical composition and dynamics of land surface temperature in the Kolkata metropolitan area: a GIS and statistical based analysis for sustainable planning." *Modeling Earth Systems and Environment* 5, no. 1 (2019): 307–329.

Grimm, Nancy B., David Foster, Peter Groffman, J. Morgan Grove, Charles S. Hopkinson, Knute J. Nadelhoffer, Diane E. Pataki, and Debra PC Peters. "The changing landscape: ecosystem responses to urbanization and pollution across climatic and societal gradients." *Frontiers in Ecology and the Environment* 6, no. 5 (2008): 264–272.

Hathway, E. A., and Steve Sharples. "The interaction of rivers and urban form in mitigating the Urban Heat Island effect: A UK case study." *Building and Environment* 58 (2012): 14–22.

Kalnay, Eugenia, and Ming Cai. "Impact of urbanization and land-use change on climate." *Nature* 423, no. 6939 (2003): 528–531.

Kayet, Narayan, Khanindra Pathak, Abhisek Chakrabarty, and Satiprasad Sahoo. "Spatial impact of land use/land cover change on surface temperature distribution in Saranda Forest, Jharkhand." *Modeling Earth Systems and Environment* 2, no. 3 (2016): 1–10.

Kumari, Madhuri, Noyingbeni Kikon, Ambrina Sardar Khan, and Prateek Srivastava. "A GIS based study of urbanization impact on land surface temperature in greater Noida, India." *Int. J. Res. Appl. Sci. Eng. Technol.* 5, no. 9 (2017): 608–615.

Lawrence, Alfred, and Nishat Fatima. "Urban air pollution & its assessment in Lucknow City – the second largest city of North India." *Science of the total environment* 488 (2014): 447–455.

Lillesand, Thomas, Ralph W. Kiefer, and Jonathan Chipman. *Remote sensing and image interpretation*. John Wiley & Sons, 2015.

Liu, Guilin, Qian Zhang, Guangyu Li, and Domenico M. Doronzo. "Response of land cover types to land surface temperature derived from Landsat-5 TM in Nanjing Metropolitan Region, China." *Environmental Earth Sciences* 75, no. 20 (2016): 1–12.

Majra, J. P., and A. Gur. "Climate change and health: Why should India be concerned?" *Indian journal of occupational and environmental medicine* 13, no. 1 (2009): 11.

Mumford, Lewis. *The city in history: Its origins, its transformations, and its prospects*. Vol. 67. Houghton Mifflin Harcourt, 1961.

Munda, Giuseppe. "Social multi-criteria evaluation for urban sustainability policies." *Land use policy* 23, no. 1 (2006): 86–94.

Proptiger (2018) 6 Urban Development Schemes You Should Know About. Retrieved from www.proptiger.com/.

Ramachandra, T. V., Bharath Setturu, K. S. Rajan, and MD Subash Chandran. "Stimulus of developmental projects to landscape dynamics in Uttara Kannada, Central Western Ghats." *The Egyptian Journal of Remote Sensing and Space Science* 19, no. 2 (2016): 175–193.

Rembold, Felix, Stefano Carnicelli, Michele Nori, and Giovanni A. Ferrari. "Use of aerial photographs, Landsat TM imagery and multidisciplinary field survey for land-cover change analysis in the lakes region (Ethiopia)." *International Journal of Applied Earth Observation and Geoinformation* 2, no. 3–4 (2000): 181–189.

Rojas, Carolina, Joan Pino, Corina Basnou, and Mauricio Vivanco. "Assessing land-use and-cover changes in relation to geographic factors and urban planning in the metropolitan area of Concepción (Chile). Implications for biodiversity conservation." *Applied Geography* 39 (2013): 93–103.

Rosenzweig, C., Solecki, W. D., Romero-Lankao, P., Mehrotra, S., Dhakal, S., & Ibrahim, S. A. (Eds.). (2018). *Climate change and cities: Second assessment report of the urban climate change research network*. Cambridge University Press.

Saikawa, Eri, Marcus Trail, Min Zhong, Qianru Wu, Cindy L. Young, Greet Janssens-Maenhout, Zbigniew Klimont et al. "Uncertainties in emissions estimates of greenhouse gases and air pollutants in India and their impacts on regional air quality." *Environmental Research Letters* 12, no. 6 (2017): 065002.

Sankhala, Sunil, and B. Singh. "Evaluation of urban sprawl and land use land cover change using remote sensing and GIS techniques: a case study of Jaipur City, India." *International Journal of Emerging Technology and Advanced Engineering* 4, no. 1 (2014): 66–72.

Singh, Prafull, Noyingbeni Kikon, and Pradipika Verma. "Impact of land use change and urbanization on urban heat island in Lucknow city, Central India. A remote sensing based estimate." *Sustainable cities and society* 32 (2017): 100–114.

Somvanshi, Shivangi S., Oshin Bhalla, Phool Kunwar, Madhulika Singh, and Prafull Singh. "Monitoring spatial LULC changes and its growth prediction based on statistical models and earth observation datasets of Gautam Budh Nagar, Uttar Pradesh, India." *Environment, Development and Sustainability* 22, no. 2 (2020): 1073–1091.

Taubenböck, Hannes, Michael Wiesner, Andreas Felbier, Mattia Marconcini, Thomas Esch, and Stefan Dech. "New dimensions of urban landscapes: The spatio-temporal evolution from a polynuclei area to a mega-region based on remote sensing data." *Applied Geography* 47 (2014): 137–153.

Tian, Hanqin, Kamaljit Banger, Tao Bo, and Vinay K. Dadhwal. "History of land use in India during 1880–2010: Large-scale land transformations reconstructed from satellite data and historical archives." *Global and Planetary Change* 121 (2014): 78–88.

Wang, L., W. P. Sousa, and P. Gong. "Integration of object-based and pixel-based classification for mapping mangroves with IKONOS imagery." *International Journal of Remote Sensing* 25, no. 24 (2004): 5655–5668.

Xiong, Yongzhu, Shaopeng Huang, Feng Chen, Hong Ye, Cuiping Wang, and Changbai Zhu. "The impacts of rapid urbanization on the thermal environment: A remote sensing study of Guangzhou, South China." *Remote sensing* 4, no. 7 (2012): 2033–2056.

Xu, Hanqiu. "Modification of normalised difference water index (NDWI) to enhance open water features in remotely sensed imagery." *International journal of remote sensing* 27, no. 14 (2006): 3025–3033.

Zha, Yong, Jay Gao, and Shaoxiang Ni. "Use of normalized difference built-up index in automatically mapping urban areas from TM imagery." *International journal of remote sensing* 24, no. 3 (2003): 583–594.

7 Analysis of Urban Growth and Examination of a Master Plan through Geospatial Techniques
A Case Study of Ranchi, Jharkhand, India

*Kamal Bisht, Shubham Kumar Sanu,
Vishwa Raj Sharma, and Priya Sharma*

CONTENTS

7.1 INTRODUCTION

The 21st century has witnessed the rapid development of India, which has led to the social and economic transformation of the society but, on the other hand [1], it has come with unexpected population growth and rural-to-urban migration [2]. It leads to extensive urbanisation, almost 3% annually, as well as its growth, having a nega-tive impact on the LULC pattern and its ecosystem [3]. In the prediction of urban LULC change, GIS and remote sensing techniques are highly preferred, especially in multi-temporal datasets [4]. Landsat satellite data are one of the recommended sources for this prediction [5]. Geospatial techniques and remote sensing imagery become a foundation to analyse spatio-temporal LULC patterns as well as change detection [6]. This study was carried out to investigate the master plan of Ranchi, India, using geospatial techniques to demonstrate the ability of GIS and remote sensing to find out whether a master plan can regulate urban growth. The study examines the role of GIS and remote sensing in mapping LULC in Ranchi and its surrounding area. The study database consists of Landsat 8 imagery of 2016 and the map of the proposed master plan by the Ranchi development authority site with the help of Arc GIS software. Geo-referencing of the Ranchi master plan was carried out and created classes in different layers which showed the proposed master plan for 2037. Also the built-up density along the national highways which pass through Ranchi was studied. The census data of India were used to discover the growth of the population in Ranchi city at different times and also the urban density in Ranchi. This study used Landsat 8 OLI_TIRS image which has 11 spectral bands, and the path and row of the study area are 140 and 44, respectively, after the layer stack of the image. An unsupervised classification for the 2016 period of Landsat image was produced which has a spatial resolution of 30 metres of the different bands of Landsat 8, such as bands no. 1–8. In Arc GIS Geo-reference, the map of Ranchi and then the creation of a personal database to store different layers were carried out. This study highlights that there is a requirement for improvement and proper implementation of the master plan of the city for proper management and regulation of the overall development of Ranchi.

7.2 STUDY AREA

Jharkhand is a tribal-dominant state with a 40% tribal population, and a rural state with 75% of the population living in villages. Jharkhand's capital city, Ranchi, is very well known for its mineral deposits, forest, dams, lakes, streams, rivers, and waterfalls [7]. The geographical extent is latitude 23.35'N and longitude 82.23'E (Figure 7.1), the total geographical area is 5231 sq. km, and the height above sea level is 2140 feet. Ranchi is the most populous city in the state. Lohardanga was the earlier name of the Ranchi district [8]. Ranchi and Bundu are the main two divisions of Ranchi, which are further segmented into blocks, panchayats, and villages. There is a total of 18 blocks and 303 panchayats. With a growth rate of more than 85% in the last four decades, Ranchi has been one of the fastest-growing cities in India and worked as the growth pole of Jharkhand. The total urban populations of Ranchi district and city are 1,257,340 and 1,073,440, respectively. The city has grown with a growth rate of 26%. Table 7.1 shows the general information about Ranchi. The population density has increased considerably from 4840 to 6133 per sq. km in 2011, mainly due to limited land resources in the city.

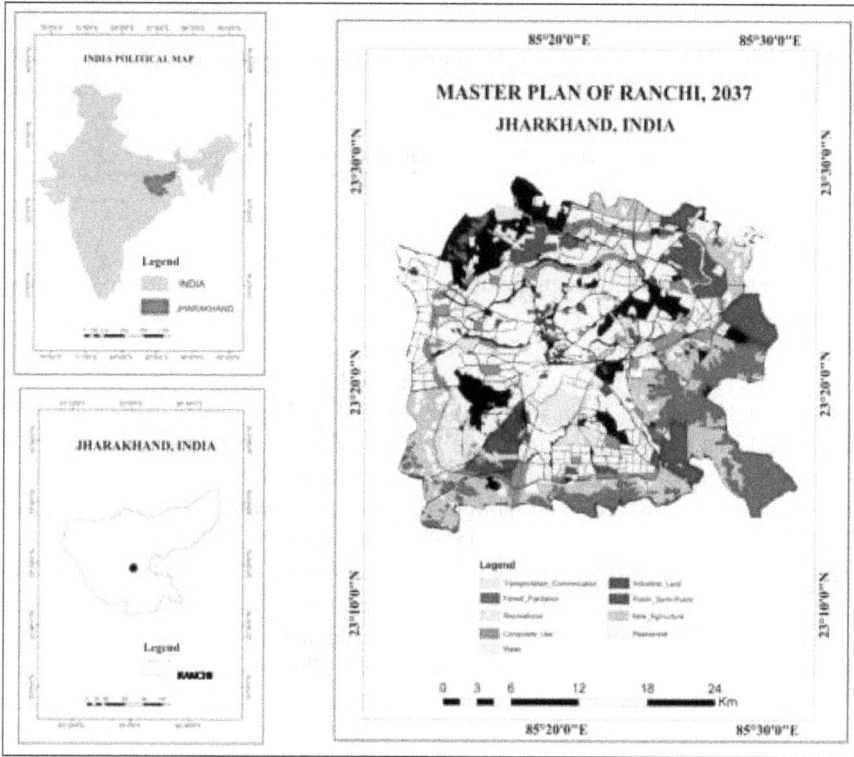

FIGURE 7.1 Study area.

Source: Prepared by the authors.

TABLE 7.1
General Information about Ranchi, Jharkhand

Indicators	Jharkhand	Ranchi district	Ranchi city
Urban population	7,929,292	1257,340	1,073,440
Sex ratio	947	950	920
Literacy (%)	67.33	77.33	88.49
Density (sq. km)	414	557	6129

Source: [9] Census 2011.

7.3 OBJECTIVES OF THE STUDY

(a) To highlight LU/LC change to understand the driving forces of urbanisation.
(b) To look at the master plan as a tool of the controller and regulator of urban growth in Ranchi.
(c) To observe the impact of LU/LC change on the master plan.

7.4 METHODOLOGY

A mixed-method approach with the inclusion of qualitative and quantitative methods has been used in this research work (Figure 7.2). The study attempted to understand population growth analysis, actual development, growth of the city, and comparison of the master plan through built-up density and urban expansion. Population change has been analysed and compared with the LULC. These two aspects have been highlighted as a tool to understand the efficiency of the master plan in the regulation and management of the city. Furthermore, the land use and master plan have been examined and their association highlighted. Superimposition of the classified image of 2016 on the reclassified zoning map of the Ranchi master plan has been done to check the similarity or dissimilarity between the proposed master plan and the situation on the ground. For these aspects, data were collected from satellite imagery, the master plan of Ranchi city, the census of India, and the City Development Plan of Ranchi. LULC have been analysed since 2016. This satellite period was chosen

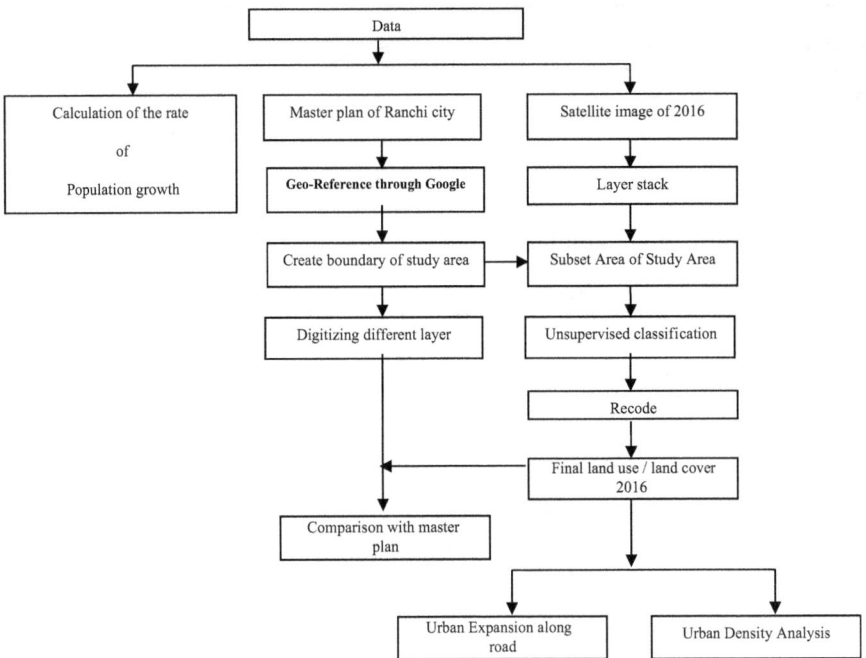

FIGURE 7.2 Data methodology chart.

because it shows the latest development of Ranchi LULC, and it helped us to examine the proposed master plan. The satellite image was acquired from Landsat 8 and the scenes used were wrs_path = 140, wrs_row = 44 and the size of the pixel was 30.

7.5 RESULTS AND DISCUSSION

7.5.1 Population Growth Rate and Status of Urbanisation in Ranchi City

In 2011, the growth rate of the urban population experienced exceptional growth as compared to 1980 to 1990 and Ranchi become the second million-plus city in Jharkhand. In 1991, the population of Ranchi city was about 5.9 lakh which has doubled in just two decades and reached 10.7 lakh in 2011 (Table 7.2). The population density of the city increased from 4840 to 6133 per sq. km in 2011, mainly due to limited land resources in the city. The new master plan included the idea of physical expansion of the city to reduce pressure on the existing land of the city. Urbanisation is the sign of transformation of a city from primary to secondary as well as tertiary activities that also lead to urban problems and their related issues [10].

From Table 7.2 it can be seen that the urban population and total population of the city have grown considerably. However, the urban population has increased very quickly and is one of the dominant causes of urbanisation. Ranchi's population was 25,970 in 1901, and it crossed the one-lakh limit in 1951. The population of the Ranchi Municipal Corporation Area in 2011 was 1,073,440. Based on past trends and the growth of the population, the urban area and rural villages of Ranchi have been divided into high, medium, and low by the Ranchi Planning Area (RPA) from 2012 to 2037. It was estimated that the population of RPA (combination of rural and urban population) will vary between 21.22 lakhs to 36.94 lakhs in 2037 (RMP). In the 2001 census data of India of Ranchi municipal ward, the total urban population was 847,093, with 450,727 males and 396,366 females, and census of 2011 shows the growth of the urban population to 1,073,427, with 558,872 males and 514,555 females. Due to urbanisation, there are many issues that create both positive and negative impacts such as sprawl, deforestation, increasing the value of land, etc. [11].

TABLE 7.2
Rate of Population Growth

Year	Ranchi district		Ranchi city	
	Population	Decadal growth (%)	Population	Decadal growth (%)
1991	22,14,048		599,306	22.4
2001	2,785,064	25.79	846,454	41.24
2011	2,912,022	4.59	1,073,440	26.72

Source: Census of India.

7.5.2 URBAN EXPANSION IN RANCHI MUNICIPAL AREA ALONG THE MAJOR ROADS

To uncover the urban expansion in Ranchi Municipal Area along the major roads, first the major roads which connect Ranchi to other states as a well intra-city connection were identified. Figure 7.3 illustrates the major roads and their routes. After the identification of major roads they were digitised in an arc map and then a multiple-ring buffer created. Each buffer was between 100 and 500 metres wide. After obtaining the Ranchi municipal area from the classified image of Ranchi and converting the Ranchi municipal raster layer to a vector using the conversion tool in arc map, we used the raster to polygon command to get a vector layer that showed all types of land use, such as agriculture, water, settlements, forests, and open space, along with the area in square kilometres. After this the multiple ring area was clipped from the municipal area of Ranchi (vector format) through the analysis tool in Arc GIS by extracting using the clip tool. Then, to reveal the urban expansion of each ring in multiple buffers each ring buffer was exported one by one, from 100 metres to 500 metres, and the settlement area calculated using the geometric calculator in sq. km. The Ranchi municipal area has 55 wards presently but, in the past, there were only 33 wards. With the expansion of wards, population growth was a major factor and the road network has played a very crucial role in the attraction of people to the city. As a well-developed road network provides easy mobility of goods and services, more opportunities for jobs were created, which attracted a greater number of people to Ranchi City. In this process, huge immigration to the city led to the construction of settlements along the roads and highways, and hotels and restaurants also flourished along the highways. Therefore, it is crucial to develop a balance between population demands and growth. There is a requirement for linkages between economic, social, and environmental planning with regional or spatial planning so that a comprehensive strategy for regional planning can be developed [11].

7.5.3 DENSITY GRADIENT ANALYSIS

It is observed from Figure 7.3 that urban development usually occurs around city centres and along major roads. The buffer analysis of GIS has been used to examine the impacts of locational functions on the spatial pattern of land development. Around the city centre and major roads of the city, a buffer zone was created to obtain the density of land development in each buffer zones. The Neighborhood Function in ERDAS was used to examine the built-up density on the spatial pattern of land development. Also, to analyse more effectively we categorised the density of built-up area in Arc GIS Density, which is found in ERDAS. Due to rapid urbanisation and the designation of Ranchi as the capital city of Jharkhand, it has experienced tremendous physical growth. Due to a lack of planning interventions, haphazard settlements have spread even beyond the municipal limit of the city (Ranchi master plan). The built-up area density is very high in the centre area of municipal wards 18, 19, 23, 25, 26, 27, 29, 30 and 31. These wards are connected by major roads such as NH 75, Circular Road, Bhonda, and Hinoo Road (Figure 7.4).

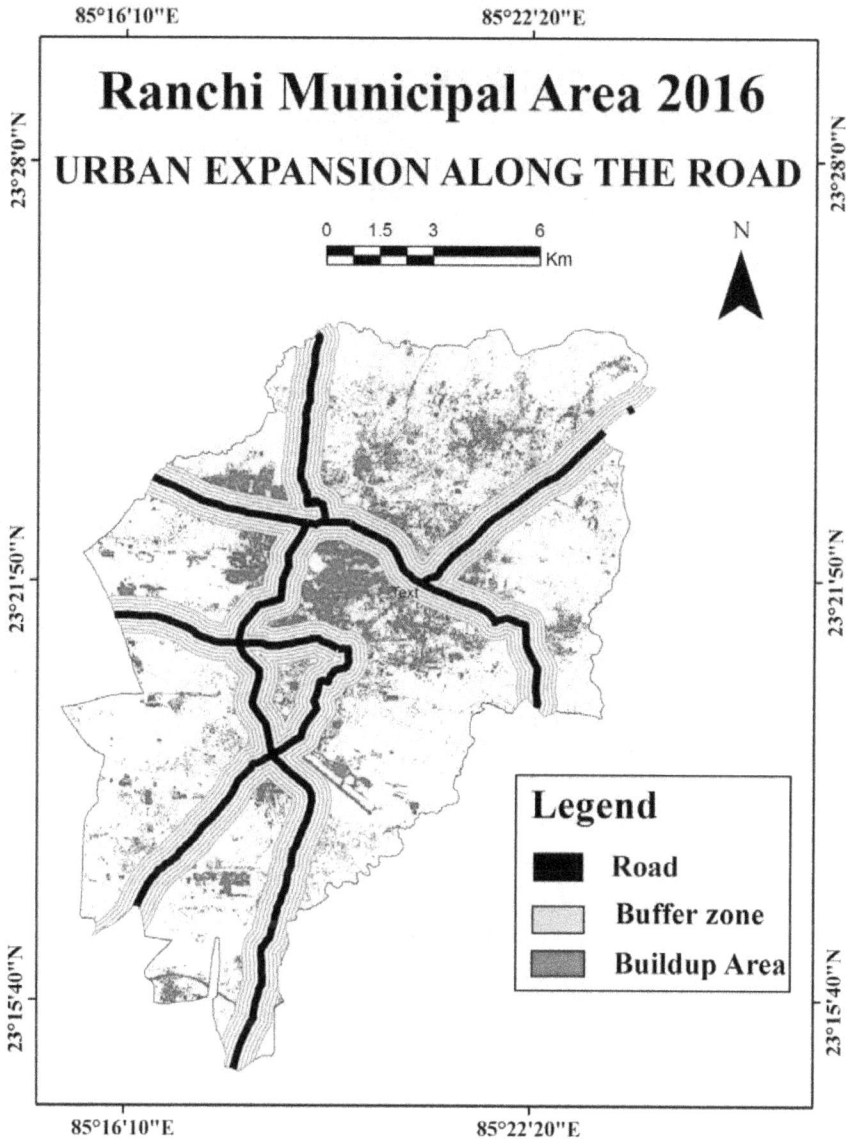

FIGURE 7.3 Urban expansion along the roadside in Ranchi.

Source: Prepared by the authors.

It has been observed that highly dense centres are very well connected with roads and that road networks played one of the most significant roles in the urbanisation of Ranchi city. The majority of the high-density areas (such as 10, 13, 14, 15, 16, 17, 20, 21, 22, 24, and 32) are located close to the very high-density area. Moderate-density wards include 6, 8, 12, 28, 46, 47, 33, 34, and 36. These wards all share a boundary

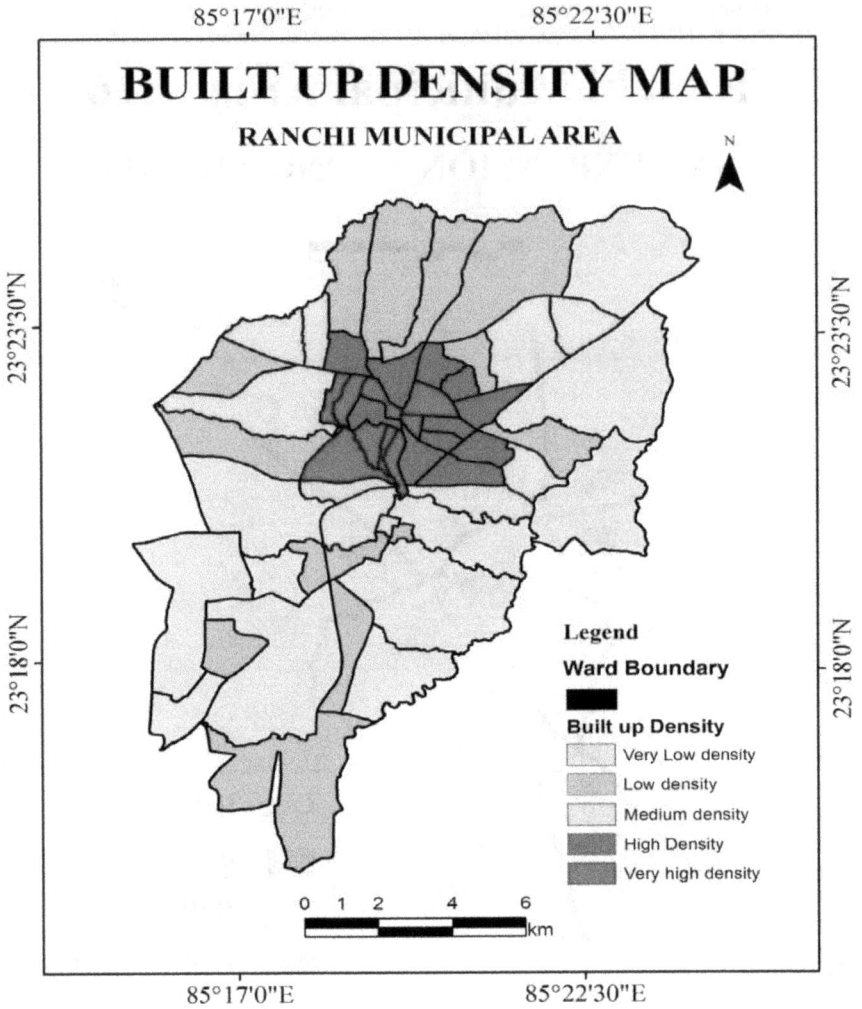

FIGURE 7.4 Built-up density map of Ranchi.

Source: Prepared by the authors.

with high- and very high-density wards, but they are outer wards of the Ranchi municipal area. Areas with lower density wards are 1, 2, 3, 4, 9, 11, 35, 37, 41, 44, 45, 51,
53, and 55, because those are outer wards of Ranchi. To understand LULC, its types,
and causative factors, its association with land use change and urban growth and for
impact assessment such as density gradient analysis is important. Furthermore, it is
helpful in the planning and implementation of developmental activities [3].

7.5.4   COMPARISON OF THE MASTER PLAN WITH LAND USE AND LAND COVER OF RANCHI (FIGURE 7.5)

The widening of cities is not uncontrollable but guided and shaped by human inter-action and by physical infrastructure, and its analysis is pivotal [12]. For this purpose, the Ranchi city LULC was analysed and compared with the proposed master plan. The results of this research work indicate that a large part of the land which was demarcated as forest and agricultural land has been transformed into built-up areas and the open space has changed its status presently in some places, and the forest has been changed into agricultural land. As can be seen in Table 7.3 lands purposed as forest areas were changed into built-up areas (2.143 sq. km) and land purposed as agricultural was changed into settlements (2.468 sq. km). This study highlights that proper implementation of the master plan would face many challenges in implications of the action on the ground, as even 2 years after its approval it is not visible on the ground. The actual land uses highly deviate from the proposed master plan of the city.

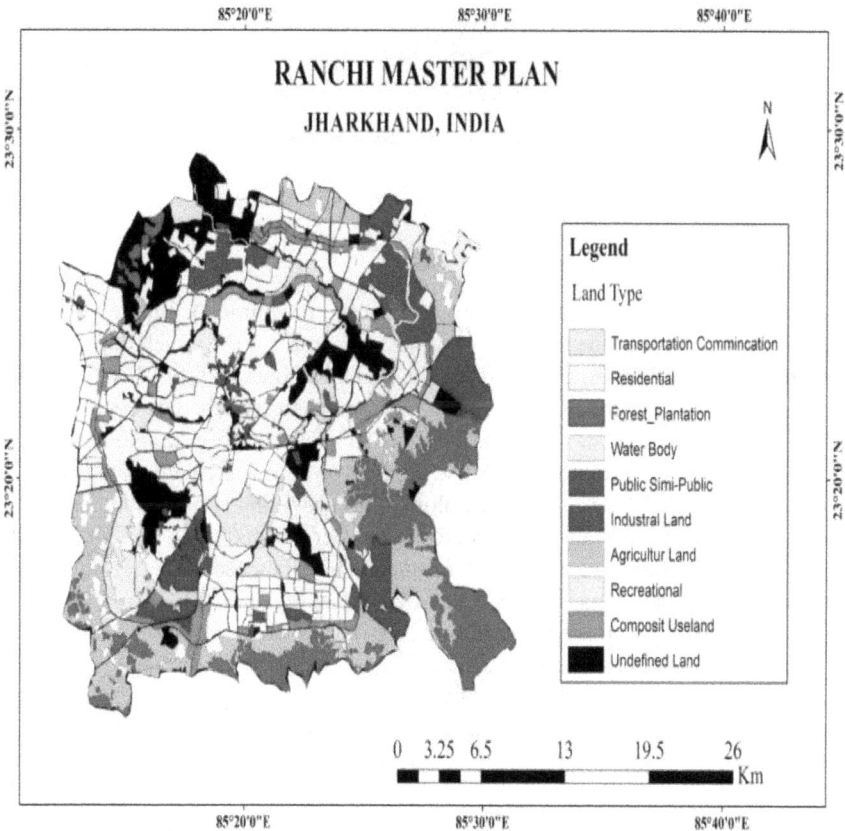

FIGURE 7.5   (A) Ranchi land use and land cover and (B) Ranchi master plan.

Source: Prepared by the authors.

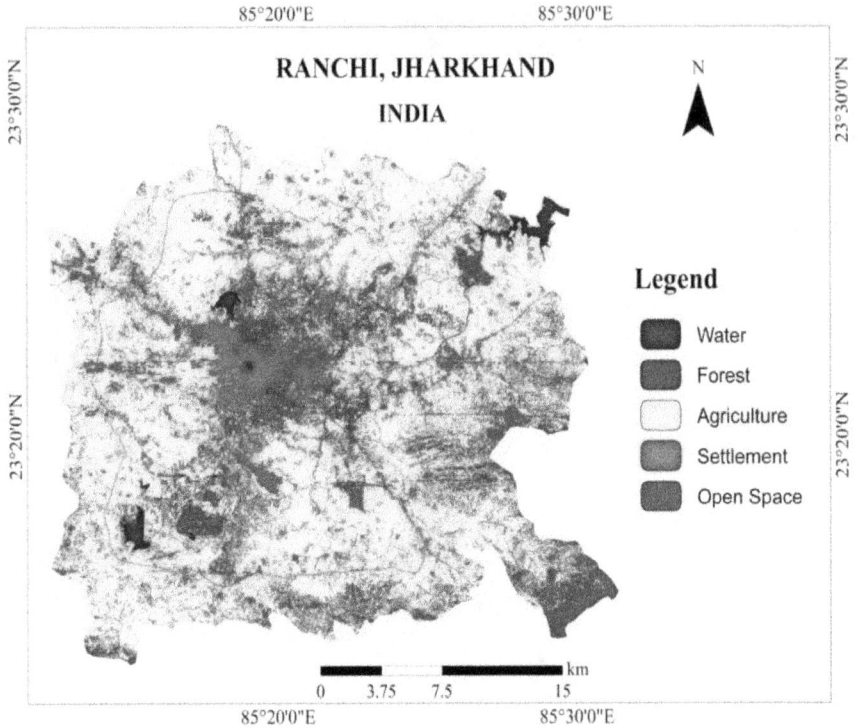

FIGURE 7.5   (Continued)

Also, the development pattern is different from the recommended master plan. The changed scenario shows us the proposed master plan for 2037 which does not overcome the current situation, and the resultant data are shown in Table 7.3.

This study has described the land use in 2016 and assessed the degree of closeness with the zoning specifications of the master plan. Furthermore, it also looks at the feasibility of a master plan to standardise urban growth in Ranchi city and to identify diverse driving forces of urbanisation, along with providing some recommendations to ensure better management of urban growth. As per the study of LULC of 2016, the results indicate a pattern of roadside development and increased cases of rapid deforestation due to urban expansion in Ranchi city. The results indicate a failure in the implementation of the master plan in bringing out the changes that it intended with the proposed regulations. Although the first master plan was developed for Ranchi city in 1965 it did not yield results as it was left unapproved due to some unavoidable circumstances which resulted in the development of the city randomly. Furthermore, in 1985, the first master plan of the city was revised to develop the city in a planned manner. This effort was one of the first comprehensive efforts of the Ranchi Regional Development Authority (RRDA) for the planned growth of the city. This plan was further revised and termed the Comprehensive Master Plan (CMP) for Ranchi and it was proposed to achieve its targets of the sustainable and planned growth of the city by 2037. For inclusive development, the Ranchi Municipal Corporation (with

TABLE 7.3
Types of Conversion of LULC in Ranchi

Type of conversion	Area (sq. km)	Percentage
Planned as a Water now Forest Land	1.744	0.307
Planned as a Water now Agriculture	4.850	0.854
Planned as a Water now Settlement	0.616	0.109
Planned as a Water now Open space	1.011	0.178
Planned as a Forest now Water	0.021	0.004
Planned as a Forest now Agriculture	30.509	5.373
Planned as a Forest now Settlement	2.143	0.377
Planned as a Forest now Open space	9.591	1.689
Planned as an Agriculture now Water	0.718	0.127
Planned as Agriculture now Forest	13.513	2.380
Planned as Agriculture now Settlement	2.468	0.435
Planned as an Agriculture now Open space	9.313	1.640
Planned as a Built-up now Water	0.895	0.158
Planned as a Built-up now Forest	52.406	9.230
Planned as a Built-up now Agriculture	232.851	41.009
Planned as Built-up now Open space	38.075	6.706

Source: Prepared by the authors.

55 wards), Kanke, Tati, and 116 revenue villages around the city have been covered under the Ranchi Planning Area (RPA). Two major factors that have exceptionally increased the pressure on Ranchi city and its resources are the continuous increase in the population of the city and the tendency of the population to live in separate houses. The population of the city has mainly increased due to the immigration of the population with the desire for a better life, jobs, and education in the capital city. Due to that, the city has experienced great changes in land use and road networks in the municipal areas [13].

7.6 CONCLUSIONS

The study of land use patterns and analysis of their changes over time have a very significant role in the proper and sustainable management of developing cities. This study has highlighted that major settlement of the city has developed along the roads in a linear pattern. Rapid urbanisation has caused the exploitation of many valuable natural resources. The uncontrolled growth of the city lacks commitment, with the absence of a combined land management system and fragmented governance being the major causative factors. One of the most important findings is that the master plan has failed to govern and regulate planned urban growth in Ranchi city. In general, a well-structured and developed master plan is required for the proper development of any city as the experiences of many cities have shown, but it is also crucial to properly implement it on the ground. Not only this, but it also has been interpreted from this study that there is an urgent need to preserve and stop the exploitation of natural resources to carry on the process of implementing urban city planning strategies alongside boosting the process of India's economic growth. Overall, it is crucial to properly implement the master plan of the city on the grassroots level, as without its proper implementation the goal of a sustainable city cannot be achieved. The aims of further studies should give greater focus to the social facets of urbanisation by finding techniques to accelerate the use of remote sensing techniques to better understand the relationship that exists between land use change and socioeconomic transformation.

REFERENCES

[1] Sharma, V. R., Bisht, K., Sanu, S. K., Arora, K., & Rajput, S. (2021). *Appraisal of Urban Growth Dynamics and Water Quality Assessment along River Yamuna in Delhi, India.*

[2] Sanu, S. K., Sharma, V. R., Rani, U., & Upadhyay, S. (2022). Some Aspects on Migration Profile: A Case Study of Kurukshetra, Haryana. *The Horizon–A Journal of Social Sciences, Volume-XIII*(January 2022), 101–110.

[3] Singh, N., & Kumar, J. (2012). *Urban growth and its impact on cityscape: A geospatial analysis of Rohtak City, India.*

[4] Singh, A. (1989). Review Article Digital change detection techniques using remotely-sensed data. *International Journal of Remote Sensing, 10*(6), 989–1003. https://doi.org/10.1080/01431168908903939

[5] Carlson, T. (2003). Preface – Applications of remote sensing to urban problems. *Remote Sensing of Environment, 86*, 273–274. https://doi.org/10.1016/S0034-4257(03)00073-7

[6] Zhou, Q., Li, B., & Zhoub, C. (2004). Detecting and Modelling Dynamic Landuse Change Using Multitemporal and Multi-Sensor Imagery. www.semanticscholar.org/paper/DETECTING-AND-MODELLING-DYNAMIC-LANDUSE-CHANGE-AND-Zhou-Li/fd477b1579ecef9d314e20bd29a9fcd77dd8e7a2

[7] Krishna, R., Iqbal, J., Gorai, A. K., Pathak, G., Tuluri, F., & Tchounwou, P. B. (2015). Groundwater vulnerability to pollution mapping of Ranchi district using GIS. *Applied Water Science, 5*(4), 345–358.

[8] Ghosh, A. (1995). A brief note on the mythology, pre-history and history of Ranchi district. *Indian Anthropologist, 25*(1), 91–96.

[9] *Census of India Website: Office of the Registrar General & Census Commissioner, India.* (n.d.). Retrieved April 11, 2022, from https://censusindia.gov.in/

[10] Datta, P. (2006). *Urbanization in India, regional and sub-regional population dynamic population process in urban areas, European Population Conference.*

[11] Mundhe, N. N., & Jaybhaye, R. G. (2014). Impact of urbanization on land use/land covers change using Geo-spatial techniques. *International Journal of Geomatics and Geosciences, 5*(1), 50–60.

[12] Castro, G. G. H., & Rocha, W. P. (2015). Change analysis of land use and urban growth in the municipalities of Culiacan and Navolato, Sinaloa, Mexico using statistical techniques and GIS. *Journal of Geographic Information System, 7*(06), 620.

[13] Gupta, R., & Sen, A. (2008). Monitoring physical growth of Ranchi City by using geoinformatics techniques. *ITPI Journal, 5*(4), 38–48.

8 A Multi-Criteria Evaluation of Urban Flood Vulnerability and Perception in Osun River Basin, Southwest Nigeria

*Oladeji Quazeem Muhammed and
Adebayo Oluwole Eludoyin*

CONTENTS

DOI: 10.1201/9781003331001-8

8.1 INTRODUCTION

Urban growth, an indication of a society's economic condition and a process of urbanisation, alongside changes in population by migration, is typically associated with population pressure in previously protected ecosystems and habitats, especially within river basins (Munoz et al., 2021). A river or drainage basin is a geographically defined land area that is drained by the river and its tributaries (Eludoyin & Adewole, 2020; Adewole et al., 2020). The river basin comprises biotic – animals, plants, and people – and abiotic factors within the environment, and these are likely to be threatened as urban growth intensifies and population increases towards the basin area. In many regions of the world, development control rules, called setback rules, are often mandated to protect against intrusion into the basins. For example, Agbola et al. (2012) argued that a government agency can gazette setback rules which can also be updated along the line of climate change and variability. In Oyo State, Nigeria, Agbola et al. (2012) showed that setbacks varied from 15–45 m, based on the river's size and response in their area of study. The setback rule is typical of what is obtainable across the states in Nigeria, and in the present study area, a report of the United Nations Human Settlements Programme (UN-Habitat, 2014) recognised the 15 m setback rule.

Despite the awareness of the setback to the river rule and other popularised interventions – including dredging and relocation of vulnerable communities – by governmental and non-governmental organisations in the study area, and many parts of Nigeria, the urban flood has reportedly become a perennial occurrence. Ologunorisa et al. (2022a) hypothesised that many communities within major river basins often consider flooding an unavoidable problem (see also Olajuyigbe et al., 2012). Studies have also associated flood occurrences and associated disasters with hydrological explanations of runoff mechanisms [infiltration-excess overland flow or saturation-excess overland flow, waste management factor and dam failure (Ologunorisa et al., 2022b)]. The infiltration-excess overland flow becomes dominant in areas that are largely characterised by impervious surfaces, such as urban areas, where less water due to rainfall is allowed to infiltrate into the soil due to the impervious landscape, while saturation-excess flow becomes dominant only during high rainfall intensity, when the soil has become super-saturated, and excess water floods the environment. The waste management factor links flooding with blockage of drains with different waste types, while many cases of flood disasters have equally been recorded due to dam or reservoir failure (see Ologunorisa et al., 2022b). In all, fatalities due to flooding have increased in Nigeria despite evidence that many fatality cases were not documented (Ologunorisa et al., 2022b). Consequently, it makes sense to underscore the importance of capacity building and improved methodologies for understanding flood control systems and urban growth sustenance as described in the literature (e.g. Li et al., 2021).

8.2 RESEARCH PROBLEM

A huge amount of attention has been focused on the flooding problem across the world, probably because of its increasing number of victims and the negative implications for the economy. Studies have justified concerns for urbanisation-related flood risks in communities in coastal regions and river basins (Ologunorisa, 2004; Onanuga et al., 2022) and the effects of extreme climatic events and climate change (Padi et al., 2011). Existing studies have also suggested a link between proximity to major rivers and fatalities, with variations owing to the presence or absence of emergency planning plans, properly or poorly monitored urban growth, and the level of adherence to planning regulations and rules, especially in riverine communities (see Salami et al., 2017; Ologunorisa et al., 2022b).

The Nigerian Hydrological Services Agency (NIHSA) issued a national early warning on disturbing flood events across Nigeria in 2019, flagging settlements in the southern parts of the country as well as the Niger-Benue trough. All across the country, the impact of the incidence of floods has been devastating, with estimated losses worth billions of naira, lives lost, and the economies of many truncated. The 2019 floods in Osun state, especially in Osogbo, wreaked havoc on many communities. Gasu et al. (2019) reported that churches, bridges, schools, farmlands, and markets were flooded in at least 10 communities in the area. An editorial of a national newspaper, *Punch*, on September 13, 2016, reported a loss of millions of naira in many communities in the area.

Furthermore, many studies have evaluated flood vulnerability in different parts of Nigeria from the perspective of either sole spatial analysis of satellite imageries (e.g. Oloukoi et al., 2014; Ouma & Tateishi, 2014; Aguda & Adegboyega, 2013; Adegboyega et al., 2018; Owolabi, 2019) or social surveys (e.g. Esan & Babatola, 2015; Fatusin, 2015). The majority of these studies have evaluated urban growth and concluded based on their evaluation of satellites without consideration of the heterogeneity of their study area because their approaches were not based on basin or catchment systems. Information is also scarce about urban growth in the Osun river basin area, the vulnerability of the residents of the area to flood and the perception of residents about floods in typical urban sections of a river basin, in a growing administrative capital in southwestern Nigeria. Consequently, this study has adopted a mix-method approach of multi-date imagery analysis and social surveys within the scope of a geographical information system to provide answers to the nature and rate of growth of urban features within the river basin areas as well as the adherence to setback rules to major rivers; perception of residents on flood disasters and coping/resilience strategies in times of flood occurrences; and inform on the spatial and temporal variability of the vulnerability of the residents and properties in the study area to floods. Specific objectives are to examine the growth of built-up areas, other land uses/cover, within Osogbo, Osun river basin; investigate the perception of the residents of the area about vulnerability to flooding disaster and their coping strategies, and assess the vulnerability of the study area to flood. This study is justified by the need to monitor the effect of urban growth on flood occurrences at the basin scale.

Osogbo in Nigeria, where the study area is located in a rapidly going administrative capital of Osun state is one of the 36 state capital which is characterised by a significant amount of pull factors for social and economic growth which will lead to an influx of immigrants and an attendant increase in population size. The increase in population size has been linked to both unintended and intended environmental problems ranging from land use transformations, fragmentation of fragile ecosystems, degradation of the water quality of rivers, increased flooding, and disturbance in the natural river basin ecology which poses a serious threat to human lives and properties in the city (Shukla et al., 2013). This study, therefore, fits well into the targets of SDG 11 (make cities and human settlements inclusive, safe, resilient, and sustainable), 13 (take urgent action to combat climate change and its impacts) and 17 (strengthen the means of implementation and revitalise the global partnership for sustainable development).

8.3   STUDY AREA

8.3.1   LOCATION

The study area, the Osun river basin (Figure 8.1), is situated in the southern part of Osogbo, the administrative capital of Osun state in southwestern Nigeria – the state was created in 1991. The basin is characterised by a perimeter of 49.47 km and is 152.73 km² in area, most of which is dominated by urban built-up and transport features. Egunjobi (1995) documented that the settlement has grown from the status of a provincial headquarters in the Western region to that of an administrative capital city in 1991, and has thus witnessed a significant level of urban growth and population increase through the years. From a cultural centre whose outwards growth from the Oja–Oba (King's market) and palace that is similar to the concentric urban setting, the settlement grew into development similar to Hoyt's sectoral model of urban development, such that growth and development have become localised. Many parts of the settlements have grown into the flood plains around the Osun river basin, such that the hitherto well-developed riparian vegetation is now preserved for farming; especially market gardening and cultural activities. In some other parts of the region, farming activities appeared to have given way to built-up areas, and areas of the rivers used for cultural activities (Osun festival) have reduced to a relatively protected section of the basin (Osun Groove).

8.3.2   CLIMATE, GEOLOGY, AND VEGETATION

The study area is characterised by wet (March–October) and dry (November–February), being within the Humid Tropical Climatic Zone that is known to exhibit double (in July and October) maximal/peak rainfall (UN-Habitat, 2014). The 38-year (1982–2020) mean values showed that the average peak occurred in September and that the wet season may vary between March and April.

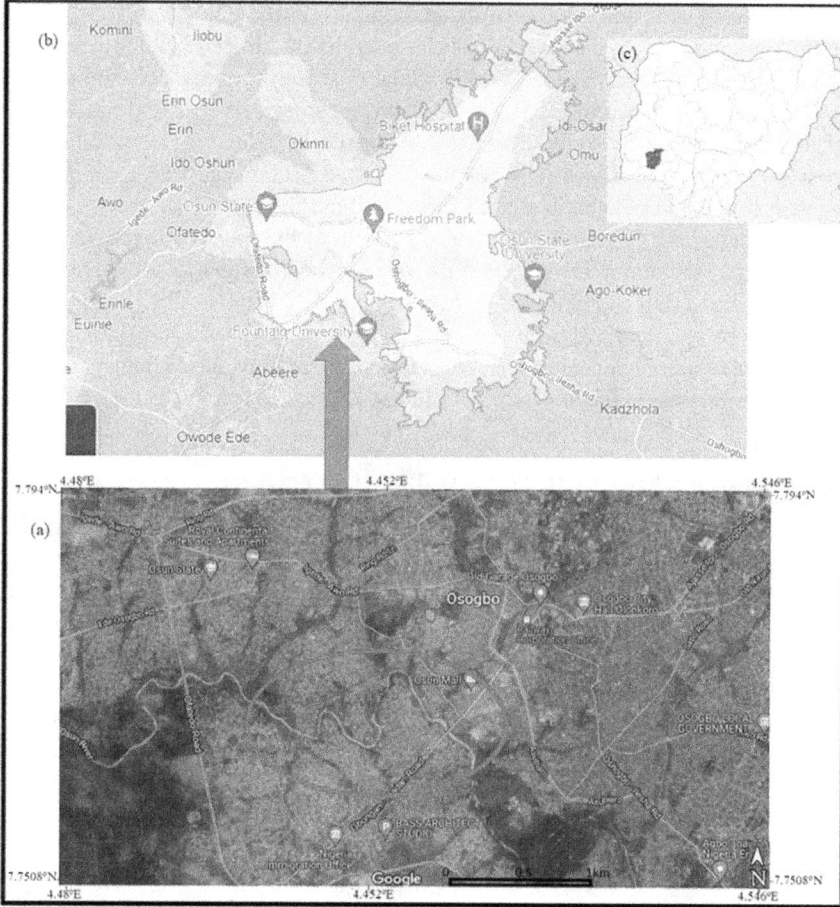

FIGURE 8.1 The study area is the Osun river basin in Osogbo, southwestern Nigeria.

Source: The image was drawn by the authors using imagery from Google Earth.

Variations in wet and dry seasons are influenced by the periodic movements of the South-West and North-East trade winds in the region (e.g., Eludoyin et al., 2009). The average humidity in the area is about 80%, and annual rainfall varies from 1200 mm to 1500 mm. The mean annual rainfall is about 1460 mm. The average annual temperature is about 26°C and varies between 21°C and 31°C.

The Osun drainage basin is generally (about 73%) underlain by metamorphic rocks of schist complex and granitic (igneous) rocks (27%), both of which exhibit varied grain size and mineral composition (Akinwumiju, 2016). The soils, on the other hand, are generally of two types: clayey and sandy soils. Whereas the clayey soils are formed on the upper slopes of the basin area, the sandy hill-wash soils are common at the lower slopes. Both soil types belong to ferruginous tropical red soils associated with Basement Complex rocks (Ogunsanwo, 1980) whose derivatives of mostly sandy clay and lateritic soils are common in the tropical regions of West Africa. The sandy soil of the lower slopes is generally considered rich for agricultural cultivation (Adejuwon & Ekanade, 1998; Salami & Sangoyomi, 2013). In addition, the study area exists within 305–418 m above mean sea level (Figure 8.2), while the northern part of the study area is characterised by numerous domed hills and occasional flat-topped ridges (UN-Habitat, 2014). The drainage system is made up of several small river channels that are seasonal, but the main Osun river is perennial with seasonally fluctuating volume. In addition, the most common vegetation is tree species, including *Terminalia Superba*, *Lophira alata*, *Khaya ivorensis*, and *Antiaris africana* (UN-Habitat, 2014). According to UN-Habitat (2014), many of the economic tree species have been removed as the town expands and its population grows, and the natural vegetation becomes degraded, mainly due to human activities and other anthropogenic factors (Salami, 1999).

8.3.3 Urban Growth and Population

Since it became an administrative capital city and seat of the Osun State Government in 1991, Osogbo has been transformed as the home to many private and public institutions and has thus received an influx of migrants primarily for economic reasons. The increase in population and urbanisation has been linked with infrastructure development that has transformed into 'leapfrog development' and 'urban sprawl' (Aguda and Adegboyega, 2013; Adedotun, 2015). According to Hovinen (1977), leapfrog development or scattered development often results in continuously built-up, low-density development of many suburbs, and this may be a causal factor of many urban sprawls. The population of the Osogbo local government area within which the Osun river basin is largely situated in Osogbo grew from 155,507 in 2006 to an estimated 217,544 in 2020 according to Geo-Referenced Infrastructure and Demographic Data for Development (GRID[3]) Nigeria Project (www.grid3.gov.ng), making an annual growth rate of 3.99%.

8.4 MATERIALS AND METHODS

8.4.1 Data

The data used were from both primary and secondary sources. The primary data included the coordinates (x,y,z) of conspicuous landmarks and these were obtained through the use of a handheld global positioning system (Magellan version; ±5 m accuracy;). The coordinates (now referred to as ground control points [GCP]) for the study area were used during pre-processing of the images that were also used for the study (discussed in the next subsection) for correction of radiometric and geometric errors. In addition, 18 purposively selected residents of the study area were selected as key informants (KI) based on their experience of flood in the area. The conditions for selection were that at least 10-year residence time in the area and they were mainly community leaders in the area. Also, a senior officer of the State Ministry of Environment, Flood Control Unit, was purposively selected for information on the role of government in the prevention and control of flood disasters in the area as well as his overall view on flood events. In all, 18 community heads were interviewed. The interview was conducted using the Open Data Kit (ODK) also known as ODK Collect. The ODK is a modern mobile phone-based, open source android operating system device that has been found useful in recent studies (see Gebreyesus et al., 2013; Raja et al., 2014). According to Gebreyesus et al. (2013), the ODK Collect implements a programmed questionnaire and immediately digitises the data for analysis, allowing remote monitoring of the collection progress, and therefore facilitates the gathering of data, eliminating the need for paper surveys, and therefore significantly reducing survey times.

Raja et al. (2014) reported that the ODK is an open-source suite of tools that were designed to help users build information services for developing nations. The ODK tools – *ODK build, collect* and *aggregate* – are designed as open standards that enable users to build services to collect and distribute information with ease and therefore support data integration (including coordinates from GPS, images, codes, sound bites, and videos). The ODK ensures smarter data gathering and management in developing countries where mobile phone usage is rapidly increasing with the expansion of service coverage, albeit with possibilities of limitations due to unreliable electricity supply, battery life, and the cost of android smartphones. Figure 8.2 describes the configuration of the *ODK collect* data collection system (Gebreyesus et al., 2013). Using this system, the *ODK collect* program was installed on an android-based phone and a set of harmonised questions in a questionnaire format were formatted with *xml* format and saved to the phone's device memory. The information

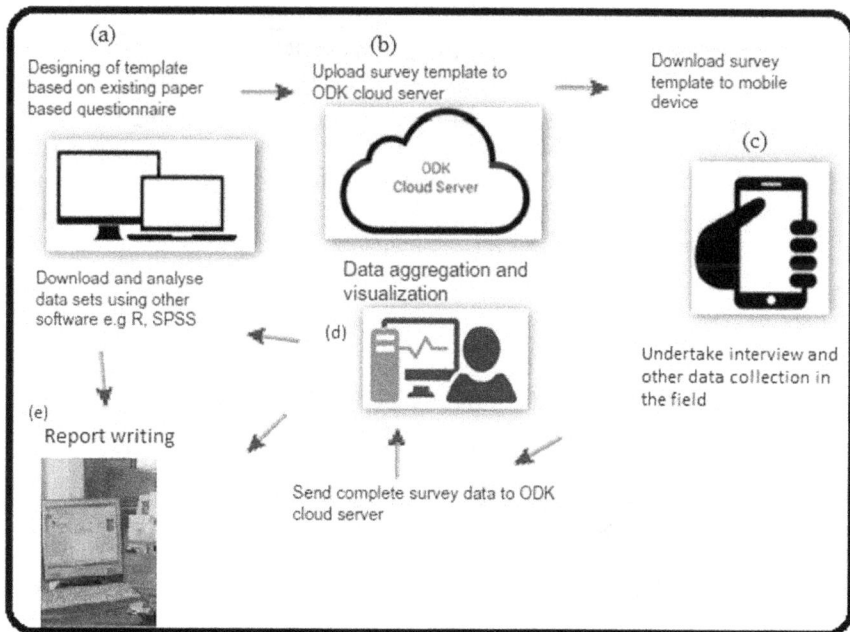

FIGURE 8.2   The basic procedure for use of ODK collects data gathering.

stored consequently became rendered into a sequence of inputs which prompt the user through accomplishing the task of data gathering and saving them.

In this study 77% of the respondents were male, 83% of whom were aged above 40 years. Only 17% had received no formal education, while 27% and 56% had secondary and tertiary education, respectively. In addition, 80% had lived in the study area for at least 10 years, before the date of the interview.

In terms of secondary datasets, the *shapefiles* of the study area (Osogbo) and the freely available multi-date Landsat (MSS, 1986; TM, 1996; ETM, 2006; OLI, 2020) images were obtained and investigated. While the *shapefiles* were obtained from the office of the Physical Planning Department of the Osun State Capital Territory Development Authority in Osogbo, Osun State, the Landsat images were downloaded from the archive of the United States Geological Survey (USGS; https://earthexplorer.usgs.gov/) through 190/55 path/row. Spatial resolution is 30 m for each image. The main characteristics of the Landsat images used are reported in Table 8.1.

The Landsat images were selected for the study because of their temporal and cost advantages over other available remotely sensed images. The Landsat data were the known image data that spanned between 1986 and 2020, and their free accessibility, as well as medium spatial resolution (see Zhang and Weng, 2013; Yu et al., 2017) made them a better candidate for a monitoring project in a developing country like Nigeria (see also Eludoyin and Iyanda, 2018). In addition, the 12.5 m Alos Palsar Digital Elevation Model, which was sourced from the archive of the Alaska Satellite Facilities (www.search.asf.alaska.gov), was used to delineate the drainage characteristics of the Osun drainage basin, and drainage variables, including elevation, slope, flow accumulation, nearness to drainage, and drainage density were derived using appropriate methods.

Furthermore, daily rainfall records for 39 years (1982–2020) were obtained from Prediction of Worldwide Energy Resource (POWER), an arm of the United States National Aeronautics and Space Administration (NASA); while daily discharge or runoff data (for 1991, 1993, and 1994) and water level/stage record (for 2001–2019) – the selected years which the data were only available at Apoje station (the closest station for which such records were available) of River Osun – were obtained from the office of the Ogun-Osun River Basin Development Authority (OORBDA) in Osogbo.

TABLE 8.1
Sensors, Resolutions, and Date of Acquisition of Acquired Landsat Data

Sensor	Spatial resolution	Spectral resolution (number of bands)	Date of acquisition
Multispectral scanner (MSS)	30 m	5	17-12-1986
Thematic mapper (TM)	30 m	7	16-11-1996
Enhanced thematic mapper (TM)	30 m	8	12-01-2010
Operation land imager/thermal infrared sensor (OLI/TIRS)	30 m	11	14-02-2020

8.4.2 Data Analysis

8.4.2.1 Image Quality Improvement and Assessment

The software used for the study is Quantum GIS (also known as QGIS) which was preferred because it is open access and for novel contribution to knowledge because QGIS had enjoyed rather minimal use by researchers for LULC classification at the time of this research. First, the Landsat images were corrected for geometric and radiometric errors in the QGIS software environment. This means that the images were first georeferenced using the previously acquired coordinates of established landmarks on the field as described by Eludoyin & Iyanda (2018) following the guide from the QGIS manual (see also Lillesand et al., 2008; Congedo, 2017). Also, the process of radiometric enhancement includes pan-sharpening and other image-rectifying processes as described in Semi-Automatic Classification Plugin Documentation (Congedo, 2017). Using Semi-Automatic Classification Plugin (SCP) in QGIS, the Pre-processing tab was opened and Landsat was selected. The directory containing Landsat bands and the corresponding metadata were specified. DOS1-atmospheric correction and pan-sharpening were selected and the result was saved as a band set containing the pre-processed imageries. From the pre-processed imageries, a false colour band combination that displays the near-infrared (NIR) and parts of the visible (red and green) bands of the images was derived to enhance visual interpretation, with the assumption that the red colouration indicates vegetation, cyan represents built-up areas, dirty ash represents bare surfaces, and black depicts water bodies, as described by Eludoyin and Iyanda (2018). The images were classified into different land uses as adapted from a modified version of Anderson et al.'s (1976) classification schema (Table 8.2). Based on the model, a training set of pixels was selected for a supervised classification using a maximum likelihood (ML) classification algorithm for five land (built-up, bare surface, farmland, vegetation, waterbody) classes, after which the LULC classification was derived.

Finally, the results of the supervised classification were assessed for accuracy using an error or confusion matrix (Jensen and Cowen, 1996; Congalton and Green, 2008). Error matrix is often used to assess the accuracy of the land cover classification. For this study, accuracy assessment under the Post-processing tab of the *SCP dock* was

---

TABLE 8.2
Classification Schema of Land Use/Cover

Code	Land use/cover	Description
1	Bare surface	Land devoid of vegetation; land with exposed soil
2	Built-up	Land used for residential, commercial, industrial, etc. purposes
3	Farmland/agricultural land	Land used for farming (plantation, cropland)
4	Vegetation/natural forest	Land covered with natural vegetation
5	Waterbodies	Reservoirs, rivers, streams, lakes

Source: Anderson et al., 1976, Eludoyin and Iyanda, 2018.

selected, and both classified data and ground-truth data (reference) were overlaid to obtain the error matrix, the *overall accuracy, Producer's accuracy, User's accuracy,* and *Kappa coefficient* that were later exported in *xlsx* format for better visualisation.

8.4.3 ANALYSIS

Sen's slope and Mann-Kendall (MK) trend tests were used to estimate the rainfall trend and its magnitude, respectively. The MK trend test (S) is a non-parametric test that has been found useful in determining the monotonic trend in a data series (Mondal et al., 2012; Aswad et al., 2020). The MK test is insensitive to the normal distribution of data time series and outliers and appropriate for identifying patterns in time series. It compares each data value in the time series with all subsequent values, and S is assumed to be zero except when a data value in subsequent periods is higher than a data value in the previous period that the S increases by 1, or vice versa. The final S is considered negative for a negative trend, zero for no trend, and positive for an increasing trend (equation 8.1; Yue et al., 2002)

$$S = \sum_{i-1}^{n-1} \sum_{j-i+1}^{n} sign(xj - xi)$$ (8.1)

where:

$$sign(xj - xi) = 1 \quad if \ (xj - xi) > 0;$$
$$= 0, if \ (xj - xi) = 0;$$
$$= -1, if \ (xj - xi) < 0;$$

Xi and Xj = values in years i and j, respectively.

Sen's slope estimator is also a non-parametric test procedure that is often used to determine the gradient of slope in the MK trend test (Sen, 1968; Gilbert, 1987). The Sen slope estimates the magnitude of the trend in the MK test. Like the MK test, Sen's slope is also insensitive to single errors or outliers. Sen's slope (Q) is typically determined as described in equation 8.2.

$$Q = \frac{xj}{j} - \frac{xk}{k}$$ (8.2)

where: Q = Sen's slope; xj and xk = values at time j and k, respectively.

In this study, the MK test statistics were determined using Paleontological Statistical (PAST) software (version 4.08), while Sen's slope was determined using *MakeSens* software (version 1.0) developed by the Finnish Meteorological Institute.

Furthermore, the standardised precipitation index (SPI) of the rainfall data was determined using Drought Indices Calculator (DrinC, http://drought-software.com/). The SPI identifies rainfall anomalies based on total precipitation amounts for some

TABLE 8.3
Category of Standardised Precipitation Index (SPI)

Anomaly	SPI range	Precipitation regime	Probability (%)
Positive	≥ 1.0	Moderately wet–extremely wet	9.2–2.3
None	1.0	Normal	68.2
Negative	≤ 1.0	Moderately dry–extremely dry	9.2–2.3

Source: WMO, 2012

time (McKee et al., 1993). Positive SPI indicates wetness, while drought is indicated by a negative trend (Table 8.3). It was obtained in this study using equation 8.3.

$$SPI = \frac{X - \bar{x}}{SD} \tag{8.3}$$

where:
$(X - \bar{x})$ = anomaly; SD = standard deviation of mean

8.4.4 MULTI-CRITERIA ANALYSIS (MCA) FOR DELINEATION OF THE FLOOD RISK ZONE

Factors to which criteria weight were given included distance to drainage (1), elevation (2), slope (3), land cover type (4), flow accumulation (5), topographic wetness index (TWI) (6), drainage density (7), soil type (8), geology (9), and normalised difference water index (NDWI) (10) as described in the literature (e.g. Ouma and Tateishi, 2014; Das, 2019). To achieve the MCA, each theme of features was reclassified using the reclassify tool in ArcMap, and unique values (ranging from 1–9) were assigned to each class based on the severity of their influence on humans and the ecosystem (see Mwangi, 2016). Classes with the value '9' possess the severest risk' while it is lowest with '1' (Table 8.4).

8.5 RESULTS

8.5.1 CHANGES IN LAND USE/COVER OVER OSUN RIVER BASIN (1986–2020)

Analysis of the multi-date (1986, 1999, 2006' and 2020) Landsat images show comparative changes in the classified land use/cover of the Osun basin area over the selected study period (Table 8.5). As of 1986, the built-up area was 6.48 km² accounting for 6.4% of the entire area, while bare land surface/open land space was 26.08 km² or 26.6% of the total area. Areas used as farmland or short vegetation (including shrubs) occupied 34.5 km² or 37.9%, while tree and tall grasses vegetation was 26.7 km² or 26.2%. The waterbody was 4.1 km², occupying (4.02%) the least proportion (when compared with other land use/cover) of the 101.8 km² total Osun basin area in Osogbo.

TABLE 8.4
Weights Assigned to Each Parameter for Weighted Overlay

Thematic layer	Fuzzy weight (w_j)	Weight (%)
Distance to drainage	0.142	14
Elevation	0,131	13
Drainage density	0.122	12
Slope	0.121	1
Flow accumulation	0.100	10
TWI	0.085	9
LULC	0.083	8
Geology	0.083	8
Soil	0.072	7
NDWI	0.070	7
Σ	1	100

Table 8.5 also shows that while built-up areas and bare land/open land surfaces were on the increase within the study area, areas covered by waterbody and tree vegetation/tall grasses have declined. Areas occupied by farmland/short grasses/shrubs first declined between 1986 and 2006 but have recently (2006–2020) experienced a growth of 7.8% of what existed between 1996 and 2006. The area has witnessed increases in the population through immigration, and accompanying accommodation and facilities, following the status of Osogbo as an administrative state capital. The increase in population and built-up areas would understandably be linked with the reduction in open land surface, tree vegetation, and waterbodies, especially as there is evidence of construction in areas that hitherto were riparian vegetation zones. In general, the distribution of the land area occupied by waterbody and tree vegetation/tall grasses are most lost land cover in the study area in 2020 (Figure 8.3).

8.5.2 Patterns of Land Use/Cover Change and Growth in the Built-Up Area

The results of the analysis of the spatio-temporal variations in the land use/cover indicated that built-up areas have spread increasingly since 1986 and that the spread appeared to have commenced in the northern part of the basin in 1996 before it steadily grew towards the southwestern part in 2020 (Figure 8.4). Figure 8.4 also shows that the majority of the green areas (farmland/short grasses/shrubs and tree vegetation/tall trees) have been converted to bare land/open land surface over the years.

When examined using the hierarchical classification method with a dendrogram plot, the plot (Figure 8.5) indicates the closest level of association between the two vegetation classes (that is, farmland/short grasses/shrubs and tree vegetation/tall trees), followed by the association between built-up area and bare land/open land surface.

The green (farmland/short grasses/shrubs and tree vegetation/tall trees) classes were associated with the water body before they became linked with the built-up

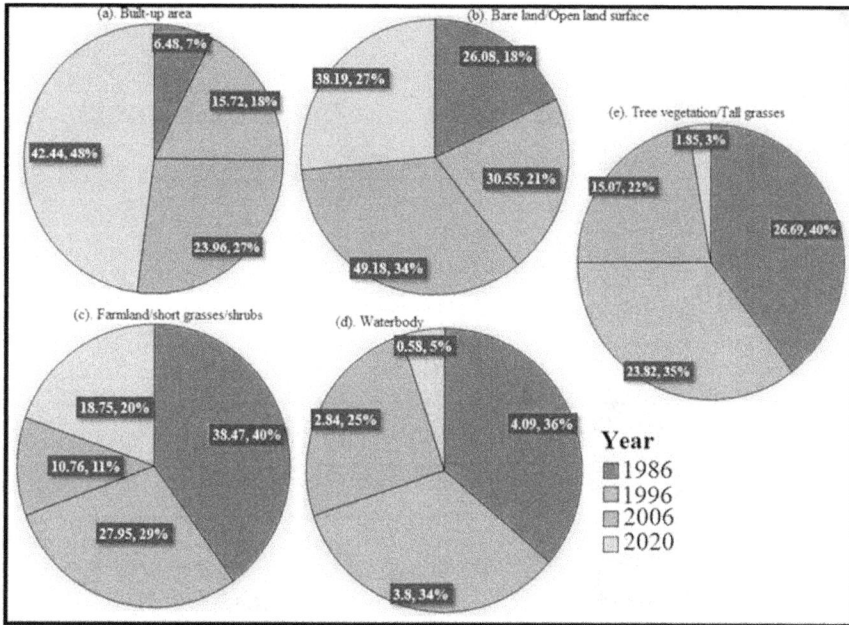

FIGURE 8.3 Distribution of different land use/cover (area in km², percentage of total area) in the different Landsat images of the Osun river basin in Osogbo.

Source: The image was drawn by the authors.

area with bare land/open land surface class. The level of association represents the level of interaction and thus appears to suggest that the previously tree vegetation/ tall grasses had changed to farmland/short grasses/shrubs with associated effects on the waterbody, on the one hand. On the other hand, the strong association between built-up areas and bare land/open land surface suggests that open spaces have been attached to an increase in built-up areas. In either case, it is easy to link the changes that occurred to anthropogenic causes, which may have been prompted by population increase and urbanisation.

Furthermore, the result of the stepwise regression that was used to examine the relationship of the growth in a built-up area with other land use/cover indicated that the tree vegetation/tall grasses were more impacted than other land use/cover (equation 8.4). The partial correlation of bare land/open land surface, farmland/short grasses/ shrubs, and waterbody with the built-up area were however weak ($r = -0.0026, -0.23$, and 0.09, respectively) and they contributed very insignificantly ($p \geq 0.85$) to the direct explanation of the changes in a built-up area in the basin.

$$y = -0.989x + 45.01 \left(F = 92.148, p = 0.011 \right) \qquad (8.4)$$

where; y = built-up area (km²), x = tree vegetation/tall grasses

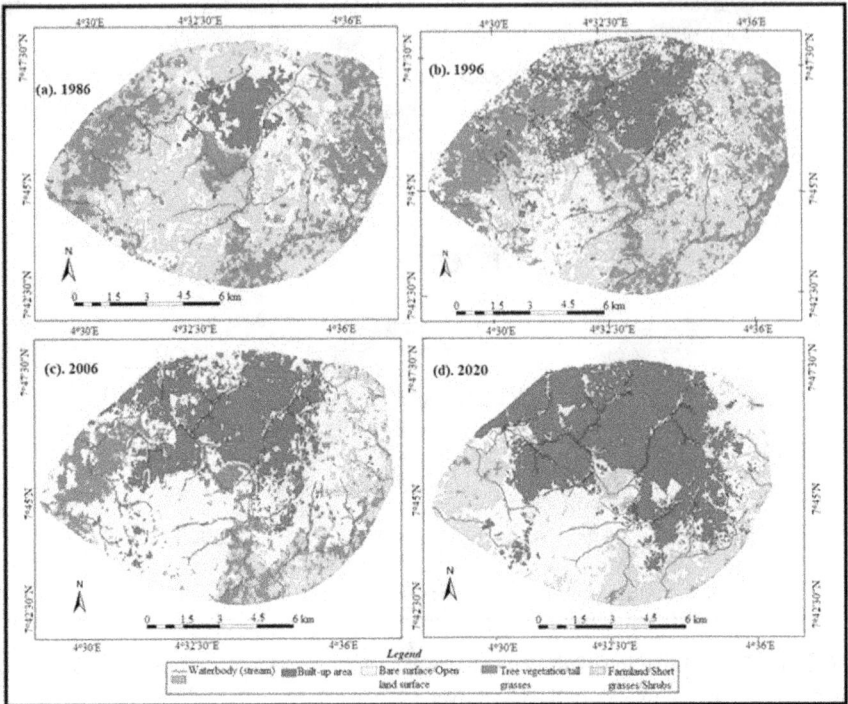

FIGURE 8.4 Pattern and direction of temporal changes in the classified land use/cover over the study area in 1986, 1996, 2006, and 2020, respectively.

Source: Image was drawn by the authors.

8.5.3 VARIABILITY IN RAINFALL PATTERNS OVER THE OSUN RIVER BASIN

Results of the mean monthly variations in rainfall and the trends are presented in Table 8.6. The peak in the 39 years (1988–2020) occurred in October, while the highest mean value occurred in September. Mean total rainfall values were generally low (less than 25 mm) for November, December, January, and February, while over 100 mm of rainfall was recorded for April through October. The monthly rainfall distribution exhibited high ($\geq 78\%$) variability throughout the year but the variability from November to April exceeded 100%. In terms of trends, January, February, April, May, and August exhibited a general decline in rainfall between 1982 and 2020 although the coefficient of determination was generally low (R^2 being less than 0.3 or 30% for the monthly trends), indicating that change in the year may not be sufficient to explain the observed monthly trends.

The annual rainfall trend between 1982 and 2020 shows fluctuating (alternating increase and decrease) patterns with extreme conditions in June, September, and October 2019 (Figure 8.6).

TABLE 8.5
Land Use/Cover Change (Area in km², and % Change in Parentheses) between 1986 and 2020

Land use/cover type	Area covered in 1986 (km²)	Change from 1986 to 1996 km² (%)	Change from 1996 to 2006	Change from 2006 to 2020
Built-up	6.5	9.2 (9.1)	8.2 (8.1)	18.48 (18.2)
Bare land/open land surface	26.18	4.5 (4.4)	18.6 (18.3)	11.0 (10.8)
Farmland/short grasses/ shrubs	38.5	−10.5 (−10.3)	−17.2 (−16.9)	8.0 (7.8)
Tree vegetation/tall grasses	26.7	−2.9 (−2.8)	−8.75 (−8.6)	−13.2 (−13.0)
Waterbody	4.1	−0.3 (−0.3)	−1.0 (−0.9)	−2.3 (−2.2)

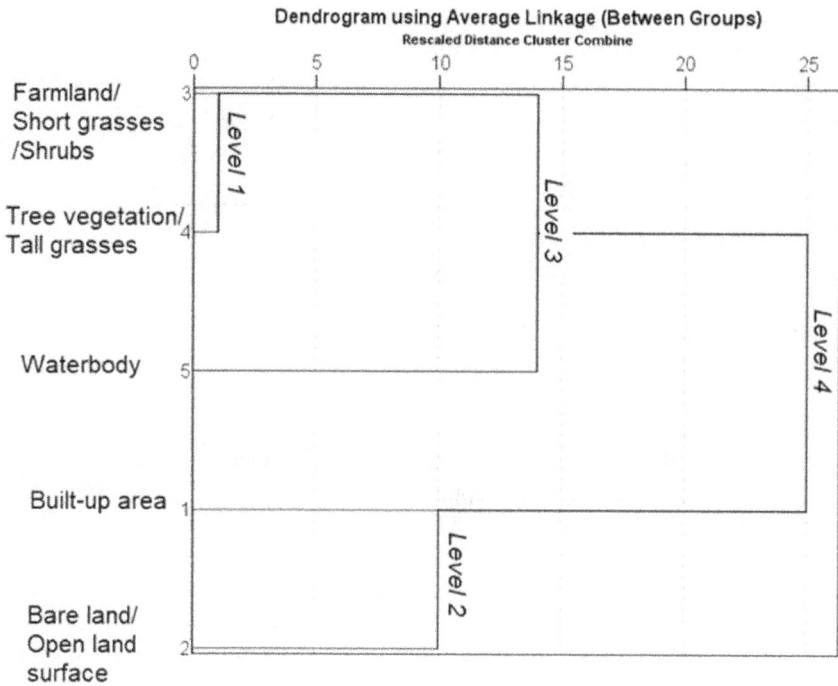

FIGURE 8.5 Dendrogram plot of the hierarchical classification of the classified land use/ cover areas in 1986, 1996, 2006, and 2020.

Source: The image was drawn by the authors.

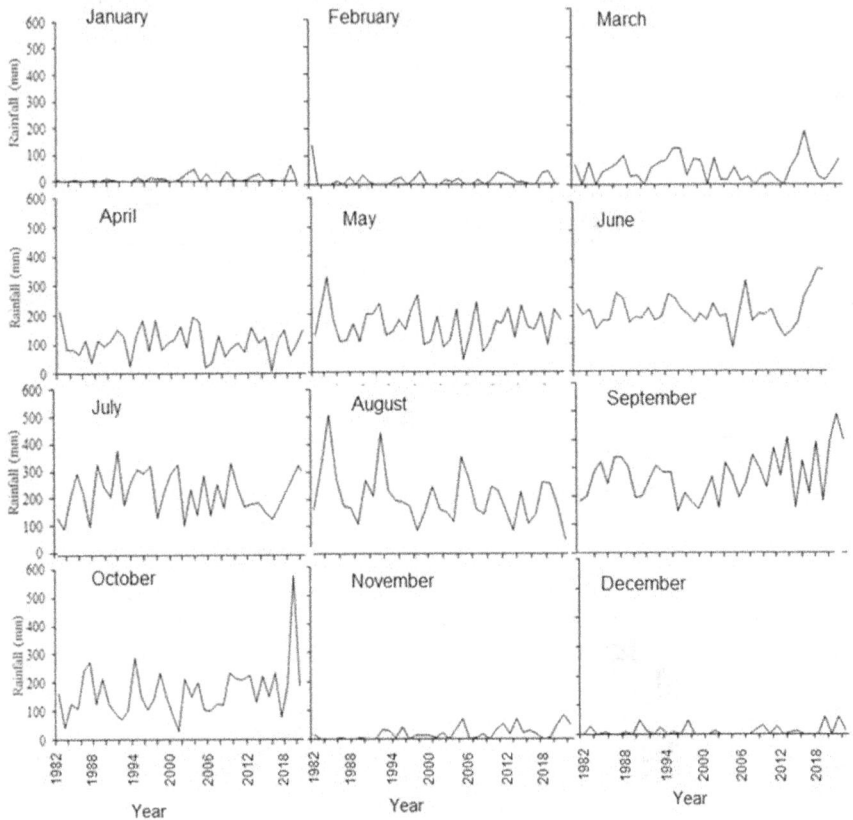

FIGURE 8.6 The pattern of rainfall trend for each month in 1982–2020.

Source: The image was drawn by the authors.

In all, rainfall steadily fluctuated without an obvious increase in most of the wet months, except September, which appeared to have increased since 2000. When compared, the results of the Mann-Kendall test and Sen slope for each of the 1982–2000 and 2001–2020 climatic periods indicate a comparative change in January, April, May, August, and September (albeit insignificantly; $p > 0.05$, except in January) (Table 8.7). Whereas rainfall in January, April, and August exhibited increased rainfall trends in the 1982–2000 period, rainfall in the same months in the 2001–2020 period showed decreasing trends. For May and September, rainfall exhibited an increasing trend in 2001–2020 in contrast to what was obtained in the same month in the 1982–2000 period.

8.5.4 VULNERABILITY OF THE OSUN RIVER BASIN AREA TO FLOODING EVENTS

The vulnerability of the study area to flooding events was assessed using the SPI and rainfall–discharge relationship. For the SPI, the results showed that 76.92% of the

TABLE 8.6
Thirty–nine (1982–2020) Monthly Rainfall Characteristics in the Study Area

Months	Mean (mm)	Peak	Coefficient of variation (%)	Trend (a ± bx)	R^2
Jan	9.48	10.6	384.21	2.5–0.06x	0.05
Feb	21.0	142.4	252.38	18.73–0.17x	0.006
Mar	66.2	179.3	156.80	47.41 + 0.17x	0.002
Apr	143.1	216.2	113.42	114.72–0.26x	0.004
May	214.0	327.0	89.58	172.99–0.51x	0.009
Jun	280.3	362.7	83.05	195.05 + 0.99x	0.04
Jul	274.8	374.4	96.32	237.84–0.94x	0.01
Aug	259.3	506.2	99.58	263.14–3.12x	0.11
Sept	347.5	491.1	78.22	221.07 + 2.50x	0.12
Oct	203.1	579.2	107.83	119.27 + 2.49x	0.09
Nov	24.6	80.2	218.29	-0.67 + 0.96	0.25
Dec	19.6	59.6	198.47	4.78 + 0.28x	0.03

TABLE 8.7
Result of Man-Kendal (MK) Test of Change in Monthly Trend and Magnitude of Rainfall for 1982–2000 and 2001–2020 Periodic Slices in the Study Area (the Trend is Significant at $p \leq 0.05$)

Month	1982–2000 MK test	Sen's slope (Q)	2001–2020 MK test	Sen's slope (Q)	p value
Jan	1.50	0.00	−0.26	−0.07	0.02
Feb	0.46	0.00	1.17	0.05	0.49
Mar	2.04	3.30	1.57	2.09	0.54
Apr	0.84	1.41	−0.23	−0.40	0.40
May	−0.28	−0.81	1.04	2.88	0.56
Jun	0.00	0.00	1.10	4.27	0.59
Jul	1.89	7.907	1.40	4.75	0.65
Aug	0.25	1.04	0.25	−1.37	0.17
Sept	−1.30	−4.06	1.98	7.82	0.07
Oct	0.00	0.00	1.43	4.632	0.34
Nov	1.51	0.40	1.11	1.32	0.08
Dec	−0.28	0.00	1.18	0.00	0.19

entire annual precipitation regimes were 'normal', indicating non-threatening rainfall, provided the drainage systems were adequate and water flow was not blocked, while there was only a case of extreme wetness in 2019 (Figure 8.7). Consequently, given the rare occurrence of the extremely wet conditions in the study area, flooding events are not expected if there is adequate urban structure and if the drainage systems are adequate.

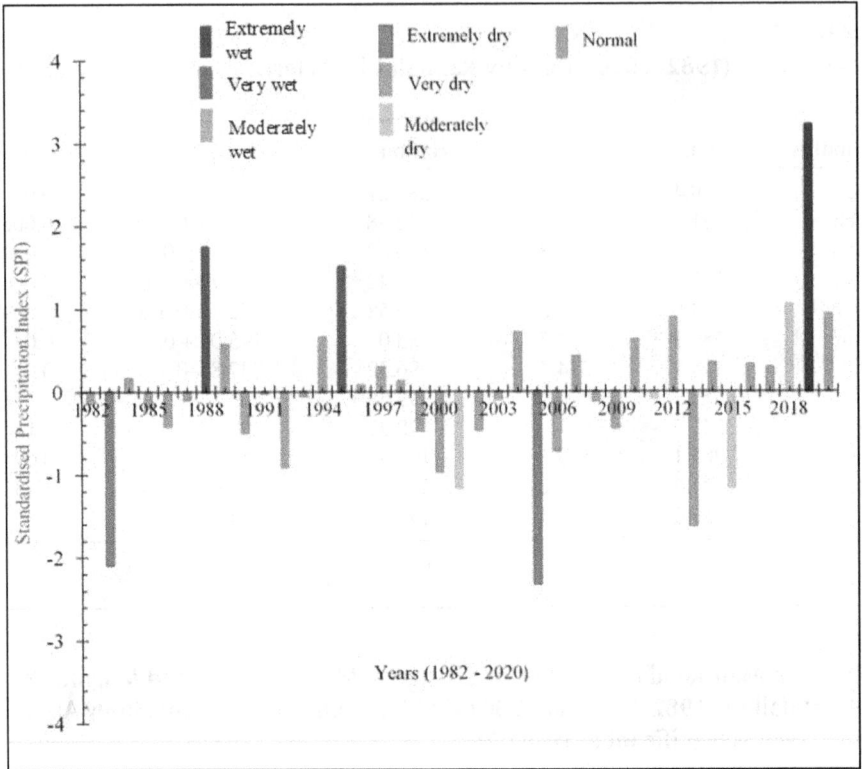

FIGURE 8.7 Annual variations in the standardised precipitation index over the Osun river basin in Osogbo (1982–2020).

Source: The image was drawn by the authors.

Since the study area is in the near river zone, the rainfall–discharge relationship was examined to understand the response of the stream flow to rainfall input and the potential impact on the immediate environment. The results indicate conspicuous differences in the hydrographs for the selected years. In 1991, the rainfall–discharge relationship was strong and positive ($r = 0.76$), and rainfall accounted for 57% of the change in discharge. By 2001, the rainfall–discharge relationship had become weak and insignificant ($r = 0.02$, $R^2 = 0.004$, $p > 0.05$), indicating the presence of other factors that substantially influence the stream flow other than rainfall input. In addition, in 2019, the relationship was weak and rainfall accounted for only 5% of the total variance in streamflow in the area (Figure 8.8). In reality, several factors, including increased built-up areas, removal of vegetation, and dredging of the river channel have occurred between 1991 and 2019 that may impact the rainfall–discharge relationship in the area.

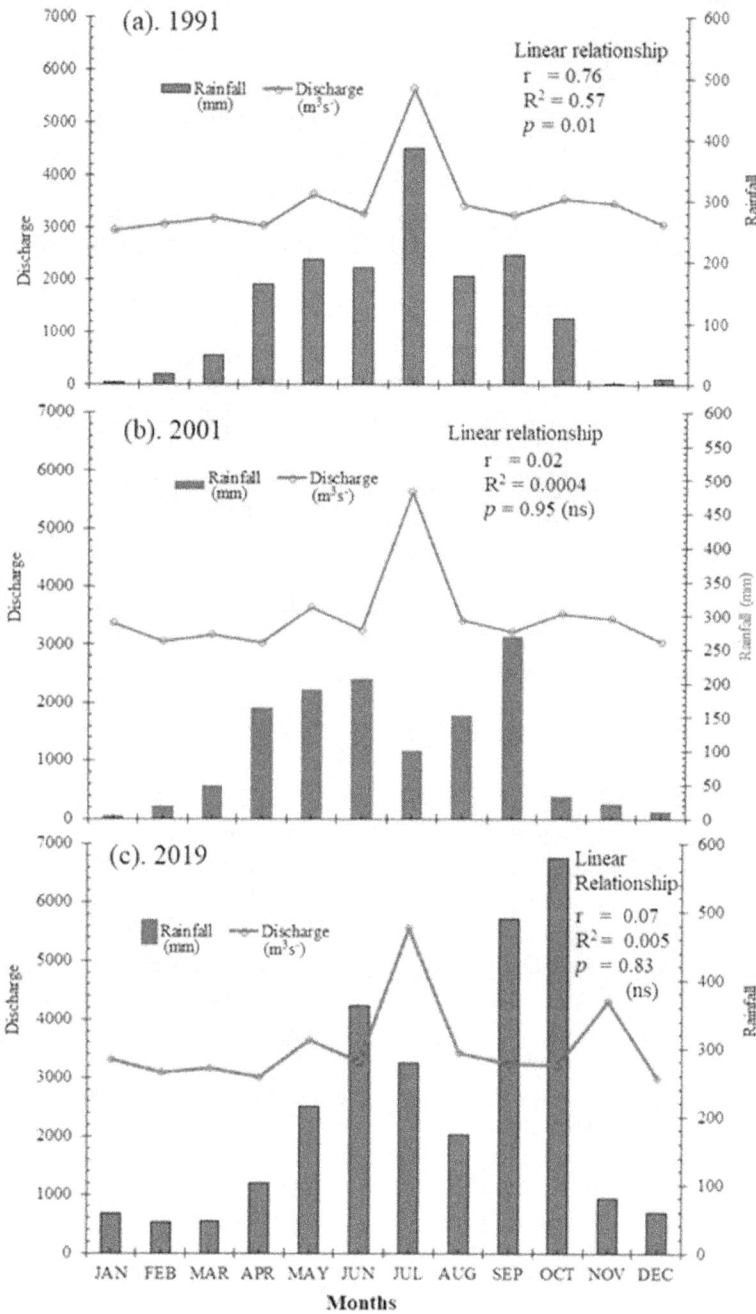

FIGURE 8.8 Rainfall–discharge relationship in selected years (1991, 2001, and 2019; based on data availability) at a measuring station in the Osun river basin.

Source: Images were drawn by the authors.

8.5.5 FLOOD RISK MAPPING OF THE RIVER BASIN

Results of the multi-criteria analysis used to determine the vulnerability of the study area to flood risks identified five major risk zones based on the level of vulnerability. Table 8.8 provides the results of the list of the average and range (as appropriate for proximity to the Osun river or major tributaries, drainage density, elevation, dominant land use/cover distance, elevation, topographic wetness index [TWI] normalised difference water index [NDWI], geology, and soil class/group) of the criteria used. The characteristics of each of the zones are described in the sub-sections below.

8.5.5.1 Zone 1: the Very High Flood-Risk Zone

This is the most vulnerable region to flood incidence in the study area. Important settlements in the zone include Alekunwodo, Oke-fia, Old garage, Fakunle, Olorunsogo, and Oke-Abesu (Figure 8.9). The area covers about 4.5 km². The zone is characterised by 0–728.5 km around the Osun river or its tributaries, drainage density in the zone is 0.5–5.2, and elevation above mean sea level varied between 307 m and 357 m with a 0–8.5 degree slope. The NDWI varied from–0.6 to 0.4 while the TWI in this zone was 4.6–19.7 in the region; flow accumulation varied from zero to 439,577 units. Major land uses in the zone are built-up areas, bare land/open land surfaces, and farmland/short grasses/shrubs. Tree vegetation/tall grasses are essentially lacking in most parts of this region. The geological underlain is essentially blended gneiss and schist and the soil composition is fluvisol and lixisols. The geological underlain and soils as well as land use/cover are common to zones 2 and 3 which experienced comparatively high and moderate risks.

8.5.5.2 Zone 2: The High Flood-Risk Zone

This zone is highly vulnerable to flooding. Important settlements in the zone include Okefia, Ilupeju, Onward, Atelewo, Tanisi, Olugunna, Oke Osun, Osunjela, Gbonmi, Gbodofon, Famson, Ajegunle, Oke Onitea, Oke Ijetu, Hallelujah Estate, and Olorunsogo (Figure 8.9). The area covers around 22.4 km². The zone is characterised by 0–924.1 km around the Osun river or its tributaries, drainage density in the zone is 0.5–5.19, and elevation above mean sea level varied between 305 m and 369 m with a 0–14.7 degree slope. The NDWI varied from –0.75 to 0.89 while the TWI in this zone was 4.14–19.07 in the region; flow accumulation varied from zero to 439,673 units. Major land uses in the zone are built-up areas, bare land/open land surfaces, and farmland/short grasses/shrubs. Tree vegetation/tall grasses are essentially lacking in most of this region. The geological underlain is essentially banded gneiss and schist and the soil composition is fluvisol and lixisols. The geological underlain and soils as well as land use/cover are common to zones 2 and 3 which experienced comparatively high and moderate risks

8.5.5.3 Zone 3: The Moderate Flood-Risk Zone

The risk of flooding in this zone is moderate except when triggered by an excess of anthropogenic and natural factors such as rainfall, blocked drainage, etc. Important settlements in the zone include Ogo Oluwa, Fagbewesa, Oluode, Kolabalogun, Testing

TABLE 8.8
Identified Attributes of the Different Classes of Flood Risk Zone in the Osun River Basin in Osogbo, Osun State

Classified vulnerable zone	Dominant land use	Proximity to Osun river/tributary (km)	Drainage density	Elevation (m)	Slope (degree)	NDWI (no unit)	TWI (no unit)	Flow Acc.	Geology	Soil
Very High	• Built-up,	0–728.5	0.5–5.2	307–357	0–8.5	−0.6–0.4	4.6–19.7	0–439577	Banded gneiss, Schist	Fluvisols, Lixisols
High	• Bare land/open surface	0–924.07	0–5.19	305–369	0–14.70	−0.75–0.89	4.14–19.07	0–439673		
Moderate	• Farmland/short grasses/shrubs	0–1573.03	0–3.83	307–384	0–18.88	−0.78–0.76	3.74–15.68	0–36315		
Low	• Tree vegetation/tall grasses	12.6–1941.7	0–2.18	309–405	0–18.66	−0.79–0.43	4.14–19.07	0–4737		Lixisols
Very Low	• Built-up, • Bare land/open surface • Farmland/short grasses/shrubs	328.3–1945.4	0–0.71	326–418	0–23.21	−0.70–0.01	4.60–19.13	0–956		

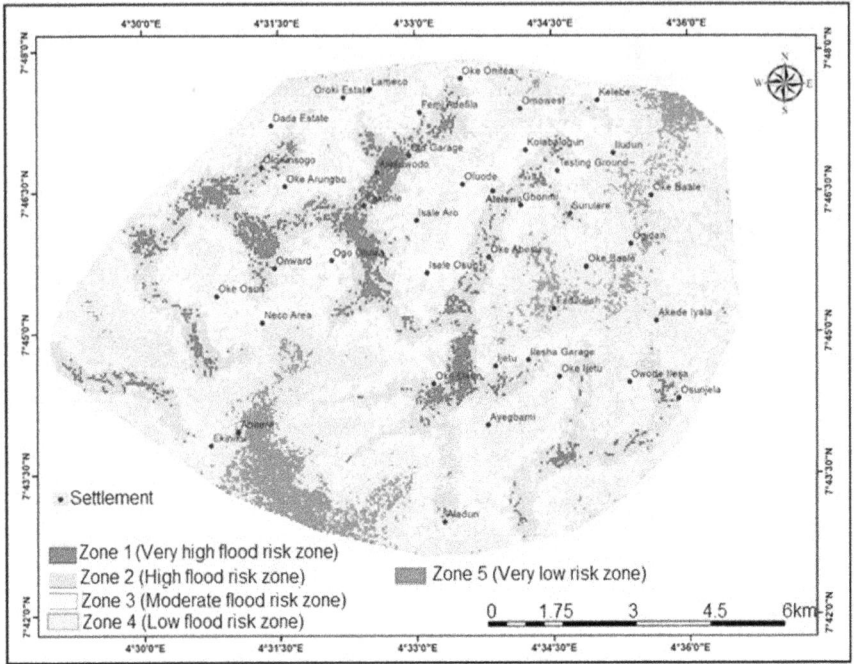

FIGURE 8.9   Classified flood-risk zones based on selected multi-criteria weights.

Source: The image was drawn by the authors.

Ground, Oke Baale, Oroki Estate, Lameco, Ilesa garage, Oluode, Oke Arungbo, Isale Osun, and Osogbo Olorunda (Figure 8.9). The area covers around 39.6 km² The zone is characterised by 0–1573.0 km around the Osun river or its tributaries, drainage density in the zone is 0.5–3.8, and elevation above mean sea level varied between 307 m and 384 m with a 0–18.9° slope. The NDWI varied from –0.78 to 0.76 while the TWI in this zone was 3.74–15.68 in the region; flow accumulation varied from zero to 36,315 units. Major land uses in the zone are built-up areas, bare land/open land surfaces, and farmland/short grasses/shrubs. Tree vegetation/tall grasses are essentially lacking in most parts of the region. The geological underlain is essentially banded gneiss and schist and the soil composition is fluvisol and lixisols. The geological underlain and soils as well as land use/cover are common to zones 2 and 3 which experienced comparatively high and moderate risks.

8.5.5.4   Zones 4 and 5: The Very Low and Low Flood-Risk Zones

These zones are the least vulnerable to flooding incidence in the study area. Important settlements in these zones include Isale Aro, Omo West, Ago Wande, Fadillulahi, Akede, and Oloruntedo (Figure 8.9). The area covers about 33.5 km². The zone is characterised by 12.6–1945.4 km around the Osun river or its tributaries, drainage density in the zone is 0–0.71, and elevation above mean sea level varied between 309 m and 418 m with a 0–23.2° slope. The NDWI varied from –0.79 to 0.01 while the TWI in this zone was 4.14–19.13 in the region; flow accumulation varied from zero to 956 units. Major land uses in the zone are built-up areas, bare land/open land surfaces, tree vegetation/tall grasses, and farmland/short grasses/shrubs. The geological underlain is essentially banded gneiss and schist and the soil composition is majorly lixisols (see Figure 8.9).

8.5.6 Perception of Stakeholders

The participants linked torrential rainfall, river overflow, poor waste management, illegal structures across drainage channels, inadequate drainage channels, blocked drainage channels, and closeness to river channels with flooding events in the river basin (Table 8.9). The officer of the Flood Control Unit of the State Ministry of Environment and Sanitation argued that floods occur any time rainfall exceeds the typical normal and the river channel is either blocked or narrowed due to sedimentation. They noted that:

> *Dredging of the river channel, followed by removal of the silt (desilting) which was done before 2011 drastically reduced the occurrences of flood events that became prominent in 2019.*

TABLE 8.9
Responses on Causes and Effects of the Flood in the Study Area

Variables	Variables	Percentage (%)
Factors associated with flooding	Torrential rain	82
	Illegal structures across drainage channels	62
	Inadequate drainage channels	72
	Collapsed dam	10
	Poor waste management	62
	River overflow	64
	Blockage of drainage channels	68
	Closeness to stream channel	60
	Supernatural forces	16
	Climate change	26
	Lack of disciplinary action towards defiant citizens	48
	Population increase and urban expansion	32
	Nonchalant attitude and indiscipline	58
Nature of loss to the flood	Building collapse	34
	Damage to vehicle	60
	Death of livestock	40
	Destruction of shops and goods	78
	Loss of properties (gadget/document)	84
	Death of relative	6
	Destruction of farmland	8
	Outbreak of disease	10

The participants blamed poor finance allocation for annual dredging and 'desilting' for continuous flooding events in the area. Other factors that were linked to flooding problems in the area included blockage of river channels and drains with refuse, blockage of river channels through the planting of plantain/banana stems, poor and inadequate drains, and the increasing number of poorly planned buildings and other structures along the river course. A key informant disclosed that increased population and poor urban planning procedures were major problems in the area:

> The flooding problem is amplified by haphazard town planning; such that access roads and line drains have been replaced with residential/permanent structures in many places. Buildings have also been erected too close to streams in other places, and setbacks have been encroached with structures.

In terms of strategies for resilience and coping or preparedness, 70% of the respondents claimed to clear their blocked drainages, while 46% constructed drainage channels and flood steps. About 52% decided to create a safe for their valuables in shelves and 24% claimed to plant trees/protective lawns. If caught by flood, 24% claimed to transfer their valuables to their neighbours, especially to those living upstairs in storey buildings, while 68% claimed to clear blocked drainages to aid the flow of the flood water, 40% placed sandbags in flooded areas and constructed makeshift channels to aid the flow of flood water, and about 30% claimed to relocate to other places. Many of the respondents have equally suffered many losses due to floods (see also Table 8.9).

8.6   CONCLUSIONS AND RECOMMENDATIONS

This study exemplifies the use of GIS as a data-coordinating approach to environmental studies, which allows information from different sources to be complementarily evaluated. In this study, results from satellite data analysis and ODK responses made inferences on flood events possible. The study, therefore, concluded on the effectiveness of GIS to provide a better flood decision-support system than single-based, standalone models. The study is ongoing, and further results are envisaged. Recommendations for a similar study of urban flood at catchment – or basin – scale are encouraged to be able to protect the basin region and avoid the depletion of basin resources.

REFERENCES

Adedotun, S.B., 2015. A Study of Urban Transportation System in Osogbo, Osun State, Nigeria. *European Journal of Sustainable Development*, 4(3), 93–93.

Adegboyega, S.A., Onuoha, O.C., Komolafe, A.A., Olajuyigbe, A.E., Adebola, A.O. and Ibitoye, M.O., 2018. An integrated approach to modelling of flood hazards in the rapidly growing city of Osogbo, Osun State, Nigeria. *Am J Space Sci*, 4, 1–15.

Adejuwon, J.O. and Ekanade, O., 1988. A comparison of soil properties under different landuse types in a part of the Nigerian cocoa belt. *Catena*, 15(3–4), 319–331.

Adewole, A.O., Ike, F. and Eludoyin, A.O., 2020. A multi-sensor-based evaluation of the morphometric characteristics of Opa river basin in Southwest Nigeria. *International Journal of Image and Data Fusion*, 11(2), 185–200.

Agbola, B.S., Ajayi, O., Taiwo, O.J. and Wahab, B.W., 2012. The August 2011 flood in Ibadan, Nigeria: Anthropogenic causes and consequences. *International Journal of Disaster Risk Science, 3*(4), 207–217.

Aguda, A.S. and Adegboyega, S.A., 2013. Evaluation of spatio-temporal dynamics of urban sprawl in Osogbo, Nigeria using satellite imagery & GIS techniques. *International Journal of Multidisciplinary and Current Research, 1*(1), 39–51.

Akinwumiju, A.S. 2016. Morphometric and Land Use Analysis: Implication on Flood Hazards in Ilesa and Osogbo Metropolis, Osun State, Nigeria. *Ethiopian Journal of Environmental Studies and Management,* 10(2), 229–240.

Anderson, J. R., Hardy, E. E., Roach, J. T. and Witmer, R.E. 1976. A Land Use and Land Cover Classification System for Use with Remote Sensor Data. *Geological Survey Professional Paper No. 964, U.S Government Printing Office, Washington DC.* 28.

Congalton, R. G. and Green, K. 2008. *Assessing the Accuracy of Remotely Sensed Data Principles and Practices.* 2nd ed. Boca Raton, FL CRC Press, Taylor and Francis Group.

Congedo, L. (2017). Semi-Automatic Classification Plugin Documentation.

Das, S. (2019). Geospatial Mapping of Flood Susceptibility and Hydro-Geomorphic Response to The Floods in Ulhas Basin, India. *Remote Sensing Applications: Society and Environment,* 14: 60–74.

Egunjobi, L. 1995. Aspects of urbanization, physical planning and development Osogbo in model of growing African Town. C.O. Adepegba, *Institute of African Studies, University of Ibadan, Nigeria* 13–28.

Eludoyin, A. O., and Iyanda, O. O. 2018. Land cover change and forest management strategies in Ife nature reserve, Nigeria. *GeoJournal, 84*(6), 1531–1548.

Eludoyin, A.O. and Adewole, A.O., 2020. A remote sensing-based evaluation of an ungauged drainage basin in Southwestern Nigeria. *International Journal of River Basin Management, 18*(3), 307–319.

Eludoyin, A.O., Eludoyin, O.M. and Oyinloye, M.A., 2009. Monthly variations in the 1985–1994 and 1995–2004 rainfall distributions over five synoptic stations in western Nigeria. *J Meteorol Clim Sci, 7*, 11–22.

Esan, A. L., and Babatola, E. B. 2015. Urbanization and its environmental impacts in Nigeria: Implications for sustainable development (A case study of Ado-Ekiti). *Pyrex Journal of Research in Environmental Studies, 2*(3), 27–34.

Fatusin, A. F. 2015. Environmental quality perception and management in industries in Ondo state, Nigeria. *Economic and Environmental Studies, 15*(4 (36)), 349–361.

Gasu, M. B., Olaiyiwola, O. and Ezekiel, A. O. 2019. Appraisal of flooding and drainage conditions in Osogbo, Osun State, Nigeria. *1st International Conference on Engineering and Environmental Sciences* (p. 907).

Gebreyesus, G., Dessie, T., Wamalwa, M., Agaba, M., Benor, S. and Mwai, O., 2013. Harnessing 'ODK collect' on smartphones for on-farm data collection in Africa: The ILRI-BecA goat project. *ILRI APM.*

Gilbert, R.O., 1987. *Statistical methods for environmental pollution monitoring.* New York: John Wiley & Sons.

Hovinen, G. R. (1977). Leapfrog developments in Lancaster county: a study of residents' perceptions and attitudes. *The Professional Geographer, 29*(2), 194–199.

Jensen, J. R., and Cowen, D. C. (1999). Remote sensing of urban/suburban infrastructure and socio-economic attributes. *Photogrammetric Engineering and Remote Sensing, 65*, 611–622.Li, H., Ishidaira, H., Souma, K. and Magome, J., 2021. The Impact of Climate Change and Urbanization on Flood Control Capacity of Sponge City. *Journal of Japan Society of Civil Engineers, Ser. G (Environmental Research), 77*(5), 117–125.

Lillesand, T.M., Kiefer, R.W. and Chipman, J.W. (2008). *Remote Sensing and Image Interpretation,* 6th Edition, New York: John Wiley and Sons.

McKee, T.B., N.J. Doesken and J. Kleist. (1993). The relationship of drought frequency and duration to time scale: Proceedings of the Conference Climatology, Anaheim, California, 17–22 January 1993. Boston, American Meteorological Society. 179–184.

Mondal, A., Kundu, S., and Mukhopadhyay, A. (2012). Rainfall trend analysis by Mann-Kendall test: A case study of north-eastern part of Cuttack district, Orissa. *International Journal of Geology, Earth and Environmental Sciences,* 2(1): 70–78.

Muñoz-Sabater, J., Dutra, E., Agustí-Panareda, A., Albergel, C., Arduini, G., Balsamo, G., Boussetta, S., Choulga, M., Harrigan, S., Hersbach, H. and Martens, B., 2021. ERA5-Land: A state-of-the-art global reanalysis dataset for land applications. *Earth System Science Data,* 13(9), 4349–4383.

Mwangi, M. P. (2016). Department of Environmental Studies and Community Development, Kenyatta University. The Role of Land Use and Land Cover Changes and GIS in Flood Risk Mapping in Kilifi County, Kenya. Unpublished M.Sc Thesis.

Ogunsanwo, O. (1980). Geotechnical investigation of some soils from S.W. Nigeria for use as mineral seal in waste disposal landfills. *Bulletin of the International Association of Engineering Geology,* 22(1): 119–123.

Olajuyigbe, A. E., Rotowa, O. O., and Durojaye, E. (2012). An assessment of flood hazard in Nigeria: The case of mile 12, Lagos. *Mediterranean Journal of Social Sciences,* 3(2), 367–367.

Ologunorisa, T.E., 2004. An assessment of flood vulnerability zones in the Niger Delta, Nigeria. *International Journal of Environmental Studies,* 61(1), 31–38.

Ologunorisa, T.E., Eludoyin, A.O. and Lateef, B., 2022a. An evaluation of flood fatalities in Nigeria. *Weather, climate, and society,* 14(3), 709–720.

Ologunorisa, T.E., Obioma, O. and Eludoyin, A.O., 2022b. Urban flood event and associated damage in the Benue valley, Nigeria. *Natural Hazards,* 111(1), 261–282.

Oloukoi, J., Oyinloye, R. O., and Yadjemi, H. (2014). Geospatial analysis of urban sprawl in Ile-Ife city, Nigeria. *South African Journal of Geomatics,* 3(2), 128–144.

Onanuga, M.Y., Eludoyin, A.O. and Ofoezie, I.E., 2022. Urbanization and its effects on land and water resources in Ijebuland, southwestern Nigeria. *Environment, Development and Sustainability,* 24(1), 592–616.

Ouma, Y. O. and Tateishi, R. (2014). Urban Flood Vulnerability and Risk Mapping Using Integrated Multi-Parametric AHP and GIS: Methodological Overview and Case Study Assessment. *Water,* 6, 1515–1545.

Owolabi, J. T. (2019). GIS as a Tool in Analyzing Flood Occurrence and Its Impact on Ikere Ekiti, Ekiti State Nigeria. *Journal of Geographic Information System,* 11(05), 595.

Padi, P.T., Di Baldassarre, G. and Castellarin, A., 2011. Floodplain management in Africa: Large scale analysis of flood data. *Physics and Chemistry of the Earth, Parts A/B/C,* 36(7–8), 292–298.

Raja, A. Tridane, A. Gaffar, T. Lindquist and K. Pribadi (2014) Android and ODK based data collection framework to aid in epidemiological analysis, *Online Journal of Public Health Informatics,* 5(3), e228–255.

Salami, A.T., 1999. Vegetation dynamics on the fringes of lowland humid tropical rainforest of south-western Nigeria an assessment of environmental change with air photos and Landsat TM. *International Journal of Remote Sensing,* 20(6), 1169–1181.

Salami, B.T. and Sangoyomi, T.E. (2013). Soil fertility status of cassava fields in South Western Nigeria. *American Journal of Experimental Agriculture.* 3(1), 124–152.

Salami, R.O., Giggins, H. and Von Meding, J.K., 2017. Urban settlements' vulnerability to flood risks in African cities: A conceptual framework. *Jàmbá: Journal of Disaster Risk*

Studies, *9*(1), 1–9.Sen, P.K. (1968). Estimates of the regression coefficient based on Kendall's tau. *Journal of American Statistical Association.* 39, 1379–1389.

Shukla, S., Khire, M.V. and Gedam, S.S., 2013. Effects of increasing urbanization on river basins-state of art. *International Journal of Engineering Research & Technology, ISSN*, pp. 2278–0181.

UN-Habitat (2014). Structure Plan for Osogbo Capital Territory and Environs (2014–2033) State of Osun Structure Plans for Projects. Pub Ministry of Lands, Physical Planning and Urban Development.

Yu, S. S., Sun, Z. C., Guo, H. D., Zhao, X. W., Sun, L. and Wu, M. F. (2017). Monitoring and analysing the spatial dynamics and patterns of megacities along the Maritime Silk Road. *Journal of Remote Sensing, 21*, 1993–2002.

Yue S, Ouarda TB, Bobée B, Legendre P, Bruneau P (2002) Approach for describing statistical properties of food hydrograph. *J Hydrol Eng,* 7(2), 147–153.

Zhang, L. and Weng, Q. H. (2013). Annual dynamics of impervious surface in the Pearl River Delta, China, from 1998 to 2013, using time series Landsat imagery. *ISPRS Journal of Photogrammetry, 113*, 86–93.

9 Site Suitability for an Information Technology Corridor in Tirunelveli District Using GIS
A Restriction Model Approach

Venkatesh Baskaran, M.A.M. Mannar Thippu Sulthan, E. Aswin Raj, and A. Krishna Kumar

CONTENTS

9.1 INTRODUCTION

The growth of the IT sector has been awe-inspiring during the 21st century in India. This industry makes a significant impact on other sector economies by improving their productivity and accelerating growth. The tech sector alone contributes 8% to India's GDP (gross domestic product). As digital transformation is the biggest priority for global corporations, this info-tech industry consistently performs and hits 12% annual growth. This sector comprises software products, IT services, hardware supporting products, IT-enabled services, and e-commerce (Raman and Chadee, 2011). The

DOI: 10.1201/9781003331001-9

145

government has taken up and executed various initiatives such as establishing special economic zones (SEZs), software technology parks (STPs), and industrial complexes to achieve a commanding position in the global IT industry (Saraswathi, 2015)

India has several IT hubs which are distributed spatially around the nation. A few notable hubs are Bangalore, Hyderabad, Chennai, Mumbai, Visakhapatnam, Trivandrum, Kochi, Mysore, Indore, Jaipur, Mangalore, Lucknow, Coimbatore, Chandigarh, Bhubaneswar Vadodara, and Patna. Many state governments have recast their existing information and communications technology (ICT) policies to attract software industries to Tier-II cities within their states, either for expansion or modernisation. This may address the needs and employment opportunities of low-income graduates and local communities (Kramer, Jenkins, and Katz, 2007).

Site suitability analysis helps in filtering suitable sites to satisfy defined criteria or restrictions. The manual search for a suitable site by considering various aspects by collecting required data through field visits is time-consuming. Even after the collection of data, their interpretation and identification are very tedious (Kumar and Shaikh, 2013). The development and construction of a new IT sector is a significant long-term investment, and in this sense choosing the location is a crucial point on the road to the success or failure of that entire sector (Rikalovic, Cosic, and Lazarevic, 2014). To overcome these time constraints and financial risks, an advanced overlay technique that follows mapping data on the natural and human-made features and their attributes around the environment of a study area called a geographic information system (GIS) is used (Malczewski, 2004). With the development of GIS and high-end computer technology, site selection and land suitability are becoming easier for planners by using an advanced and scientific method (Ayele, Wondim, and Abebe, 2018; Baskaran and Velkennedy, 2022). In the developing stage, the planning agencies and local governments used GIS in general by promoting the merits of this technological tool. As the cost of introducing, learning, and maintaining a GIS is an expense, a cost–benefit analysis is performed to justify the acquisition of the technology. The researchers, practitioners, and academicians in civil engineering have witnessed the advantages of GIS in the earlier stage (Nedović-Budić, 1998; Miles and Ho, 1999). GIS is a powerful technique to investigate the high precision of multiple geospatial data with the flexibility to perform land suitability analysis (Pramanik, 2016; Hama, Al-Suhili, and Ghafour, 2019). While planners try to design and develop a project that benefits economic development and maintains ecological balance in systems and quality locations, they face various challenges (Srinivasan, Haque, and Rao, 2012). It will not allow them to decide depending on diverse behaviour in all aspects. The development of GIS analysis and its decision-making tools reduces the complexity by estimating future growth patterns with suitable assumptions (Patten and Arafat, 2010). The methods to analyse site suitability can simple or complex and elaborate. It includes graduated screening, weighted factors, pass/fail screening, penalty point assignment, composite rating, weighted composite rating, and direct assignment (Anderson, 1987; Banai-Kashani, 1989).

This chapter performs an analysis of suitable sites based on the pass/fail screen method, also known as the restriction model approach – this method screens only the locations that satisfy all aspects. Here, selecting alternatives may not result in the best alternative selection since the relationship within the criteria is not considered (Banai-Kashani, 1989).

The critical factors are location-based criteria; either presence or absence directly answers the site suitability in the restriction model (Liang and Wang, 1991). The critical factors alone are enough for this approach. Criteria such as the cost of land and occurrence details come under objective factors. Subjective factors deal with assigning weightage or priorities between the factors (Chou, 2010). Physical, ecological, infrastructure, and social criteria were considered globally for a suitability study on industrial developments (Fernando, Sangasumana, and Edussuriya, 2015; Johar, Jain, and Garg, 2013). The criteria considered for the IT corridor land suitability analysis are rivers and waterbodies, roadways, city municipal corporation limit, land use – bare land, slope, and area continuity (MD, 2019; Gong, Liu, and Chen, 2012). A suitable site is selected that does not create any impact on the surrounding environment (Bunruamkaew and Murayama, 2011). Therefore, the IT corridor location is planned beyond the buffer zones of rivers and waterbodies. Steps carried out to screen for the suitable location with a spatial map and supporting discussion are presented in this chapter.

9.2 METHODOLOGY AND STUDY AREA

9.2.1 STUDY AREA

The area selected for the study to perform a suitability analysis for the IT corridor is Tirunelveli District, situated in the southern part of Tamil Nadu. Tirunelveli is the only district in the state which contains five different landforms as stated in Tamil literature, including mountain region (Kurinji), forest (mullai), cropland (marutham), seashore (neithal), and dryland (palai). This district lies between 77° 10' E–77° 58' E and 08° 08' N–08° 56' N, over an an area of about 407,602 hectares, which extends to a boundary of 415 kilometres. Thamirabarani, a perennial river with a subbasin coverage of 2055 km², is also located in this district (Magesh and Chandrasekar, 2014). The location of our study area is shown in Figure 9.1. Most of the IT corridors

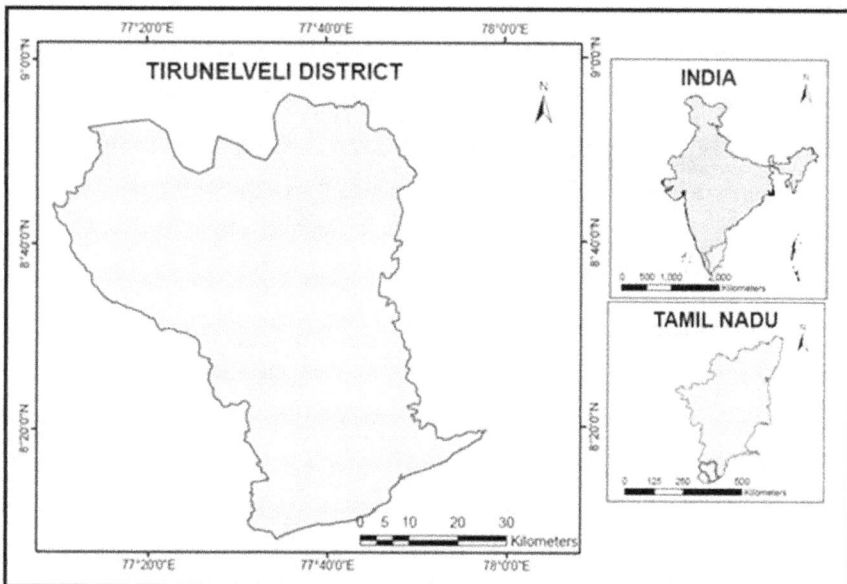

FIGURE 9.1   Site suitability analysis – study area.

in India were planned and developed in the vicinity of metropolitan cities (Sekar and Kanchanamala, 2011).

Tirunelveli also has its city municipal corporation, where various public infrastructure development activities are carried out under the panel of Tirunelveli Smart City. They are operated as a Centrally Sponsored Scheme (CSS) by the Ministry of Housing and Urban Affairs, Government of India (Sarbeswar and Han, 2019). Before proceeding with the detailed analysis, the suitable site for the corridor needs to be be located around the corporation's boundaries where the city is expanding.

9.2.2 SELECTION AND PREPARATION OF THE CRITERIA MAP

The suitable site selection for a specific purpose depends on a set of selected criteria. Certain aspects are given priority when screening for the best location for a defined facility. Our selected criteria should satisfy the norm and policies listed by the government body (Kumar and Shaikh, 2013; Kumar and Bansal, 2016). The detailed sequence of activities carried out is shown in Figure 9.2.

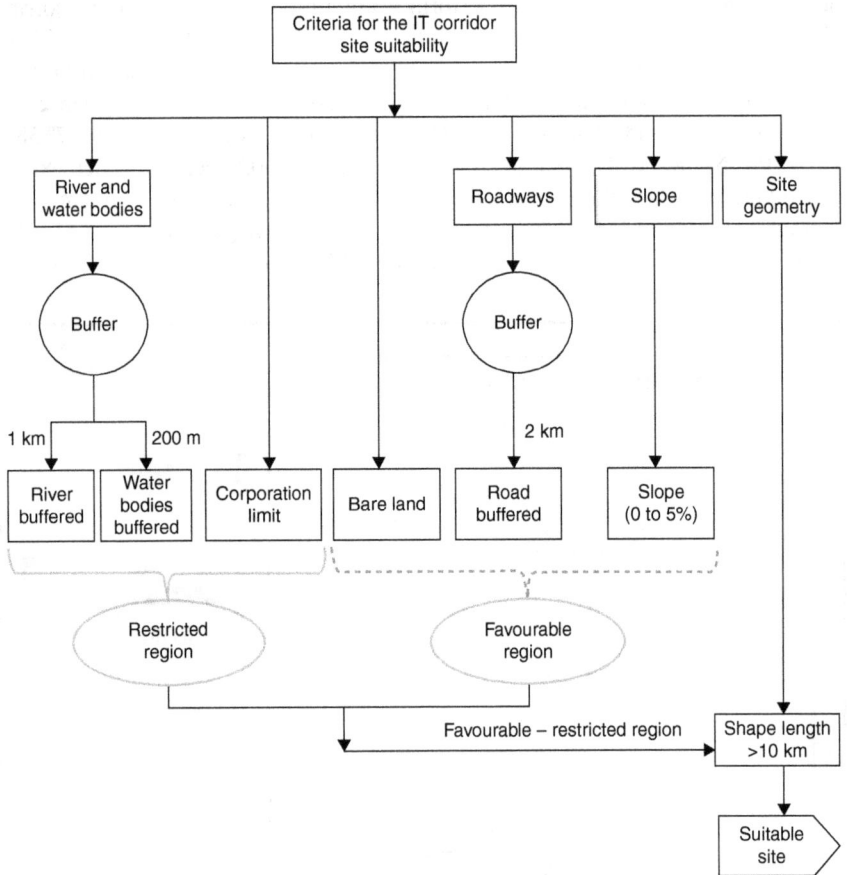

FIGURE 9.2 Methodology flowchart for the IT corridor site suitability.

TABLE 9.1
Source Details of IT Corridor Parameter

S. no.	Parameters	Source details
1	Rivers and waterbodies	Landsat 8:
2	Bare land	(Scene ID = LC81430542021038LGN00) • Spatial Resolution: 30 metres • Datum: WGS84 • Access Portal: USGS Earth Explorer (https://earthexplorer.usgs.gov/)
3	Slope	Cartosat-1: (CartoDEM) • Spatial Resolution: 2.50 meters • Datum: WGS84 • Access Portal: Bhuvan by NRSC (https://bhuvan.nrsc.gov.in/)
4	Roadways	Open street map (www.openstreetmap.org/)
5	City municipal corporation limit	Tirunelveli Corporation Master Plan (Digitisation)

The study considered six criteria: rivers and waterbodies, slope, bare land, road, corporation limits, and site geometry. The preparation steps for each criteria map are discussed in the corresponding heads below. The criteria maps were prepared by performing the various operations in a GIS platform called ArcMap. This allows us to create and interact with meaningful maps, and also, one can view, edit, and analyse the created map spatially (Ormsby et al., 2010; Kumar and Bansal, 2016). The sources of the IT corridor suitability parameters are listed in Table 9.1.

9.2.2.1 Rivers and Waterbodies

River Thamirabarani flows across the Tirunelveli district from the western ghats (Agastyarkoodam peak of Pothigai hills) to Thoothukudi, with various subbasins. The site should not be located near the river basin so as to not disturb the river ecosystem. The settlements and other encroachments are at high risk when a low-probability high-impact flood occurs. Urbanisation leads to a higher chance of flood risks in settlements (Basnet, 2020). Urban encroachment has materialised as a warning to the existence of any river. Trade and commerce, housing, industry, religion, culture, and agriculture are the possible land use categories that provide a risk to the river ecology (Hossain, 2017; Ashraful et al., 2015).

Similarly, many waterbodies lose their storage capacity and are polluted due to the city's rapid urban sprawl. Some opportunistic people have attempted to transform public resources into private property, resulting in a water crisis in the surrounding area (Chigurupati and Prasad, 2008).

Processing Landsat 8 satellite images provides the rivers and waterbodies data obtained from USGS earth explorer with free access. The image is masked with

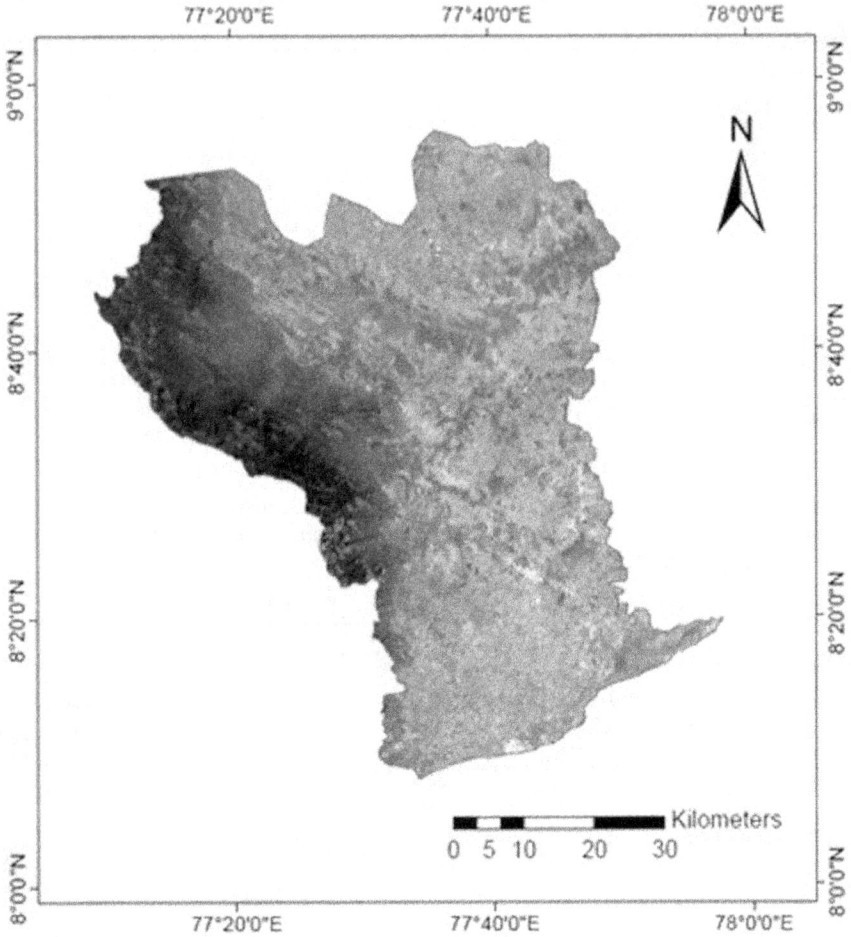

FIGURE 9.3 Satellite image of the study area.

our study area –Tirunelveli district – as shown in Figure 9.3. Supervised classification with trained samples helps derive a land use classification map with rivers
and waterbodies, mountains, vegetation, urban region, and bare land. Land use and
cover change have supported studies on impacts on ecosystems, global biogeochem-
istry, human vulnerability, land suitability, and climate change (Verburg et al., 2009).
A unique value is assigned in the supervised classification raster layer, as shown in
Figure 9.4. The derived river stream data are cross-checked with delineated watershed
information from ArcGIS.

9.2.2.2 Slope

The slope of any terrain directly decides the changes for foundation positioning and
execution of any project. In general, an area with a steeper slope is categorised as a

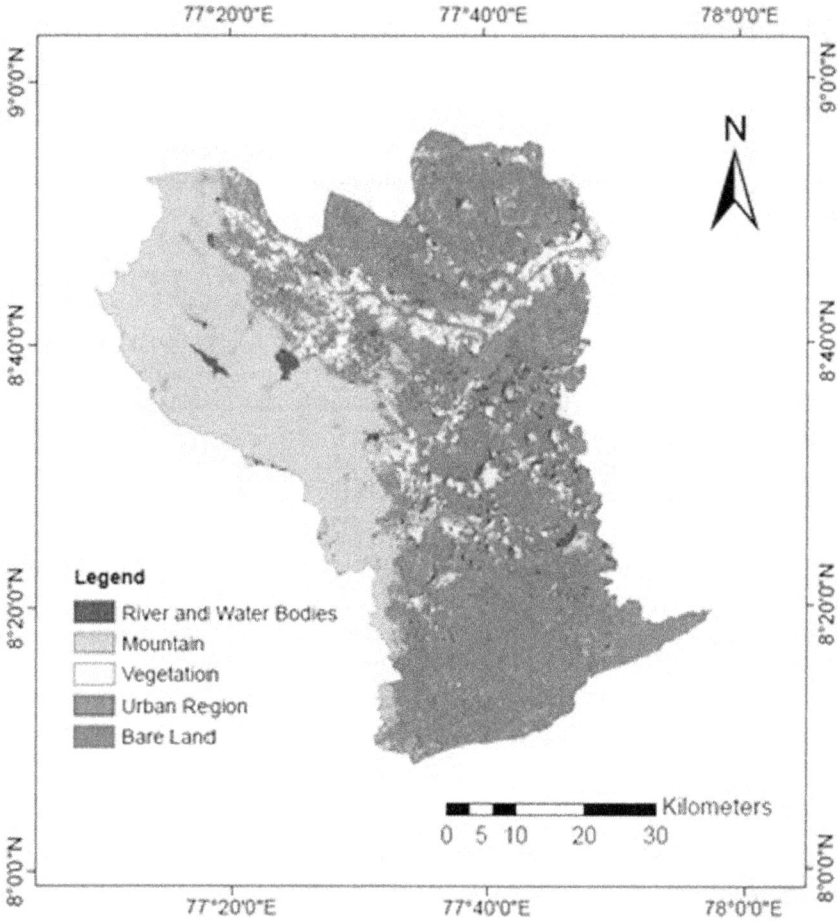

FIGURE 9.4   Supervised land use classification for the study area.

protected area (Parry, Ganaie, and Sultan Bhat, 2018). A region with diverse slopes may be preferred for the water supply scheme. Selecting a site with minimal slope decreases the variance during material handling. Slopes of 10% and above are susceptible to erosion (Hopkins, 1977). The Processing Digital Elevation Model (DEM) is an effective method to determine the slope. The slope accuracy depends on the DEM resolution (Warren et al., 2004).

   We accessed the DEM data from the SRTM data sets – USGS portal with 30-metre resolution. The masking procedure is repeated to extract the study area data alone. The map clearly states that the maximum level difference between any two points in our study area is 1828 metres. The slope layer is achieved in ArcGIS by setting the DEM raster as an input parameter in the slope tool under the surface toolset. The slope values on the result are rearranged from flat terrain to very steep terrain (Choubey and Litoria, 1990; Sobolewska-Mikulska, Krupowicz, and Sajnóg, 2014).

The final terrain slope map in percentages depicts that almost 80% of our study area is flat.

9.2.2.3 Bare Land

The undisturbed area on the Earth's surface, excluding roads, vegetation, and buildings, is classified as bare land. For a rapidly urbanising city, most of the development activities take place on bare land only. Such developments are emerging because of their most negligible impact on the environment (Li et al., 2017).

The spatial-temporal change in any bare land indicates the city expansion due to the intervention of humans. The ability to monitor dynamic changes in the urban context is a highly desirable goal for planners and decision-makers (Zhou et al., 2014). The bare land map is similar to the rivers and waterbodies map from the supervised classification image. From the pixel values summary, it is observed that bare land occupies 32% of the total area of Tirunelveli district.

9.2.2.4 Roadways

The IT corridor is just a provisional space in a long stretch that develops an opportunity for IT/ITES companies to set up their facilities (Venkatesh et al., 2022). The distance to the selected site is significant for planning this corridor. The roadways go to be heart of this site's suitability criteria. Various studies recommend a buffer distance for roads. However, there is no proper valid definition for the maximum distance from the site to roadways, as different studies have taken disparate distances for the buffer (Baseer et al., 2017). Data sets for roadways are extracted from OpenStreetMap, an open-source wiki world map. Since we decided to locate our site near major roads only, the query extended to national and state highways.

9.2.2.5 City Municipal Corporation Limit

The city municipal corporation is a legal local governing body of the city that works to develop a city with more than one million population. The municipal area is divided into many wards. The collection of property tax, water tax, drainage tax, etc., is the revenue for the corporation. The corporation limit is said to be an engine of growth. Until the city government solves the problems related to basic amenities, immigration of people from rural to urban areas is a problem for this municipal corporation (Basu Roy and Saha, 2011). Therefore, it is not wise to plan our site within this limit even if other criteria favour it. This city municipal corporation area will be a restricted criterion, like rivers and waterbodies. We obtained corporation boundaries by georeferencing the available city planning map and vectorisation.

9.2.2.6 Site Geometry

The IT corridor should be long enough to handle various economic activities and adaptable to the development phase. However, there is no stringent rule on how far the corridor should run. The final suitability map may contain numbers of smaller polygons representing the site for the corridor. To avoid such incongruous results, the minimum longitudinal length suitable for the corridor is to be set as 10 kilometres.

9.2.3 Methodology

A detailed and precise workflow is essential for any site suitability analysis. GIShas a streamlined workflow. Here the sequence of operation is planned in such a way as to obtain output flexibility. This kind of proper planning in processing, managing, and analysing data opens up an opportunity for reengineering existing flow patterns (Li and Coleman, 2005).

The workflow for our study starts with classifying the criteria into restricted and favourable. The criteria are highly unsuitable for the planning and development of the IT corridor listed under the restricted region. The criteria are also highly suitable for locating the IT corridor listed in a favourable region. Rivers and water bodies and corporation limits are restricted criteria, while bare land, roadways, and slope are favourable criteria. There is a need to modify the criteria layer to clear out the environmental impact assessment and feasibility check. The river layer alone is buffered for 1 kilometre to avoid disturbing the river and its allied ecosystem.

Similarly, waterbodies like lakes, ponds, and canals are also buffered for 200 metres (Ashraful et al., 2015). Creating a buffer gives a positive opportunity in solving water crises and groundwater table-level problems. As the landscape of city corporation limit ia already stacked with urban activists, planning a new corridor is not an engineered approach. Several land-use changes happen only on bare land. Therefore, planning a colossal corridor is very suitable in the space classified as bare land by conducting supervised classification from satellite images. The continuous stretch of development activities creates a corridor. Roadways are the influence criteria for this study (Pareta, 2013). As there is a corridor expansion possibility in the future in lateral directions, this layer is buffered for 2 kilometres. Choosing a flat terrain decreases various risks of the execution stage. The slope map is reclassified to obtain an area with a slope of 0–5%.

The layers grouped as restricted and favourable regions are combined with the help of the union-geoprocessing tool. This is followed by removing the common areas between both regions from favourable regions, giving a distributed suitability layer. The output may contain many polygons. The concern of placing geometry criteria as a final screening is well illustrated by referring to the suitability output layer. Sorting the grouped polygons by shape length results in features with a large area. A polygon with a length greater than 10 kilometres is selected and exported as a result of site suitability analysis through the minimum bounding geometry tool under data management toolsets.

9.3 RESULTS AND DISCUSSION

The appropriate criteria for the site suitability were decided and based on the restrictions; the six criteria are classified into favourable and unfavourable. Finally, by applying stretch length constraints, the suitability site map for the IT corridor was accomplished, as shown in Figure 9.5.

One can consider several criteria, but the land should be available to execute any project (Hama, Al-Suhili, and Ghafour, 2019). The criteria were carefully selected

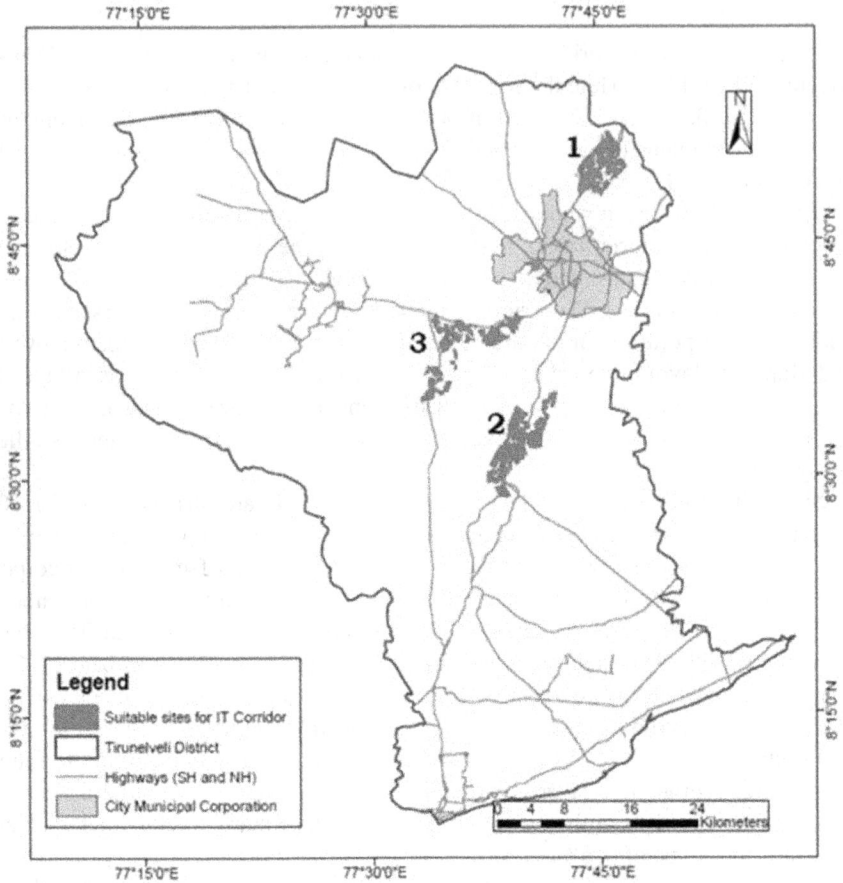

FIGURE 9.5   Site suitability map for the IT corridor in Tirunelveli district.

and studied as per the task. All the suitable sites should be located on bare land alone (Zhou et al., 2014). Supervised classification indicated that 32% of the study area is bare land. Therefore, there is not an issue of land availability in the study area. The roadways (national and state highways) have buffers of 2 kilometres created while considering accessibility. The areas which match with bare land and buffered road layers are strictly suitable (Liu et al., 2014). It is difficult to remove the common area in the raster with a single click on any geoprocessing tool. Therefore, the supervised classification image in raster format is converted into a vector file with five land-use attributes for this restriction approach (Congalton, 1997).

From the suitability map, we can decide that the three stretches are suitable sites for planning and developing the IT corridor in Tirunelveli district. Stretch 1 is located between Tirunelveli and Kovilpatti on NH 44 (the longest highway in India) and extends about 10 kilometres. This shows that our study is valid and the selected suitable area is faultless (Venkatesh and Mannar Thippu Sulthan, 2022).

TABLE 9.2
Extracted Suitable Sites for IT Corridors

Proposed corridor number	Located between	Bounded area (sq. km)	Distance b/w farther ends (km)	Roads passing through
1	Tirunelveli town and Kovilpatti	3244.55	10.15	NH 44
2	Tirunelveli town and Valliyur	4887.95	14.20	NH 44
3	Tirunelveli town and Kalakkad	2315.20	12.95	SH 40 and SH 177

This project has already started to provide employment opportunities to the surrounding population depending on their educational qualification. This is solid evidence of how the economic conditions of people rise through the development of the corridor. Stretch 2 is located between Tirunelveli and Valliyur (the southern part of Tirunelveli) on NH 44, extending about 14 kilometres. This region contributes most of the bare land from the Tirunelveli district. Finally, stretch 3 of 13 kilometres is located between Tirunelveli and Kalakkad, partially in SH 40 and SH 177. The presence of agricultural land between this stretch makes the suitable region discontinuous. Also, the slope criteria contributed to making this stretch somewhat less satisfactory. A few areas have a slope value greater than 5%. This stretch is the only suitable stretch located near mountainous regions.

Table 9.2 depicts that stretches 1 and 2 occupy an area greater than stretch 3. Due to the vast suitable land availability in these two stretches, the lateral expansion for industrial activities is offers numerous opportunities for planners and decision-makers.

9.4   CONCLUSIONS

Urban development provides a strain on the environment. As development should step towards sustainability, considering natural and environmental factors is requisite. The appropriate criteria were selected and processed to select suitable sites for the IT corridor. GIS is utilised for viewing and analysing the selected criteria. Since this restriction model approach analysis can be made only with vector layers, a few format conversions take place for flexible workflow. No provision should be given for taking up any development activities near rivers and other waterbodies. The restriction mode is very effective and facile to adopt the above condition. Integrating the selected criteria results in a suitable site for the facility. This study may be helpful for decision-makers and relevant authorities who are working in developing special economic zones in Tirunelveli district and other districts. Establishing such new IT corridors in Tirunelveli city attracts small-scale software enterprises to become recognised globally through occupying a space in this location.

REFERENCES

Anderson, L T. 1987. *Seven Methods for Calculating Land Capability/Suitability. Planning Advisory Service (PAS) Re–Port No. 402. American Planning Association.*

Ashraful, Md, Islam Chowdhury, Mohammed Amir, Hossain Bhuyain, and Mohammad Mahbub Kabir. 2015. Assessment of River Encroachment and Land-Use Patterns in Dhaka City and Its Peripheral Rivers Using GIS Techniques. *International Journal of Geomatics and Geosciences* 6 (2): 1556–1567.

Ayele, Leykun Getaneh, Yirga Kebede Wondim, and Abiyu Demessie Abebe. 2018. GIS Based Suitable Site Selection and Road-Map Preparation for Equitable Distribution of Secondary Schools of Amhara Region. *Journal of Environment and Earth Science* 8 (1): 100–113.

Banai-Kashani, Reza. 1989. A New Method for Site Suitability Analysis: The Analytic Hierarchy Process. *Environmental Management* 13 (6): 685–693. doi:10.1007/BF01868308

Baseer, M A, S Rehman, J P Meyer, and M Alam. 2017. GIS-Based Site Suitability Analysis for Wind Farm Development in Saudi Arabia. *Energy* 141 (II): 1166–1176.

Baskaran, Venkatesh, and R Velkennedy. 2022. A Systematic Review on the Role of Geographical Information Systems in Monitoring and Achieving Sustainable Development Goal 6: Clean Water and Sanitation. *Sustainable Development*, 1–9. doi:10.1002/sd.2302

Basnet, Keshav. 2020. Flood Modelling of Patu River in Tulsipur City of Nepal and Analysis of Flooding Impact on Encroached Settlement along the River. *International Journal of Advance Research*, no. September.

Basu Roy, Tamal, and Sanjoy Saha. 2011. A Study on Factors Related to Urban Growth of a Municipal Corporation and Emerging Challenges: A Case of Siliguri Municipal Corporation, West Bengal, India. *Journal of Geography and Regional Planning* 4 (14): 683–694.

Bunruamkaew, Khwanruthai, and Yuji Murayama. 2011. Site Suitability Evaluation for Ecotourism Using GIS & AHP: A Case Study of Surat Thani Province, Thailand. In *Procedia–Social and Behavioral Sciences*, 21:269–278. Elsevier Ltd. doi:10.1016/j.sbspro.2011.07.024

Chigurupati, Ramachandraiah, and Sheela Prasad. 2008. *Impact of Urban Growth on Water Bodies :The Case of Hyderabad. Centre for Economic and Social Studies.* 6th ed.

Chou, Chien Chang. 2010. An Integrated Quantitative and Qualitative FMCDM Model for Location Choices. *Soft Computing* 14 (7): 757–771. doi:10.1007/s00500-009-0463-8.

Choubey, V. D., and P. K. Litoria. 1990. Terrain Classification and Land Hazard Mapping in Kalsi-Chakrata Area (Garhwal Himalaya), India. *ITC Journal* 1990–1 (May): 58–66.

Congalton, Russell G. 1997. Exploring and Evaluating the Consequences of Vector-to-Raster and Raster-to-Vector Conversion. *Photogrammetric Engineering and Remote Sensing* 63 (4): 425–434.

Fernando, G M T S, Ven Pinnawala Sangasumana, and C H Edussuriya. 2015. A GIS Model for Site Selection of Industrial Zones in Sri Lanka (A Case Study of Kesbewa Divisional Secretariat Division in Colombo District). *International Journal of Scientific & Engineering Research* 6 (11): 172–175.

Gong, Jianzhou, Yansui Liu, and Wenli Chen. 2012. Land Suitability Evaluation for Development Using a Matter-Element Model: A Case Study in Zengcheng, Guangzhou, China. *Land Use Policy* 29 (2). Elsevier Ltd: 464–472. doi:10.1016/j.landusepol.2011.09.005

Hama, Ako Rashed, Rafea Hashim Al-Suhili, and Zeren Jamal Ghafour. 2019. A Multi-Criteria GIS Model for Suitability Analysis of Locations of Decentralized Wastewater Treatment Units: Case Study in Sulaimania, Iraq. *Heliyon* 5 (3). Elsevier Ltd: e01355. doi:10.1016/j.heliyon.2019.e01355

Hopkins, Lewis D. 1977. Methods for Generating Land Suitability Maps: A Comparative Evaluation. *Journal of the American Planning Association* 43 (4): 386–400. doi:10.1080/01944367708977903

Hossain, Muhammad Selim. 2017. Mapping Urban Encroachment in the Rivers around Dhaka City : An Example from the Turag River. *Journal of Environment and Earth Science* 7 (10): 79–88.

Johar, Amita, S S Jain, and P K Garg. 2013. Land Suitability Analysis for Industrial Development Using GIS. *Journal of Geomatics2* 7 (2): 101–106.

Kramer, W.J. Wj, Beth Jenkins, and R.S. Rs Katz. 2007. The Role of the Information and Communications Technology Sector in Expanding Economic Opportunity. *International Journal for Management Science and Technology* 2 (3): 27–32.

Kumar, Manish, and Vasim Riyasat Shaikh. 2013. Site Suitability Analysis for Urban Development Using GIS Based Multicriteria Evaluation Technique: A Case Study of Mussoorie Municipal Area, Dehradun District, Uttarakhand, India. *Journal of the Indian Society of Remote Sensing* 41 (2): 417–424. doi:10.1007/s12524-012-0221-8

Kumar, Satish, and V. K. Bansal. 2016. A GIS-Based Methodology for Safe Site Selection of a Building in a Hilly Region. *Frontiers of Architectural Research* 5 (1). Higher Education Press Limited Company: 39–51. doi:10.1016/j.foar.2016.01.001

Li, Hui, Cuizhen Wang, Cheng Zhong, Aijun Su, Chengren Xiong, Jinge Wang, and Junqi Liu. 2017. Mapping Urban Bare Land Automatically from Landsat Imagery with a Simple Index. *Remote Sensing* 9 (3). doi:10.3390/rs9030249

Li, Songnian, and David Coleman. 2005. Modeling Distributed GIS Data Production Workflow. *Computers, Environment and Urban Systems* 29 (4): 401–424. doi:10.1016/j.compenvurbsys.2003.12.002

Liang, Gin Shuh, and Mao Jiun J. Wang. 1991. A Fuzzy Multi-Criteria Decision-Making Method for Facility Site Selection. *International Journal of Production Research* 29 (11): 2313–2330. doi:10.1080/00207549108948085

Liu, Renzhi, Ke Zhang, Zhijiao Zhang, and Alistair G.L. Borthwick. 2014. Land-Use Suitability Analysis for Urban Development in Beijing. *Journal of Environmental Management* 145. Elsevier Ltd: 170–179. doi:10.1016/j.jenvman.2014.06.020

Magesh, N. S., and N. Chandrasekar. 2014. GIS Model-Based Morphometric Evaluation of Tamiraparani Subbasin, Tirunelveli District, Tamil Nadu, India. *Arabian Journal of Geosciences* 7 (1): 131–141. doi:10.1007/s12517-012-0742-z

Malczewski, Jacek. 2004. GIS-Based Land-Use Suitability Analysis: A Critical Overview. *Progress in Planning* 62 (1): 3–65. doi:10.1016/j.progress.2003.09.002

MD. 2019. Environment Policy. *State Industries Promotion Corporation of Tamilnadu Ltd (SIPCOT)*. https://sipcot.in/pages/view/typeof-Industries.

Miles, SB, and CL Ho. 1999. Applications and Issues of Gis As Tool for Civil Engineering Modelling. *Journal of Computing in Civil Engineering* 13 (July): 144–152.

Nedović-Budić, Z. 1998. The Impact of GIS Technology. *Environment and Planning B: Planning and Design* 25 (5): 681–692. doi:10.1068/b250681

Ormsby, Napoleon, Burke, Groessl, and Bowden. 2010. *Getting To Know ArcGIS Desktop*. Redlands, California: ESRI Press.

Pareta, Kuldeep. 2013. Remote Sensing and GIS Based Site Suitability Analysis for Tourism Development. *International Journal of Advanced Research in Engineering and Applied Sciences*.

Parry, Jahangeer A., Showkat A. Ganaie, and M. Sultan Bhat. 2018. GIS Based Land Suitability Analysis Using AHP Model for Urban Services Planning in Srinagar and Jammu Urban Centers of J&K, India. *Journal of Urban Management* 7 (2). Elsevier B.V.: 46–56. doi:10.1016/j.jum.2018.05.002

Patten, Iris E., and A. Amin Arafat. 2010. Site Selection and Suitability Modeling. In *ESRI International User Conference*, 1–54.

Pramanik, Malay Kumar. 2016. Site Suitability Analysis for Agricultural Land Use of Darjeeling District Using AHP and GIS Techniques. *Modeling Earth Systems and Environment* 2 (2). Springer. doi:10.1007/s40808-016-0116-8

Raman, Revti, and Doren Chadee. 2011. A Comparative Assessment of the Information Technology Services Sector in India and China. *Journal of Contemporary Asia* 41 (3): 452–469. doi:10.1080/00472336.2011.582714

Rikalovic, Aleksandar, Ilija Cosic, and Djordje Lazarevic. 2014. GIS Based Multi-Criteria Analysis for Industrial Site Selection. In *24th DAAAM International Symposium on Intelligent Manufacturing and Automation, 2013*, 69:1054–1063. Elsevier B.V. doi:10.1016/j.proeng.2014.03.090

Saraswathi, T. 2015. Emerging Trends of It Sector in India. *Shanlax International Journal of Management* 2 (3): 107–110.

Sarbeswar, and Hoon Han. 2019. Cutting through the Clutter of Smart City Definitions: A Reading into the Smart City Perceptions in India. *City, Culture and Society* 18 (April). Elsevier: 0–1. doi:10.1016/j.ccs.2019.05.005

Sekar, S.P., and S. Kanchanamala. 2011. An Analysis of Growth Dynamics in Chennai Metropolitan Area. *Institute of Town Planners, India* 8 (4): 31–57.

Sobolewska-Mikulska, Katarzyna, Wioleta Krupowicz, and Natalia Sajnóg. 2014. Methodology of Validation of Agricultural Real Properties in Poland with the Use of Geographic Information System Tools. *International Multidisciplinary Scientific GeoConference Surveying Geology and Mining Ecology Management, SGEM* 2 (2): 345–356. doi:10.5593/sgem2014/b22/s9.044

Srinivasan, S., Mainul Haque, and U. Rao. 2012. Socio, Medico and Economic Impacts on in and around Dwellers of Industrial Effluent Pretentious Areas–A Critical Survey. *Research Journal of Humanities and Social Sciences* 3 (2): 182–195.

Venkatesh, Baskaran, and M.A.M Mannar Thippu Sulthan. 2022. An Automated Geoprocessing Model for Accuracy Assessment in Various Interpolation Methods for Groundwater Quality. In *Environmental Informatics*, edited by P. K. Paul, Amitava Choudhury, Arindam Biswas, and Binod Kumar Singh, 1st ed., 285–299. Singapore: Springer Nature Singapore. doi:10.1007/978-981-19-2083-7

Venkatesh, Baskaran, C Subash Raj, R.V Subbu Sankar, and K Blessy. 2022. The Influence of CBR Value on the Cost of Optimal Flexible Pavement Design. *Sustainability, Agri, Food and Environmental Research* 12 (1). https://doi.org/10.7770/safer-V12N1-art2780.

Verburg, Peter H., Jeannette van de Steeg, A. Veldkamp, and Louise Willemen. 2009. From Land Cover Change to Land Function Dynamics: A Major Challenge to Improve Land Characterization. *Journal of Environmental Management* 90 (3). Elsevier Ltd: 1327–1335. doi:10.1016/j.jenvman.2008.08.005

Warren, S. D., M. G. Hohmann, K. Auerswald, and H. Mitasova. 2004. An Evaluation of Methods to Determine Slope Using Digital Elevation Data. *Catena* 58 (3): 215–233. doi:10.1016/j.catena.2004.05.001

Zhou, Yi, Guang Yang, Shixin Wang, Litao Wang, Futao Wang, and Xiongfei Liu. 2014. A New Index for Mapping Built-up and Bare Land Areas from Landsat-8 OLI Data. *Remote Sensing Letters* 5 (10): 862–871. doi:10.1080/2150704X.2014.973996

10 Integrating Environmental Monitoring Techniques for an Effective Healthcare System

Nupur Joshi and Ambrina Sardar Khan

CONTENTS

10.1 INTRODUCTION

A rapid spread of an infectious disease covering a large portion of the population worldwide is called a pandemic. One such ongoing pandemic, COVID-19 (coronavirus) has severely affected humankind in many ways (Figure 10.1). SARS-CoV-2 (severe acute respiratory syndrome 2) has been traced to originate from Wuhan, Central China's Hubei Province, China, as a pneumonia of unknown cause in November 2019 (Bogoch et al., 2020). In India, the first case was found on 27 January 2020 in Kerala. After facing a whole year of rising and falling of the first wave of the virus spread, it resulted in affecting 219 countries and territories, of which 118,000,000 people worldwide and 11,300,000 in India caught the virus, with global deaths reachingd 2.62 million, and in India above 158,000 deaths had been recorded by 14 March 2020. The second wave which hit India from 15 February was much more frightening, with 250,000 cases being reported daily at the time of writing. The highest single-day spike of 314,835 new cases was recorded on 22nd April 2021 and it continued to break its records daily. According to Health Ministry Data, India's daily positivity rate doubled

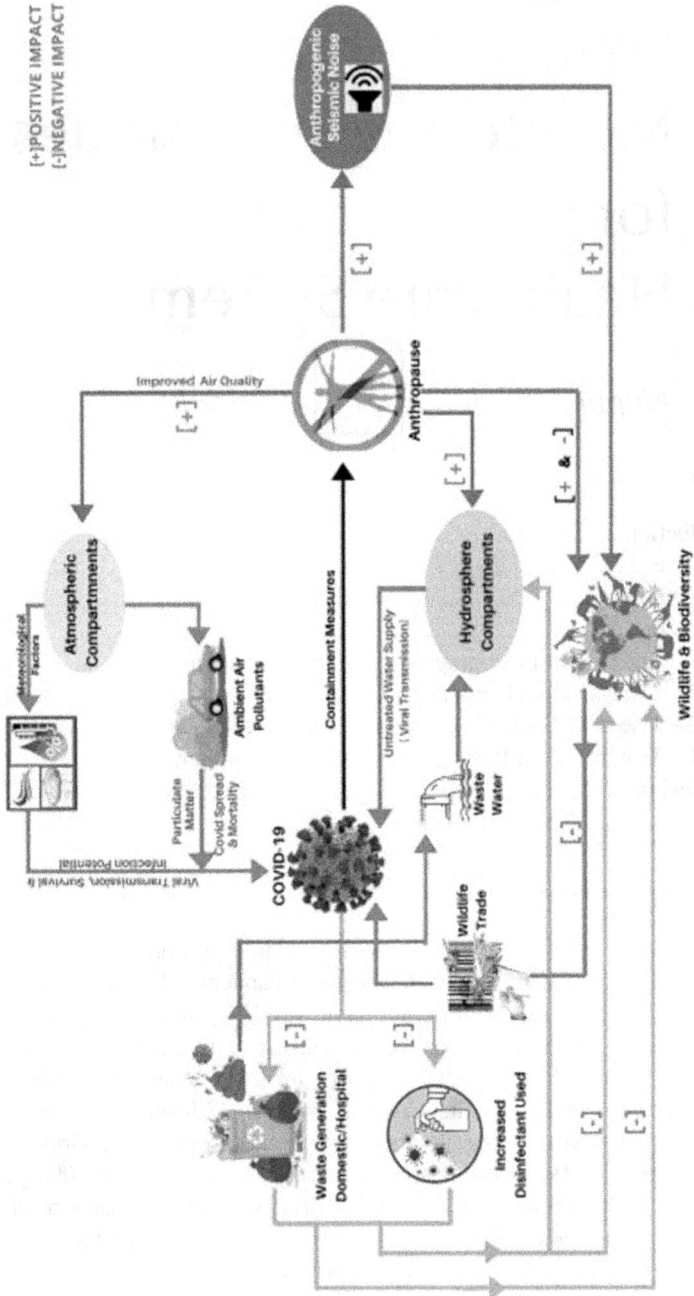

FIGURE 10.1 Impact of COVID-19 on the environment and its interlinking throughout the various spheres increasing its spread.

Source: doi: 10.1007/s42398-021-00207-4.

from 8% to 16.7% in the most recent trend observed. As per current data, 15,9000,000 cases are active in India, and more than 184,000 deaths have been reported.

According to clinical reports, one of the symptoms which COVID-2 patients exhibit is diarrhoea (Cheung et al., 2020). Some results illustrate that many symptomatic individuals along with asymptomatic patients, tend to discharge the virus and through faecal discharge, and so it ultimately reaches sewage treatment plants (Haramoto et al., 2020; Kumar et al., 2020). Surveys and research have indicated the presence of the virus in wastewater even when patients stop exhibiting respiratory symptoms. This is because RNA strains of coronavirus-2 are found in faeces for a median 22-day duration (Wu et al., 2020; Zheng et al., 2020). First proof of the detection of COVID-2 strains in faeces through wastewater surveillance in India was accomplished in July 2020 and the temporal variation of genetic material loadings was examined (Kumar et al., 2020).

A portion of the population affected by COVID 2 is asymptomatic. Because of this, they don't go for testing (this has been proven through the authors' community survey) and still they carry the virus within them which they ultimately excrete through faeces. Without knowing the problem and without getting proper treatment they stay at their residence, and they might directly or indirectly contaminate their surroundings and the people around them. The people who get affected and even those who show symptoms are also quarantined for 14 days, whereas research has shown that patients are found to excrete faeces with strains of SARS CoV-2 for another 8 days after recovering (Wu et al., 2020; Bivins et al., 2020).

One of the best approaches, i.e., wastewater-based epidemiology (WBE), is found to be a leading approach which helps in understanding the appropriate status of disease spread in certain catchment areas by monitoring the virus concentration in the wastewater as this wastewater contains excreted waste from pre-symptomatic, symptomatic, as well as asymptomatic individuals (Choi et al., 2020; Yang et al., 2015). Thus, the data also provide an estimate of the actual number of infected in the population. This is not the first time that WBE has been used as an effective tool as past studies show it also helped track the viral load during outbreaks of viruses such as poliovirus, norovirus, and hepatitis A (Hellmér et al., 2014; Ahmed et al., 2020; Kitajima et al., 2020).

Due to the lack of proper spatial reference data and the increasing population, the health situation of society and the population increasingly deteriorates. The greater the population, the more will be the spread of disease, and thus there is high pressure on healthcare services to meet the demands of growing patient lists. Thus, proper utilisation, management, and advancements in GIS can help overcome these problems, with 'health' being a basic human right that can affordably reach every individual. According to the situation, plans should be made, modified, and adjusted for the welfare of the nation and its people. Once health and family welfare are achieved through the integrated approach of GIS and health impact assessment, we can aspire to have a stabilised population and thus a sustainable healthy environment can be created for future generations.

These days, GIS is becoming a very significant device for the correct and accurate assessment of applications in urban healthcare. This comprises various practices such as managing databases, planning for healthcare systems, threat and risk mapping of

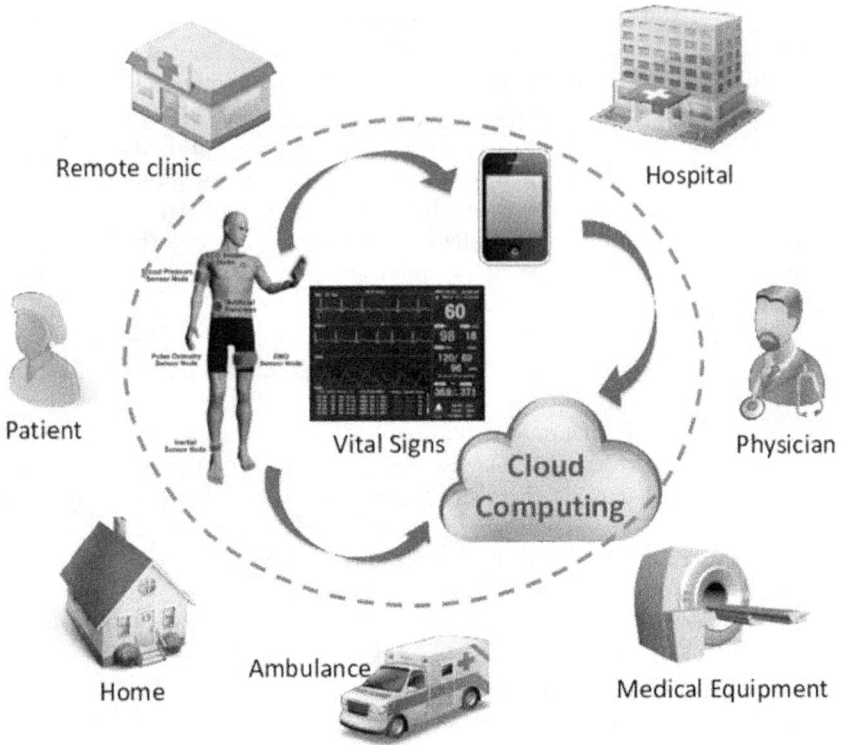

FIGURE 10.2   Use of remote sensing and GIS in healthcare systems.

Source: doi: 10.1109/NEBEC.2013.108.

infectious areas, location identification, cases of diseases, clinics, roads, channels, country boundaries, and health catchment areas, etc. The present shift in using GIS in healthcare applications is because GIS helps us to understand the consequences and to support health services in the population. These GIS agencies have become effective in monitoring, observing, evaluating, and planning strategies for urban healthcare and providing immediate assistance in contaminated zones, as shown in Figure 10.2. Over time, a great deal of exploration and technical research has been conducted using GIS for the investigation of various epidemics, pandemics, and other communicable diseases. GIS also enables the exploration and discovery of the availability of clean drinking water, contamination of water, and the assessment of disease spread. This is where and how GIS functions as an especially valuable tool.

10.2   OBJECTIVE

The purpose of this study is to highlight the importance of GIS and other spatial parameters in the field of health and healthcare systems, access to healthcare, the application of GIS in various field studies, and how the technologies which have been used for decades can be used for the betterment of community health in different

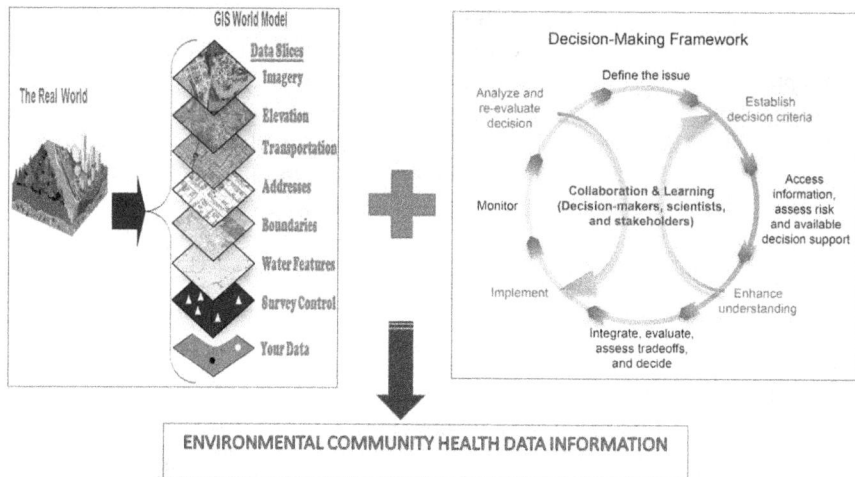

FIGURE 10.3   Use of remote sensing and GIS in developing a decision support system.

aspects. Furthermore, how GIS and other spatial technology can become beneficial decision-making tools in the near future is shown in Figure 10.3.

10.3   METHODOLOGY

For the completion of this study, the method used was going through a literature review of the past two to three decades. The purpose of this current review was focused on the health sector and how GIS plays a role in assessing or viewing environmental impacts on human health. The study material included publications and articles related to the topic.

From an initial search of the literature, from databases such as Google Scholar, Pub-med, and HEAL-LINK, around 130 review and research-based titles were collected based on terms such as 'GIS', 'human health', 'public health', 'applications of GIS', and 'environmental impacts'. Based on abstracts, titles, and conclusions, around 80 were eliminated due to lack of availability of the full text, some did not include the role of GIS in the health sector, and some had focused on GIS and not on health sectors. Based on abstracts, content, and full-text availability, around 35 articles and publications were chosen.

Of these 35 articles and publications, around 60% of the articles included GIS, 37% included health-related information, and 3% were focused on other spatial tools and extra related information.

10.4   DISCUSSION

10.4.1   Overview of Geographic Information System (GIS)

GIS is a platform which helps capture spatial data from the surface of the Earth, and computerised GIS tools help create an interphase between the user and data analysis and manipulation. Like the evolution of every species, event, and technology, the evolution of GIS over the last five decades has transformed it into a science from just a concept.

The chronology of the advancements in the field of GIS is well explained in Foresman's edited book where he involves evolution in overlapping ages: the pioneer age, the research and development age, the implementation and vendor age, the client applications age, and the local and global network age (Waters, 2016; Foresman, 1998). The year 2020 marks the 30th anniversary of the coining of the term GIS. In the 1960s, Roger Tomlison created the term CGIS for the Canada Geographic Information System (Waters, 2016). Using spatial and convergent analysis data, Tomlison was able to form the world's first true operational GIS which helped in storing, manipulating, and analysing data collected for Canada's land inventory (Waters, 2016; Samet, 1990).

Before GIS, cartographic representation data helped provide evidence through the formation of maps. In 1854, British physicians provided data tracing a cholera outbreak source to be a water- and not an airborne disease (Waters, 2016). Maps have played an important role in census operations for a long time. Thus, the development of technologies to capture data, process the data, and its distribution benefits these census operations. Thus, digital mapping and GIS are the natural evolution of the ancient cartographic methods, helping in census data dissemination in the analysis of household and population data and in predicting and tracing the events leading to some health problems.

Every object, be it in air, water, land, etc., has its absorbance and reflectance or emission values, GIS and remote sensing work on this phenomenon. The sun, being a natural source of light, has its light reflected or emitted by the objects which fall in its path. It is this emitted or reflected light which is captured by the sensors on satellites which provides a raw image, which is then processed, digitalised, and analysed for assessment. There are many simple and complex stages, depending on the situation, involved in the process of gathering geographical information through such assets.

10.4.2 OVERVIEW OF HEALTH

Over time the definition of health has been changing and the perceptions of people about the concept of health has evolved. For example, a person who is overweight but has no disease calls themself healthy and not fat, on the other hand dieticians could describe this person unhealthy due to the higher BMI. If we see health from the biomedical perspective, early definitions focused on the body's ability to function; health is normal and could be hampered from time to time by any disease. One of the examples of such a definition is: 'a state characterized by physiologic, anatomic, and psychological integrity; ability to perform personally valued family, work, and community roles; ability to deal with social stress, physical, psychological, and biological' (Stokes et al., 1982).

The World Health Organisation (WHO) provided a broader definition in 1948, relating health to well-being, in terms of 'mental, social, and physical well-being, and not just the absence of disease and infirmity' (WHO, 1958). The WHO in 1984 revised the definition of health and defined it as 'the extent to which an individual or group can realize aspirations and satisfy needs and to change or cope with the environment. Health is a resource for everyday life, not the objective of living; it is a positive concept, emphasizing social and personal resources, as well as physical capacities' (WHO, 1984). Thus, in the 21st century we can say that health is not something very specific but something on a very personal level of an individual's perception of how he/she feels for themselves and how comfortable they are with their body. However, if we consider health in relation to disease, then healthcare is an important field to work in and provide the basic rights of health and its facilities. India was the first developing country to launch a family welfare programme in 1952 and it was introduced in the state of Punjab in 1956, with its major objective focused on how to reduce the population growth rate for stabilisation. The National Family Planning Programme has undergone various stages and plan periods of adaptation in terms of approaches and strategies. Each plan period was roughly 5 years. in the first 5 years (1956–1961), two plans were introduced mainly aiming for a clinical approach. In the 9th and the 10th plan periods, population growth rate reduction became the priority objective.

10.4.3   Use of GIS and Other Spatial Techniques in Health

The health of individuals and the community is affected by their interaction with their surroundings. Spatial factors, how they change with time, what impact the changes have on humans, how and where can we have access to the healthcare facility, and human health risk assessment (can all be answered more effectively and with a little more ease if we can represent all the geographical information system and thus using GIS tools, as shown in Figure 10.4. The health of the public is hampered by lots of factors (agents, vectors, etc.) and may be distributed from one place to another through biological or environmental contaminants (Cromley, 2003). GIS has mainly been used for the prevention of disease rather than treatment, although we can find the source of causative diseases using GIS and act on improving the conditions of the area to minimise its effects. Usually, the assessment is done after the spread of a disease and then these factors are kept for future reference so the spread can be minimised. GIS has been used in the study of different health parameters ranging from assessment of the impact of the built environment on obesity (Thornton et al., 2011), examining birth outcomes (Seto et al., 2007), urban planning decisions for human health improvements (Seto et al., 2007; Poggio et al., 2009), to any other uncertainty in health impact assessment (Briggs et al., 2009).

Before GIS, using cartographic skills, John Snow (epidemiologist, and physician) was able to make a handmade map plotting individual affected areas of cholera in London in 1854. This cartographic skill of John Snow not only helped pinpoint the source area (waterborne not airborne) but also was able to relate geographically related phenomena (MacQuillan et al., 2017). Health impacted by environmental contaminants can occur through the land, water, air, food chain, etc. Many studies have been done on how these factors have caused health impacts and this chapter will cover examples of these environmental sources and the use of GIS, remote sensing, and other spatial techniques in detecting and investigating the sources of diseases.

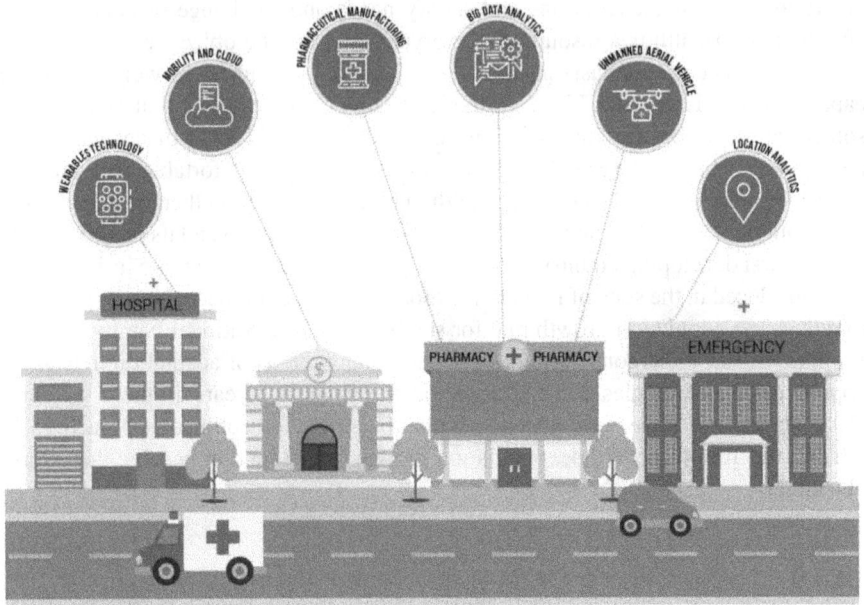

FIGURE 10.4   IoT for healthcare.

Source: www.geospatialworld.net/article/geospatial-technologies-iot-healthcare/

Models are made to study the area for the presence or absence of certain factors that can lead to the disease or any related situation.

10.4.4   Use of GIS in Water, Land, and Air Assessment

Pollution of water bodies, including surface water, marine water, or groundwater pollution, impacts a very large amount of the population worldwide. Almost 3.4 million people annually die due to water problems (scarcity, contamination) worldwide, and by the end of 2020, 135 million people were expected to die as a result, according to Peter H. Gleick in his research done in 2002 (Gleick, 2002; WHO, 2012). Thus, it is important to use methods to monitor waterbodies and their changes over time. GIS and other spatial tools have been used in many such ways in studying water-borne diseases. Measuring the domicile distance from the treatment stations as done in Brazil (Eyler, 2001) would help in tracing down the places which require more attention in the treatment of water. It can be used for the proper measurement of what and how many types of species reside in a pond ecosystem, how much arsenic content is present in that water ecosystem, and the concentration of arsenic in each species (Zeilhofer et al., 2007) which would be consumed and bio-accumulated and thus transferred to a higher trophic level (Liang, 2010). The risk of arsenic exposure and thus its risk assessment through GIS is possible by special risk assessment patterns (Liu et al., 2007). GIS is also used in tracking down the pathways of toxicity spill risk assessment (Manzurul, 2003) and thus helping in creating early warning and

emergency response systems (EWERS), which are responsible for water quality management and chemical spill responses, and they then point to the emergency decision
support system (EDSS) (Jiang et al., 2012; Camp et al., 2010). GIS works on risk
assessment, tracking and mapping, data storage, and large data retrievals and acts as
an additional help in the field of health assessment and healthcare facilities.

Models and field tests are prepared for the assessment of the presence or absence
of normal factors, which considers parameters of water example models of DO, BOD,
pH, salinity (Liou et al., 2004), turbidity and suspended particles (Liedtke et al.,
1995), and surface temperature (Bolgrien et al., 1995) mappings are made which are
then overlaid on the spatial geography of the targeted site and thus the assessment
becomes more simplified and less time-consuming. GIS also helps in storing complex
and huge data which can also help in the temporal assessment of the area so that any
changes that have occurred over a period of time can be recorded. Thus urban-use
water quality mapping can also be done to prevent diseases (Zeilhofer et al., 2007). It
is to be noted here that GIS alone cannot be used for assessment purposes but acts as
an additional important aid in the assessment.

Similarly, as assessment is done for the presence or absence of normal water
conditions, assessments for normal land, soil, vegetation cover, and forest cover are
also made using different models and different indices. Vegetation indices help in
making models which help identify land cover and the difference between healthy
and diseased crops. The indices are made in such a way that they consider many
parameters associated with the health of a crop and can differentiate between plants
having high chlorophyll contents and crops having less; plants having more chlorophyll would absorb more and reflect less light and thus after digital processing of
the image obtained through sensors crops having chlorophyll problems can be identified in dark-shaded portions of crops, indicating less or absence of chlorophyll
(Rundquist et al., 1996). Soil indices are used for the assessment of soil quality
(Bastida et al., 2008). Therefore, suppose people in an area suffer from a particular
disease and the reasons are unknown for a long time, GIS and remote sensing could
help trace temporal changes which took place and identify any change in the health
of crops being consumed which might cause disease in the people living there. Thus,
this helps in inter-relating all the circumstances leading to a problem once a field
study is done. The GIS assessment is also used for human health risk assessment
and how this risk can be minimised as done in urban green space planning (Poggio
et al., 2009).

Many respiratory diseases occur due to air pollution (Laumbach et al., 2012),
which can be both natural as well as anthropogenic pollution. Around 7 million people
die worldwide annually because of air pollution, making it the world's largest environmental health risk (Iversen, 2018; WHO, 2014). The problems might range from
acute to chronic respiratory diseases. Thus, keeping a check on the atmosphere and
its parameters becomes an important aspect of study in maintaining health systems.
GIS and remote sensing help in the detection of air quality with the help of models
prepared according to different parameters involved in the study of air quality known
as air quality indices (AQI). These air quality indices also include different indices for
particulate matter, ozone, sulphur dioxide, CO, and nitrogen dioxide (Nagendra et al.,
2007). These models would help in the temporal and spatial assessment of changes

that have occurred in the atmosphere in the past and what areas are affected by higher concentrations of these pollutants. It is then possible to take precautionary steps and minimise the adverse impacts of pollutants which could bring harm to humans if they are under long-term exposure to pollutants (Gulliver et al., 2011). Epidemics due to biological factors such as vector-borne disease can be analysed through GIS for risk assessment and to provide sustainable health (Allen et al., 2006; Palaniyandi et al., 2017).

It is important to note that GIS, remote sensing data, etc. are additional assets for assessment. Ground data information is very important in the final assessment of any situation. This review has focused on how GIS and spatial tools have been playing a role in the assessment of environmental factors leading to diseases. There is possible scope for an increase in the use of GIS and with the current situation of facing a pandemic situation like COVID-19, many more such situations might be faced in the future, and we need to be ready to face them. Thus, we need more work to be done by utilising these spatial tools and integrated methods for risk assessment, so the situations are known well before time so that we are ready with precautions.

10.5 CONCLUSIONS

Finally, IT professionals and various stakeholders such as hospitals, clinics, health service providers, and developers can employ GIS technology to collaborate and work in a more coordinated fashion with one another to address national healthcare issues, particularly epidemics and pandemics like COVID-19 (Figure 10.5).

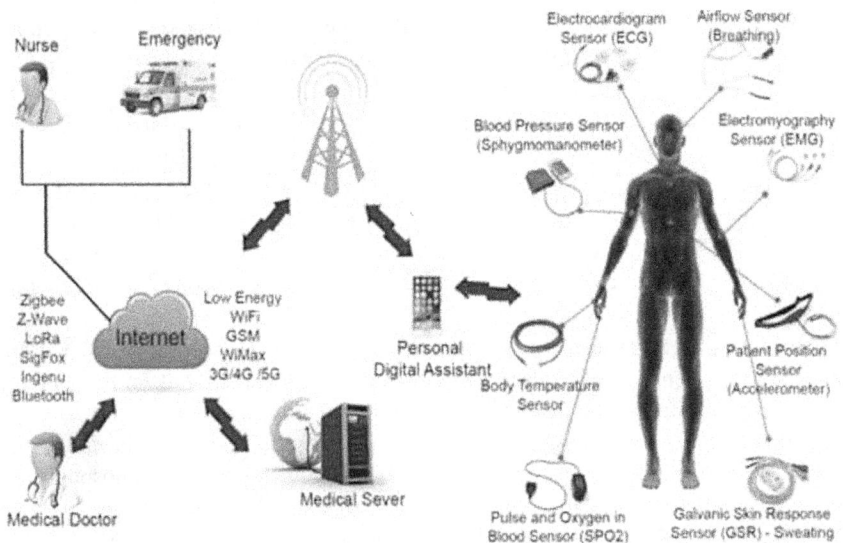

FIGURE 10.5 Representation of the use of various technologies and GIS remote sensing healthcare monitoring system.

Source: doi: 10.1109/ACCESS.2017.2789329.

This approach can help identify which areas need more health services, and care facilities for elders and children. These questions can be answered by examining patient demographic data, location, and accessibility. GIS is a strong tool that has been effectively used to solve a variety of important health challenges in urban areas, from prediction of spreading and disease management to enhance, improve, and provide better services to people and make them accessible for all. As the knowledge and understanding of this tool will gain importance and acknowledgement from the various stakeholders and professionals, more and more healthcare professionals will accept and integrate this programme into their practice. The number of benefits is expected to grow, including connections between hospitals and the communities they serve, which is the most crucial link to make for now.

REFERENCES

Allen TR, Wong DW. Exploring GIS, spatial statistics and remote sensing for risk assessment of vector-borne diseases: a West Nile virus example. International Journal of Risk Assessment and Management. 2006 Jan 1;6(4–6):253–75.

Bastida F, Zsolnay A, Hernández T, García C. Past, present and future of soil quality indices: a biological perspective. Geoderma. 2008 Oct 31;147(3–4):159–71.

Bolgrien DW, Granin NG, Levin L. Surface temperature dynamics of Lake Baikal observed from AVHRR lmages. Photogrammetric Engineering & Remote Sensing. 1995 Feb;61(2).

Briggs DJ, Sabel CE, Lee K. Uncertainty in epidemiology and health risk and impact assessment. Environmental geochemistry and health. 2009 Apr 1;31(2):189–203

Camp JS, LeBoeuf EJ, Abkowitz MD. Application of an enhanced spill management information system to inland waterways. Journal of Hazardous Materials. 2010 Mar 15;175(1–3):583–92.

Cromley EK. GIS and disease. Annual review of public health. 2003 May;24(1):7–24.

Eyler JM. The changing assessments of John Snow's and William Farr's cholera studies. Sozial-und Präventivmedizin. 2001 Jul 1;46(4):225–32

Foresman TW. The history of geographic information systems: perspectives from the pioneers. Upper Saddle River, NJ: Prentice Hall PTR; 1998.

Gleick PH. Dirty-water: estimated deaths from water-related diseases 2000–2020. Oakland: Pacific Institute for Studies in Development, Environment, and Security; 2002 Aug 15.

Gulliver J, de Hoogh K, Fecht D, Vienneau D, Briggs D. Comparative assessment of GIS-based methods and metrics for estimating long-term exposures to air pollution. Atmospheric environment. 2011 Dec 1;45(39):7072–80.

Iversen PL. Chemicals in the Environment. InMolecular Basis of Resilience 2018 (pp. 141–168). Springer, Cham.

Jiang J, Wang P, Lung WS, Guo L, Li M. A GIS-based generic real-time risk assessment framework and decision tools for chemical spills in the river basin. Journal of hazardous materials. 2012 Aug 15;227:280–91.

Laumbach RJ, Kipen HM. Respiratory health effects of air pollution: update on biomass smoke and traffic pollution. Journal of allergy and clinical immunology. 2012 Jan 1;129(1):3–11.

Liang CP, Jang CS, Liu CW, Lin KH, Lin MC. An integrated GIS-based approach in assessing carcinogenic risks via food-chain exposure in arsenic-affected groundwater areas. Environmental Toxicology: An International Journal. 2010 Apr;25(2):113–23.

Liedtke J, Roberts A, Luternauer J. Practical remote sensing of suspended sediment concentration. Photogrammetric engineering and remote sensing. 1995 Feb 1;61(2):167–75.

Liou SM, Lo SL, Wang SH. A generalized water quality index for Taiwan. Environmental Monitoring and Assessment. 2004 Aug 1;96(1–3):35–52.

Liu CW, Liang CP, Lin KH, Jang CS, Wang SW, Huang YK, Hsueh YM. Bioaccumulation of arsenic compounds in aquacultural clams (Meretrix lusoria) and assessment of potential carcinogenic risks to human health by ingestion. Chemosphere. 2007 Aug 1;69(1):128–34.

MacQuillan EL, Curtis AB, Baker KM, Paul R, Back YO. Using GIS mapping to target public health interventions: examining birth outcomes across GIS techniques. Journal of community health. 2017 Aug 1;42(4):633–8

Manzurul Hassan M, Atkins PJ, Dunn CE. The spatial pattern of risk from arsenic poisoning: a Bangladesh case study. Journal of Environmental Science and Health, Part A. 2003 Mar 1;38(1):1–24.

Nagendra SS, Venugopal K, Jones SL. Assessment of air quality near traffic intersections in Bangalore city using air quality indices. Transportation Research Part D: Transport and Environment. 2007 May 1;12(3):167–76.

Palaniyandi M, Anand PH, Pavendar T. Environmental risk factors in relation to occurrence of vector borne disease epidemics: Remote sensing and GIS for rapid assessment, picturesque, and monitoring towards sustainable health. Int J Mosq Res. 2017;4(3):9–20.

Poggio L, Vrščaj B. A GIS-based human health risk assessment for urban green space planning – An example from Grugliasco (Italy). Science of the total environment. 2009 Nov 15;407(23):5961–70.

Rundquist DC, Han L, Schalles JF, Peake JS. Remote measurement of algal chlorophyll in surface waters: the case for the first derivative of reflectance near 690 nm. Photogrammetric Engineering and Remote Sensing. 1996 Feb 1;62(2):195–200.

Samet H. The design and analysis of spatial data structures. Reading, MA: Addison-Wesley; 1990 Jun.

Seto EY, Holt A, Rivard T, Bhatia R. Spatial distribution of traffic induced noise exposures in a US city: an analytic tool for assessing the health impacts of urban planning decisions. International journal of health geographics. 2007 Dec 1;6(1):24

Stokes J, Noren J, Shindell S. Definition of terms and concepts applicable to clinical preventive medicine. Journal of Community Health. 1982 Sep 1;8(1):33–41.

Thornton LE, Pearce JR, Kavanagh AM. Using Geographic Information Systems (GIS) to assess the role of the built environment in influencing obesity: a glossary. International Journal of Behavioral Nutrition and Physical Activity. 2011 Dec 1;8(1):71.

Waters N. GIS: history. International Encyclopedia of Geography: People, the Earth, Environment and Technology. 2016 Dec 12:1–3.

World Health Organisation. 7 million premature deaths annually linked to air pollution. World Health Organisation, Geneva, Switzerland. 2014 Mar 25.

World Health Organisation. Health promotion: a discussion document on the concept and principles: summary report of the Working Group on Concept and Principles of Health Promotion, Copenhagen, 9–13 July 1984. Copenhagen: WHO Regional Office for Europe; 1984.

World Health Organisation. Rapid assessment of drinking-water quality: a handbook for implementation. World Health Organisation; 2012.

World Health Organisation. The first ten years of the World Health Organization. World Health Organization; 1958.

Zeilhofer P, Zeilhofer LV, Hardoim EL, Lima ZM, Oliveira CS. GIS applications for mapping and spatial modeling of urban-use water quality: a case study in District of Cuiabá, Mato Grosso, Brazil. Cadernos de saude publica. 2007;23:875–84.
Zeilhofer P, Zeilhofer LV, Hardoim EL, Lima ZM, Oliveira CS. GIS applications for mapping and spatial modeling of urban-use water quality: a case study in District of Cuiabá, Mato Grosso, Brazil. Cadernos de saude publica. 2007;23:875–84.

11 Advancing the Interventions of Nature-Based Solutions in Cities for Urban Climate Resilience

Harshita Jain, Renu Dhupper, and Deepak Kumar

CONTENTS

DOI: 10.1201/9781003331001-11

11.1 INTRODUCTION

Globally, urban populations are increasingly growing, placing a strain on city resources and their equitable distribution. When combined with current climate change forecasts, such as increased urban temperatures, hurricanes, extreme drought/precipitation, and other environmental hazards, urban areas are under a lot of pressure to provide conditions that sustain human health and well-being [1,2]. Most of the world's population now lives in urban towns and cities, and this is set to expand to three-quarters by 2050. In the near future, urbanisation and climate change will be dominant factors for shaping the trends of global development. On one hand, urbanisation is inevitable and is followed by a rise in per capita income, and also the GDP of the country. Urban cities have the potential to act as a key driver for propelling economic growth in developing countries and lifting people out of poverty. On the other hand, because urbanisation places enormous strain on environmental assets, global warming may undercut all of this by making resource shortages even worse and placing individuals at danger from climate-related natural disasters [1,3].

Ecological, sociological, and economic matters are becoming more prevalent in cities, and as they do, they threaten the ability of urban populations to survive and thrive [3]. Among such dangers are recurring stressors and sudden disturbances that are made worse by global warming. Ecosystem-based techniques can be combined with nature-based tactics as a notion to aid a range of social issues [4]. Nature-based approaches explicitly target and help improve urban resilience. Urban resilience and nature-based solutions (NBS) principles have made it possible to solve contemporary issues in towns in an eco-friendly and sustainably conscious way over the past 10 years. The major challenge we face globally is that cities are rapidly expanding, with the majority of this development taking place in peri-urban areas [5].

The idea of balanced urban growth is complicated, and is related to urban liveability, as well as water, food, and energy protection [6,7]. Building resilience is a crucial task for governments and organisations, as evidenced by global long-term agendas such as the Paris Climate Agreement, the New Urban Agenda 2030, and the Sendai Framework for Disaster Risk Reduction 2015–2030. The UN Agenda 2030 for Sustainable Development and other long-term global initiatives, like the Paris Agreement on Climate Change, depend on NBS to be successful. The introduction of the NBS has been promoted in European cities in recent years as a potential solution to urban problems including climate change, urban degeneration, and ageing infrastructures. In addition to a variety of other advantages, NBS harnesses natural resources to provide ecosystem services to resolve key social, environmental, and economic challenges [8]. Some examples of the latest trends which would help in the development of a general NBS impact mitigation framework are green roofs and covering shelters, green barriers, agroforestry (rural land management), peri-urban parks, natural and constructed wetlands, etc. [9,10]. Applying nature-based alternatives, though, is fundamentally challenging due to the variety of ecological systems, their interactions and the trade-offs between activities in addition to all other temporal and geographical scales.

11.2   URBAN AND PERI-URBAN AREAS RESILIENCE DEVELOPMENT

The flow of commodities and services, such as the natural continuity, connects metro-politan regions with peri-urban regions, making it impossible to build urban resilience in solitude. Urban sustainability and resilience depend on the ecosystem processes that peri-urban ecosystems and peri-urban agriculture offer for a few reasons, notably agricultural production and disaster prevention [11,12]. However, the ecosystems on which these services rely are increasingly threatened and are frequently ignored in policy and planning. The numerous ecosystem services offered by these transition zones have a direct effect on both urban and rural development sustainability [13]. The Sustainable Development Goals of the United Nations (Goal No. 11) also stress the need to make cities inclusive, sustainable, resilient, and robust through the imple-mentation of integral resource-efficiency and climate change adaptation policies and plans, which is not possible without conservation of these peri-urban areas.

Due to peri-urban areas' position adjacent to the city's edges, where many dynamic socio-economic processes exist, a peri-urban region is an ever-changing zone of both activity and change [14]. As a result, its potential for resilience is pri-marily influenced by its input–output dynamics with the nearby urban region on the one side, and with the purely rural structures on the other [15]. But, primarily, a shift from current reactive measures to the establishment and execution of prominent local planning and oversight procedures to encourage improved quality of life is required to increase urban and peri-urban resilience to natural disasters and climatic extremes (Figure 11.1). The resilience of urban and peri-urban areas depends on maintaining ecosystem health. To strengthen urban resilience, city/region planning, and policies should consider the value of peri-urban environments for functions such as disaster

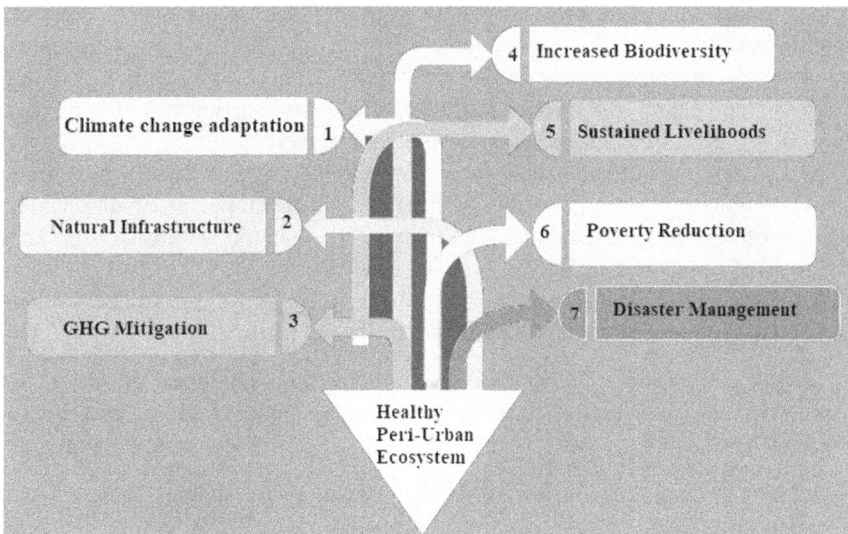

FIGURE 11.1   Peri-urban ecosystems and role in resilience.

risk management, including flood prevention, urban heat island effects reduction, air and water purification, protection of food and water, and waste management.

11.3 NATURE-BASED SOLUTIONS (NBS)

Nature-based approaches are operations that draw their motivation, reinforcement, or model from the natural world. Although they offer a large number of opportunities to be resource and energy saving in addition to environmentally resilient, they should be adapted to local circumstances in order to be effective [16]. Early attempts to save species and control the nature were centred on ecosystem-based approaches. The increased inclusion of social and economic variables indicated the necessity for deeper study of ecological systems. By this integrated solution, the contribution of environment to employment creation, in addition to bettering healthcare and well-being, was acknowledged [17]. To promote connections among environment, society, and business, NBS has been extensively embraced as a phrase and advanced in the EU Research and Innovation Policy agenda since 2013.

Biological architecture, watershed systems integration, greenways, environmental infrastructural facilities, ecosystem-based methodology, ecosystem-based adaptation/mitigation, re-naturing, and ecological processes are examples of prior ideas that NBS builds on and promotes [18]. At first look, these ideas seem to work well together, but closer examination reveals that their views, purposes, and sample selection are different. On the contrary, each of these ideas supports a strategic plan that takes into account ecosystems as a whole without neglecting humans and their effects, which result from growing populations and outweigh natural influences [19]. As per NBS, the environment serves as a source of creativity and offers long-term sustainability solutions for addressing the impacts of human activities and strengthening ecological integrity.

To deal with the cultural, ecological, and financial facets of global crises, NBS provide a multifunctional role. NBS have been cited as being essential for urban renewal and advancement, coastal adaptability, multifunctional watershed control, and biosphere preservation, as well as for increasing conservation issues and electricity use, enhancing ecosystem healthcare value, and boosting carbon sequestration [20,21]. NBS have indeed been promoted as a method for the self-sustaining use of essence to address urban societal problems such as mitigating climate change, regulating water resources, land use and development of urban areas by practitioners (in particular the International Union for Nature Conservation [IUCN]) and policy (EC) [22].

According to the Final Report 2020 on Nature-Based Solutions and Re-naturing Cities by the European Commission, the goals of NBS include:

Sustainable urbanisation: Urban regions are home to a significant portion of the world's population, which faces a variety of issues (natural resources shortage, human well-being, etc.).

Restoration of damaged ecosystems: Human activities and actions have seriously damaged a number of ecosystems (agriculture, industry, etc.).

Climate change mitigation and adaptation: This global issue affects not only the environment but also the economical and societal variables.

Risk management and resilience: Without adequate planning, a variety of dangers might cause severe losses to social and natural resources.

11.4 URBAN RESILIENCE AND NATURE-BASED REMEDIES

The resilience of a city is based on its capacity to alter and adapt. Responding to incremental change and persistent pressures, often of a socioeconomic type, as well as sudden change or acute shocks, such as natural disasters, are all part of resilience [23]. Ecological systems perform a variety of roles that support humans and the cities they live in. Nature-based remedies have arisen as a paradigm, or overarching term, encompassing ecosystem-based ways to solve concerns such as changing climate, environmental risks, food and drinking water preservation, people's health and well-being, and financial and social progress. Ecosystem services are delivered by nature-based remedies to solve these societal challenges. Provisioning, regulating, cultural, and supporting services of nature-based solutions were established. The concept of NbS was used to emphasise the provision of ecosystem services in urban areas that contribute to a city's resilience [24]. NBSs can be used to enhance sustainable urbanisation, restore degraded habitats, mitigate and adjust climate change, and manage risk and resilience. The incorporation of nature-based remedies and their related delivery of ecosystem services in urban areas enhances urban resilience [25]. With the provisioning of valuable ecosystem-promoting socioeconomic, religious, and communal health, ecosystem services can help cities thrive. Urban green spaces and nature-based remedies provide a setting for enjoyment, community development, community stability, and the promotion of wellness [20]. These services help cities to be more resilient to the chronic pressures and incremental changes that cities encounter. Ecosystem resources are increasingly being regarded as adding to resilience to sudden change, disturbances, and natural calamities. Through suppressing the effects of changing climate, like heat waves and storms, ecosystems will buffer cities and strengthen their resilience (Figure 11.2).

In general, nature-based approaches are more reliable than massive, centrally located grey infrastructure because they operate as dispersed, autonomous infrastructural service delivery systems.

11.4.1 SDGS AGENDA 2030 FOR SUSTAINABLE DEVELOPMENT

The UN 2030 Agenda for Sustainable Development, which was adopted by all UN Member States, is another action plan. It consists of 17 Sustainable Development Goals (SDGs), which address global challenges and place strain on society, the economy, and the environment [26]. Similarly, one of NBS's main priorities is to solve global issues that are specifically related to the Sustainable Development Goals. NBS contribute to a variety of UN Sustainable Development Goals, not just those relevant to biodiversity and ecosystems [27] (Figure 11.3). There are campaigns using

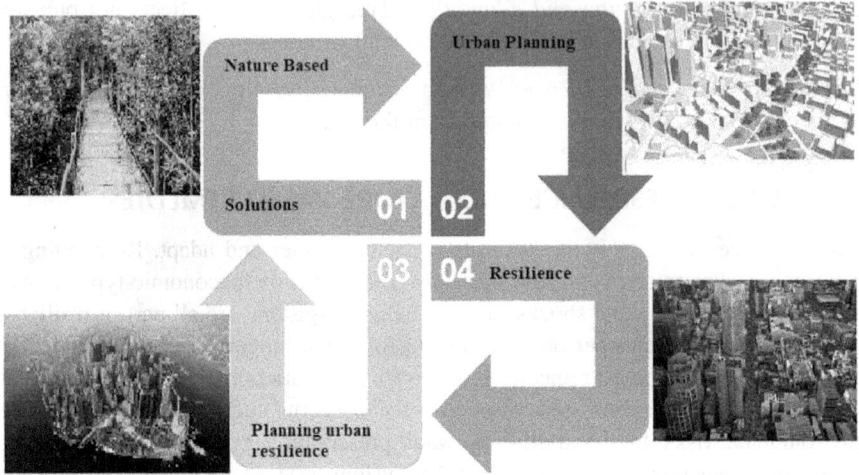

FIGURE 11.2   Relationship between nature-based solutions, urban planning, and resilience.

NBS in relation to different SDGs all over Europe. The following are some examples of interactions between various forms of NBS and SDG [28]:

- SDG 1 for poverty alleviation can be linked to green investments.
- To ensure food security and better nutrition, cities' agriculture practices are connected to SDG 2.
- SDG 3 for health and well-being is connected to urban cities' ecological zones.
- NBS-based education is related to SDG 4, which calls for inclusive and equal quality education as well as the promotion of lifelong learning.
- Initiatives for natural water retention are related to SDG 6 for water quality sustainability.
- Adaptation strategies for climate change can also be applied to SDG 7 for renewable energy.
- Both SDG 8 for sustainable economic growth and SDG 1 are related to innovative farming initiatives.
- SDG 10 for reducing inequality is related to urban restructuring with NBS (social cohesion).
- SDG 11 for healthy cities and neighbourhoods is related to vegetated roofs and pocket parks (these solutions are also to connected to SDGs 3, 10, and 13).
- Projects for urban renewal are related to SDG 12 for ensuring that resource use is sustainable (matter, energy, etc.).
- The SDG 13 goal of adapting to and combating climate change is related to urban green space planning.
- SDG 14 for the sustainable management of oceans and aquatic resources is related to natural coastal conservation measures.

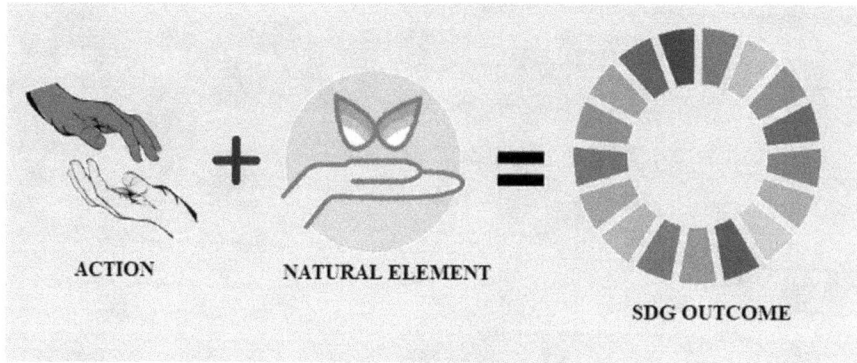

FIGURE 11.3   NbS as a three-part equation.

- Afforestation in rural areas is related to both SDG 15 and SDG 13, which seek to conserve, restore, and promote the sustainable use of terrestrial habitats.
- Residential green corridors are connected to SDG 16, which calls for the fostering of inclusive communities for long-term growth, as well as SDG 3.

11.4.2   APPLIED EXAMPLES OF NATURE-BASED SOLUTIONS

Some case studies are given below, to provide a better understanding of what NBSs are and how they function.

11.4.2.1   Case Study 1: Copenhagen Strategic Flood Masterplan

Location: Copenhagen, Denmark

In Copenhagen in 2011 a cloud burst took place. The city was inundated by 150 mm of rainfall, up to 1 metre in altitude. In order to counter potential floods that include climate adaptation options and current grey infrastructures, the town switched to NbS (blue and green master plan) [29]. The strategy to revise Lake Sankt Jørgens was a major difference between the new and old masterplans (Figure 11.4). The lake level was decreased from +5.8 m to +2.8 m, resulting in a new storage area of 40,000 m³ for cloud burst storage. The lakefront development allowed people to enter an area otherwise abandoned. A tunnel of 2.5 m diameter is designed to flow into a nearby port the water from the overflowing lake. This strategic master plan protects Copenhagen from floods and provides an urban climate of high quality.

11.4.2.2   Case Study 2: Stuttgart Green Ventilation Corridor

Location: Stuttgart, Germany

Stuttgart's location makes it vulnerable to poor air quality due to its moderate climate, low wind speeds, and heavy industrial activity. As a result, the municipality devised a new plan focused on the city's environment and population (Figure 11.5).

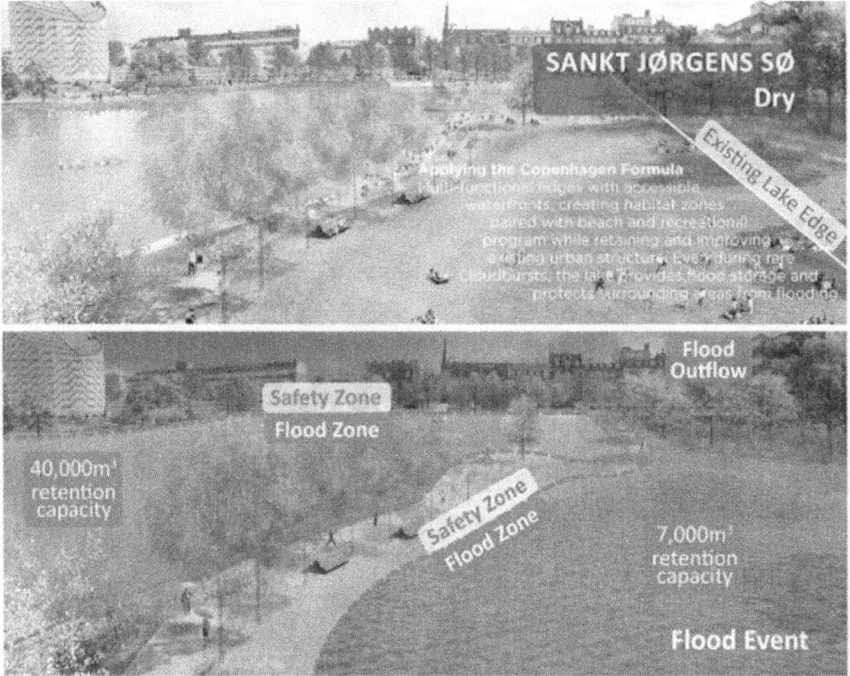

FIGURE 11.4 Lake development for flood masterplan.

FIGURE 11.5 Vegetation barrier along the developmental area.

A significant vegetation barrier was built around the development, with connected green spaces strategically positioned to allow for air exchange. Valleys that provided clean air were cordoned off to allow for free flow of air [30]. A preservation order was given for all trees with a trunk diameter of 80 cm or more. These nature-based solutions have reduced the urban heat island effect and increased biodiversity, while also sequestering carbon.

11.4.2.3 Case Study 3: Shenzen Sponge City Transition

Inside the tropical southwest of China, the Shenzhen region encounters severe floods during the monsoon and food scarcity amid dry spells. The local government has long turned to natural remedies to address this dual problem, particularly when Guangming New District was being built [31] (Figure 11.6). For instance, Guangming's People's Sports Center includes a green roof, raingardens, and permeable pavement that can collect more than 60% of the city's annual rainfall. In the subject of stormwater runoff, the region was designated as China's initial prototype village for reduced architecture in 2011. Shenzhen was selected to take part in the national Sponge City program 1; for the next 3 years, an additional 1.5 billion yuan (205 million euros) in subsidies will be given to the municipality. The action plan for the local government asks for the re-designing of an additional 256 square kilometres in 24 regions with consideration for water.

11.4.3 Urban Societal Challenge and Nature-Based Solutions

NBSs are used to tackle different urban societal challenges that societies face. NBSs that concentrate on solving issues (individually or collectively) do not tackle the challenges only, but also help to improve urban climate resilience.

FIGURE 11.6   Shenzen's peri-urban wetland.

11.5 CLIMATE RESILIENCE AND ADAPTATION

Our ecosystem and civilisation, which have evolved to and been constructed for a certain type of climate, are negatively impacted by climate change. When a culture is sensitive to or unable to handle the negative impacts of climate change, it becomes vulnerable. Increased pollution levels, habitat loss, and urban heat island effect are examples of negative impacts of climate change. Expected increases in precipitation, temperature, severe events, and elevated CO_2 levels will all have an impact on ecological functions, biology, biological relationships, and community composition [32,33]. At the COP21 in Paris, the importance of local action on carbon mitigation and greenhouse gas control was highlighted, highlighting that as the world becomes more urbanised, local action becomes increasingly important. Loss of biodiversity weakens ecosystem resources that humans depend on, decreasing our resilience.

The European Environment Agency (EEA) claims that a city's susceptibility to climate change depends on several variables. It is important to consider aspects such as the number of senior individuals that are more vulnerable to factors such as heat waves and related UHI, as well as urban growth, real estate development, and economic viability like facilities and numerous services. Cities must adapt to all these factors to become more resilient. According to the EEA, failing to adapt to these impacts will result in costs such as ill health, destroyed environment, buildings, and industry [34,35]. Flexible, adjustable solutions are required to combine insecurity with necessity, and the solutions themselves must fulfil changing needs or apply new information and must be adaptive.

11.5.1 Urban Heat Island (UHI) Effect

The dangers of extreme temperatures and their impacts on human health as well as biodiversity can be exacerbated by contemporary urban form. When temperatures in urban areas are substantially greater than those in the nearby area, this is known as an urban heat island [36,37]. The UHI effect raises the heat in a city by 5–15^{0}C. Therefore, the observed temperature increase can be attributed to the city's continuous conurbation [38]. The conversion of natural and agricultural regions into built-up areas is among the primary contributing factors to such increased temperature levels.

11.5.2 Water Management

Water resources in urban areas are under severe stress as a result of growing urban populations, pollution, and economic activities, putting a strain on their quality and quantity. As a result, sustainable water resource management is a major challenge for changing climate risk assessment and resilience in cities. Changes in rainfall patterns and temperature regimes are expected to exacerbate existing problems related to urban water resources [39]. Intense precipitation events will more frequently produce run-off volumes that exceed the capacity of urban sewerage systems, putting cities along rivers at risk of flooding, while changes in rainfall patterns in some regions will increase the risk of urban water scarcity. Because of the pollution load it transports, urban run-off water poses a threat to water quality.

11.5.3 GREEN SPACE MANAGEMENT

Blue and green regions consisting of indigenous and semi-indigenous features are called spaces, and they offer several ecological, financial, and social advantages. These include managing green and blue areas such as marshes, sports grounds, grasslands, green spaces, and parkland as well as incorporating green and blue components [40]. The management of green space is a social dilemma since urban growth typically results in a decrease in the quantity, accessibility, and quality of green space. Therefore, when green and blue areas are lost or diminished, the advantages that stem from their natural or seminatural qualities decrease.

11.5.4 AIR QUALITY

The number of pollutants and particles in the air, such as oxides of nitrogen, ozone, suspended particulate matter, oxides of sulphur, and carbon monoxide are increasing, hence air quality is deteriorating. These are primarily derived by the burning of waste and fuel products and these have direct health consequences, such as cardiovascular disease, cancer, and reduced lung function, as well as being harmful and toxic to plants and animals. From 1990, increased vehicular traffic in large and megacities has resulted in significant increases in various air pollutants. Extensive construction in a growing urban area also contributes towards the poor air quality of that area [41, 42].

Rain makes pollutants settle down, but it affects the overall air quality when emissions from other sources like industries are continuous. The fact that an area experiences patchy rains could be a factor affecting the average air quality index but it hardly reaches a good or satisfactory category.

11.5.5 REGENERATION IN URBAN CITIES

Urban regeneration aims to achieve sustainable development in urban cities by revitalizing the physical and socioeconomic conditions of a region and maximizing its economic and functional opportunities. This is achieved by enhancing various parameters such as current spatial structures, the environment, the economy, and social conditions. The goal is to improve the economic, physical, and socio-environmental conditions of a vulnerable area that has undergone negative changes and lacks resilience. Urban regeneration can cover aspects such as local business development, housing growth and improvement, community development, and environmental enhancement. However, it should also address ecological restoration on a large scale, as well as social justice issues. Given the challenges of limited available space, deprived areas, social inequity, and global environmental changes, urban regeneration offers cities new opportunities to reconsider their planning strategies.

11.5.6 PUBLIC HEALTH AND WELL-BEING

In shaping the health, well-being, and lifestyle of their inhabitants, urban environments play a significant role. A wide variety of factors, including inactive lifestyles, greater exposure to thermal and sound emissions, and stress in the air have been correlated with urban environments and increased mortality rates [43]. NBS can enhance the health and wellness of citizens living in urban regions by providing urban green area ecosystem services. Many ecosystem services under climate regulation mitigate environmental health challenges posed by urbanisation and climate change.

11.5.7 GREEN JOBS AND ECONOMICAL OPPORTUNITIES

By investing in NBSs and green infrastructure, there is an opportunity to create millions of jobs that are both low-skill, entry-level green jobs as well as high-skill, higher-paid green jobs [44]. According to Eurostat, nearly 18 million women and men were unemployed in the EU in January 2018, but by investing in NBSs, many job opportunities have opened up.

11.6 DISCUSSION

NbS are seen to bridge the gap between economic development and socio-environmental issues, providing a concrete route to a sustainable economy and urban resilience. However, scaling up and connecting and absorbing small-scale initiatives on the ground into larger and potentially more impactful interventions is a major challenge and opportunity for NbS. In this case, stronger policy coherence may be beneficial. Future research should be conducted to provide a solid evidence base for the contribution of NbS to job growth, as well as to demonstrate the economic feasibility of NbS in relation to other types of solutions on a timeline that is reasonable. Established urban planning frameworks for nature-based solutions, however, have significant gaps and omissions that must be resolved and improved if planning is to comprehensively support their implementation [45–47]. New and emerging research methods using multi-species approaches are needed for urban planning to effectively promote the implementation of nature-based solutions. The evidence for NBS benefits is inadequate, and generalisations are often made. NBSs are location-specific, which may clarify why generalisations are used. In terms of the environmental value of carbon sequestration, for example, the effects of a project of this size may not be sufficient to meet local greenhouse gas emission reduction goals. However, if applied in larger scales and quantities, a difference can be distinguishable. For sufficient documentation and to bridge information gaps, it is recommended that the results of the implemented NBSs be monitored on a regular basis.

Since NBSs are still in the early stages of growth, there are knowledge gaps in both research and practice. This can be a drawback because information gaps can lead to uncertainty on how to implement solutions. It does, however, open up possibilities for collaboration [48]. Collaboration with research projects, for example, could result in a win–win situation in which the research group is able to address their research questions while the city benefits from the implementation of NBSs in their urban environments. Our aim should be to contribute to this emerging multidisciplinary area by developing a planning system for nature-based solutions (NbS).

11.7 CONCLUSIONS

The aim of this chapter has been to provide solutions for how cities may get regenerated or built in a sustainable manner while also giving extra defence against the adverse consequences of changing climate by using NBSs. There are several urban growth drivers in cities, such as the natural increase in the human population, rural to urban migration, inclusion within the city limits of peripheral (peri-urban) cities, and reclassification of settlements from rural areas to urban cities, etc. The unprecedented urban growth is often termed as pseudo-urbanisation as this growth is exceptionally unbalanced. The impacts of urbanisation, concretisation, and land use conversion in cities are visible in the form of environment degradation. Hence, Urban restoration, climate adaptation and resilience, and green space management are the urban sustainable challenges that the city has to deal with. Blue green infrastructure, pavement shrubs and trees, roof structure, green corridors, drip irrigation, drainage channels, and public parks are all examples of green infrastructure and are some of the suggested NBSs to fix these issues. Nature-based solutions aim to meet particular needs or challenges while also maximising those environmental, social, and economic benefits. They reflect a cost-effective, resource-efficient, and adaptable strategy for achieving long-term, inclusive economic development while also enhancing human health and the natural environment. They will increase natural hazard tolerance and provide cost-effective climate change adaptation and emissions reduction solutions. Natural inspiration and encouragement will help to boost technological creativity and the economy. When combined with other more traditional sectoral approaches, nature provides solutions that support many other conservation, financial, and growth goals.

REFERENCES

[1] B. Wickenberg, K. McCormick, and J. A. Olsson, Advancing the implementation of nature-based solutions in cities: A review of frameworks, *Environmental Science & Policy,* vol. 125, pp. 44–53, 2021/11/01/ 2021.
[2] N. Frantzeskaki and T. McPhearson, Mainstream nature-based solutions for urban climate resilience, *BioScience,* vol. 72, no. 2, pp. 113–115, 2022.
[3] N. Frantzeskaki et al., Nature-based solutions for urban climate change adaptation: linking science, policy, and practice communities for evidence-based decision-making," *BioScience,* vol. 69, no. 6, pp. 455–466, 2019.
[4] S. Meerow, P. Pajouhesh, and T. R. Miller, Social equity in urban resilience planning, *Local Environment,* vol. 24, no. 9, pp. 793–808, 2019/09/02 2019.
[5] T. Emilsson and Å. Ode Sang, Impacts of climate change on urban areas and nature-based solutions for adaptation, in *Nature-Based Solutions to Climate Change Adaptation in Urban Areas: Linkages between Science, Policy and Practice*, N. Kabisch, H. Korn, J. Stadler, and A. Bonn, Eds. Cham: Springer International Publishing, 2017, pp. 15–27.
[6] S. Dhyani, M. Karki, and A. K. Gupta, Opportunities and advances to main-stream nature-based solutions in disaster risk management and climate strategy, in

Nature-based Solutions for Resilient Ecosystems and Societies, S. Dhyani, A. K. Gupta, and M. Karki, Eds. Singapore: Springer Singapore, 2020, pp. 1–26.

[7] M. C. Garg, H. Jain, N. Singh, and R. Dhupar, Chapter 15–Application of emerging nanomaterials in water and wastewater treatment, in *Current Directions in Water Scarcity Research*, vol. 6, A. L. Srivastav, S. Madhav, A. K. Bhardwaj, and E. Valsami-Jones, Eds.: Elsevier, 2022, pp. 319–340.

[8] N. Frantzeskaki, I. H. Mahmoud, and E. Morello, Nature-based solutions for resilient and thriving cities: opportunities and challenges for planning future cities, in *Nature-based Solutions for Sustainable Urban Planning: Greening Cities, Shaping Cities*, I. H. Mahmoud, E. Morello, F. Lemes de Oliveira, and D. Geneletti, Eds. Cham: Springer International Publishing, 2022, pp. 3–17.

[9] R. Antolin-Lopez and N. Garcia-de-Frutos, Nature-based solutions, in *Partnerships for the Goals*, W. Leal Filho, A. Marisa Azul, L. Brandli, A. Lange Salvia, and T. Wall, Eds. Cham: Springer International Publishing, 2021, pp. 804–814.

[10] R. C. Brears, Nature-based solutions, in *Financing Nature-Based Solutions: Exploring Public, Private, and Blended Finance Models and Case Studies*, R. C. Brears, Ed. Cham: Springer International Publishing, 2022, pp. 7–27.

[11] C. Norman, A. Surjan, and M. Booth, Making resilience a reality: the contribution of peri-urban ecosystem services (BGI) to urban resilience," in *Ecosystem-Based Disaster and Climate Resilience: Integration of Blue-Green Infrastructure in Sustainable Development*, M. Mukherjee and R. Shaw, Eds. Singapore: Springer Singapore, 2021, pp. 185–200.

[12] M. Buxton, R. Carey, and K. Phelan, The role of peri-urban land use planning in resilient urban agriculture: a case study of Melbourne, Australia, in *Balanced Urban Development: Options and Strategies for Liveable Cities*, B. Maheshwari, B. Thoradeniya, and V. P. Singh, Eds. Cham: Springer International Publishing, 2016, pp. 153–170.

[13] G. Lucertini and G. Di Giustino, Urban and peri-urban agriculture as a tool for food security and climate change mitigation and adaptation: the case of Mestre, *Sustainability,* vol. 13, no. 11, 2021.

[14] A. Galderisi, Chapter 2–The resilient city metaphor to enhance cities' capabilities to tackle complexities and uncertainties arising from current and future climate scenarios, in *Smart, Resilient and Transition Cities*, A. Galderisi and A. Colucci, Eds.: Elsevier, 2018, pp. 11–18.

[15] R. Atta ur, G. A. Parvin, R. Shaw, and A. Surjan, 3–Cities, vulnerability, and climate change, in *Urban Disasters and Resilience in Asia*, R. Shaw, R. Atta ur, A. Surjan, and G. A. Parvin, Eds.: Butterworth-Heinemann, 2016, pp. 35–47.

[16] C. Albert *et al.*, Planning nature-based solutions: Principles, steps, and insights, *Ambio,* vol. 50, no. 8, pp. 1446–1461, 2021/08/01 2021.

[17] R. Pineda-Martos and C. S. C. Calheiros, Nature-based solutions in cities—Contribution of the Portuguese National Association of Green Roofs to urban circularity, *Circular Economy and Sustainability,* vol. 1, no. 3, pp. 1019–1035, 2021/11/01 2021.

[18] N. Kabisch, N. Frantzeskaki, and R. Hansen, Principles for urban nature-based solutions, *Ambio,* vol. 51, no. 6, pp. 1388–1401, 2022/06/01 2022.

[19] M. Pineda-Pinto, N. Frantzeskaki, and C. A. Nygaard, "he potential of nature-based solutions to deliver ecologically just cities: Lessons for research and urban planning from a systematic literature review, *Ambio,* vol. 51, no. 1, pp. 167–182, 2022/01/01 2022.

[20] N. Seddon, A. Chausson, P. Berry, C. A. J. Girardin, A. Smith, and B. Turner, Understanding the value and limits of nature-based solutions to climate change and other global challenges, *Philosophical Transactions of the Royal Society B: Biological Sciences,* vol. 375, no. 1794, p. 20190120, 2020/03/16 2020.

[21] A.D. Hughes, Defining nature-based solutions within the blue economy: the example of aquaculture, Policy Brief vol. 8, 2021-July-29 2021.

[22] J. Dick et al., How are nature based solutions contributing to priority societal challenges surrounding human well-being in the United Kingdom: a systematic map protocol, *Environmental Evidence,* vol. 8, no. 1, p. 37, 2019/11/22 2019.

[23] P. B. Holden et al., Nature-based solutions in mountain catchments reduce impact of anthropogenic climate change on drought streamflow," *Communications Earth & Environment,* vol. 3, no. 1, p. 51, 2022/03/09 2022.

[24] S. Pauleit, T. Zölch, R. Hansen, T. B. Randrup, and C. Konijnendijk van den Bosch, Nature-based solutions and climate change – four shades of green," in *Nature-Based Solutions to Climate Change Adaptation in Urban Areas: Linkages between Science, Policy and Practice*, N. Kabisch, H. Korn, J. Stadler, and A. Bonn, Eds. Cham: Springer International Publishing, 2017, pp. 29–49.

[25] A. Ossola and B. B. Lin, Making nature-based solutions climate-ready for the 50 °C world, *Environmental Science & Policy,* vol. 123, pp. 151–159, 2021/09/01/ 2021.

[26] E. Gómez Martín, R. Giordano, A. Pagano, P. van der Keur, and M. Máñez Costa, Using a system thinking approach to assess the contribution of nature based solutions to sustainable development goals, *Science of The Total Environment,* vol. 738, p. 139693, 2020/10/10/ 2020.

[27] A. Haase, The contribution of nature-based solutions to socially inclusive urban development – some reflections from a social-environmental perspective, in *Nature-Based Solutions to Climate Change Adaptation in Urban Areas: Linkages between Science, Policy and Practice*, N. Kabisch, H. Korn, J. Stadler, and A. Bonn, Eds. Cham: Springer International Publishing, 2017, pp. 221–236.

[28] A.F. Giannetti, F. Agostinho, and C. M. V. B. Almeida, Chapter 2–Sustainable development and its goals, in *Assessing Progress Towards Sustainability*, C. Teodosiu, S. Fiore, and A. Hospido, Eds.: Elsevier, 2022, pp. 13–33.

[29] A. Camponeschi, Narratives of vulnerability and resilience: An investigation of the climate action plans of New York City and Copenhagen, *Geoforum,* vol. 123, pp. 78–88, 2021/07/01/ 2021.

[30] W. Wang, D. Wang, H. Chen, B. Wang, and X. Chen, Identifying urban ventilation corridors through quantitative analysis of ventilation potential and wind characteristics, *Building and Environment,* vol. 214, p. 108943, 2022/04/15/ 2022.

[31] X. Liu et al., Wind environment assessment and planning of urban natural ventilation corridors using GIS: Shenzhen as a case study, *Urban Climate,* vol. 42, p. 101091, 2022/03/01/ 2022.

[32] S. Mehryar, I. Sasson, and S. Surminski, Supporting urban adaptation to climate change: What role can resilience measurement tools play?, *Urban Climate,* vol. 41, p. 101047, 2022/01/01/ 2022.

[33] J. Leandro, K.-F. Chen, R. R. Wood, and R. Ludwig, A scalable flood-resilience-index for measuring climate change adaptation: Munich city, *Water Research,* vol. 173, p. 115502, 2020/04/15/ 2020.

[34] H. Kim, D. W. Marcouiller, and K. M. Woosnam, Coordinated planning effort as multilevel climate governance: Insights from coastal resilience and climate adaptation, *Geoforum,* vol. 114, pp. 77–88, 2020/08/01/ 2020.

[35] E. Petridou, J. Sparf, and K. Pihl, Chapter 4.1–Resilience work in Swedish local governance: Evidence from the areas of climate change adaptation, migration, and violent extremism, in *Understanding Disaster Risk*, P. P. Santos, K. Chmutina, J. Von Meding, and E. Raju, Eds.: Elsevier, 2021, pp. 225–238.

[36] H. Zhang, J.-j. Han, R. Zhou, A.-l. Zhao, X. Zhao, and M.-y. Kang, Quantifying the relationship between land parcel design attributes and intra-urban surface heat island effect via the estimated sensible heat flux, *Urban Climate,* vol. 41, p. 101030, 2022/01/01/ 2022.

[37] L. Zhang, X. Yang, Y. Fan, and J. Zhang, Utilizing the theory of planned behavior to predict willingness to pay for urban heat island effect mitigation, *Building and Environment,* vol. 204, p. 108136, 2021/10/15/ 2021.

[38] X. Huang, J. Song, C. Wang, T. F. M. Chui, and P. W. Chan, The synergistic effect of urban heat and moisture islands in a compact high-rise city, *Building and Environment,* vol. 205, p. 108274, 2021/11/01/ 2021.

[39] R. Ray Biswas, R. Sharma, and Y. Gyasi-Agyei, Adaptation to climate change: A study on regional urban water management and planning practice, *Journal of Cleaner Production,* vol. 355, p. 131643, 2022/06/25/ 2022.

[40] M. Graça, S. Cruz, A. Monteiro, and T.-S. Neset, Designing urban green spaces for climate adaptation: A critical review of research outputs, *Urban Climate,* vol. 42, p. 101126, 2022/03/01/ 2022.

[41] M. L. Melamed, J. Schmale, and E. von Schneidemesser, Sustainable policy—key considerations for air quality and climate change, *Current Opinion in Environmental Sustainability,* vol. 23, pp. 85–91, 2016/12/01/ 2016.

[42] R. McArdle, Intersectional climate urbanism: Towards the inclusion of marginalised voices, *Geoforum,* vol. 126, pp. 302–305, 2021/11/01/ 2021.

[43] J. C. Semenza, Lateral public health: Advancing systemic resilience to climate change, *The Lancet Regional Health–Europe,* vol. 9, p. 100231, 2021/10/01/ 2021.

[44] J. Lefèvre, T. Le Gallic, P. Fragkos, J.-F. Mercure, Y. Simsek, and L. Paroussos, Global socio-economic and climate change mitigation scenarios through the lens of structural change, *Global Environmental Change,* vol. 74, p. 102510, 2022/05/01/ 2022.

[45] Z. Vojinovic et al., Effectiveness of small- and large-scale nature-based solutions for flood mitigation: The case of Ayutthaya, Thailand, *Science of The Total Environment,* vol. 789, p. 147725, 2021/10/01/ 2021.

[46] N. Srivastava and R. Shaw, "8–Enhancing City Resilience Through Urban-Rural Linkages," in *Urban Disasters and Resilience in Asia*, R. Shaw, R. Atta ur, A. Surjan, and G. A. Parvin, Eds.: Butterworth-Heinemann, 2016, pp. 113–122.

[47] G. A. Parvin, A. Surjan, R. Atta ur, and R. Shaw, "2–Urban Risk, City Government, and Resilience," in *Urban Disasters and Resilience in Asia*, R. Shaw, R. Atta ur, A. Surjan, and G. A. Parvin, Eds.: Butterworth-Heinemann, 2016, pp. 21–34.

[48] A.-u. Rahman, R. Shaw, A. Surjan, and G. A. Parvin, "1–Urban Disasters and Approaches to Resilience," in *Urban Disasters and Resilience in Asia*, R. Shaw, R. Atta ur, A. Surjan, and G. A. Parvin, Eds.: Butterworth-Heinemann, 2016, pp. 1–19.

12 Management of Flood Disasters in Peri-Urban Ecosystems of Noida Using Remote Sensing and Geographical Information Systems

Renu Dhupper, Harshita Jain, Anil K. Gupta, Pritha Acharya, and Deepak Kumar

CONTENTS

DOI: 10.1201/9781003331001-12

12.1   INTRODUCTION

Urbanisation is a necessary aspect of growth at all levels, but rapid urbanisation is increasing the region's vulnerability. Due to their capacity for coping with disasters being limited, cities in developing countries are more prone to flood damage [1]. Surface roughness is diminished because of urban land use, which reduces the time it takes for streamflow to reach the stream channel [2,3]. Changes in land use and land cover are influenced by the spread of human settlements and associated anthropogenic activities, particularly rapid urbanisation, which results in variations in biological methods and water resources from global to local scales [4,5]. Unhindered grassland as well as sandy regions, on the other hand, create less runoff owing to variables like rainfall reception by the forest canopy, a thick web of roots that enhances infiltration rate, soil properties, and increased runoff. Clearly, places having a greater amount of agricultural or urban land use produce greater storm drainage than grasslands with identical soils and terrain [4]. For understanding the variety of problems associated with urban flooding and its consequences on the surrounding environment, it is necessary to consult admissible local, national, and international literature [6]. Despite a significant body of work on urbanisation and flooding in large cities, the comprehensive response to flooding risk in recently constructed small and intermediary urban areas in India is mostly unknown [7]. A venture to conduct such a work could represent a foundation for assessing the risk of urban flash floods in emerging countries' recently urbanised regions. Many satellite towns have sprouted up around India's megacities to facilitate industrial development [8]. The Indian economy grew rapidly in the 1990s, and extra relocation around Delhi far outstripped expectations. A significant area of land was recuperated from the Yamuna flood plain on the outskirts of Delhi to build a region for nearly half a million people called the New Okhla Industrial Development Area (NOIDA) in the Gautam Buddha Nagar district of Uttar Pradesh [9].

The architecture and services of Noida were meticulously built out as part of a planned growth. It was built to accommodate Delhi's future population expansion for the next 25–30 years. However, due to the high migration of people to Delhi, it began to overcrowd in less than 16 years. Noida is among Asia's greatest designed industrial townships, and it represents India's balance of human settlement and industry. This is a significant outpost of Delhi and it has been identified as a flood-risk area, despite the construction of an embankment [10]. The ecosystem of this location is unsuitable for industrial growth. Because the site was once a river catchment area, it is a shallow area and hence prone to flooding, with liquid waste discharge and logging issues. Noida has had multiple severe floods in the last 30 years, the devastating repercussions of which highlight the area's fragility in relation to drainage. The number of reported flash-flood incidents in Noida over the last 30 years has similarly shown a positive trend. While an increase in impermeable areas because of urbanisation can be attributed to some of the increased flash flood hazard, it poses significant impact on water balance in local areas.

Using remote sensing techniques and evaluating pertinent meteorological and hydrological data, this study is going to link the impervious area and flow of the river [11,12]. The land use change related with runoff of storm water and other emissions to rivers in urban catchments will be highlighted [13]. The outcomes can be directly applied to the creation of land use planning and future urbanisation policies stabilised by conservation measures [14,15]. If the impact of urbanisation on river flowrate has been quantified, forecasting future flooding trends will be easier, allowing for measures to be taken to meet rising needs for utilities without putting at risk the increased storm water intensity and extent during rainy seasons [16]. The findings of this study demonstrate that urbanisations in flood-prone areas are the primary cause of increased flood danger. Forecasting and simulating floods is consequently critical for civil protection planning and operation, as well as early flood warning. As a result of these outcomes, emergency management and local planning authorities should consider land use planning as an efficient flood risk mitigation option. The sections that follow provide some insights into the study region, its topographical location, and soil analysis considering urbanisation and flooding, as well as a review of current outcomes and some concluding thoughts.

12.2 METHODOLOGY

12.2.1 STUDY AREA

Noida township's latitude and longitude are 28° 34'N and 77° 18'E. Noida is a planning region which is wholly included within Gautam Buddha Nagar, a newly established district. It has over 80 revenue villages, 65 of which are in Tehsil Dadri (previously in Ghaziabad district) and 16 in Tehsil Sikandrabad [17]. It is situated alongside the eastern and south-eastern borders of Delhi, adjacent to the metropolis of Delhi (Figure 12.1). Noida is classified as a Delhi Metropolitan Area under the framework of the National Capital Region's regional plan because it is located within the DMA town. The city is considered a part of the NCR's Uttar Pradesh region. Noida has about 203 square kilometres total area. It is a major industrial and satellite town, located on the Delhi–Uttar Pradesh border on the sides of the rivers Yamuna and Hindon (TCPO, 1992). In 1982, 37,500 people lived in the Noida Notified Area, which is part of the National Capital Region (NCR) and spans 21,116 hectares (including 81 villages) (Census of India, 1982). By 1992, it had been designated as a Census Town, and its population had expanded fourfold to 180,600 people. Noida's population grew from 297,998 in 2001 to 809,987 in 2021. As the metropolis has grown, it has encompassed all 81 settlements.

12.2.2 SOIL TYPE

The area's geography is mostly flat, approximately 200 metres above sea level (MSL), and it has a moderate slope ranging from 0.1–0.2% from northeast to southwest. The highest point in the area is 204 metres above sea level, close to the northeast region in Parthala Khanjarpur village, and the lowest point is 196 metres above sea level, close to Garhi village in the southwest. The site's ground level is lower than the Yamuna's

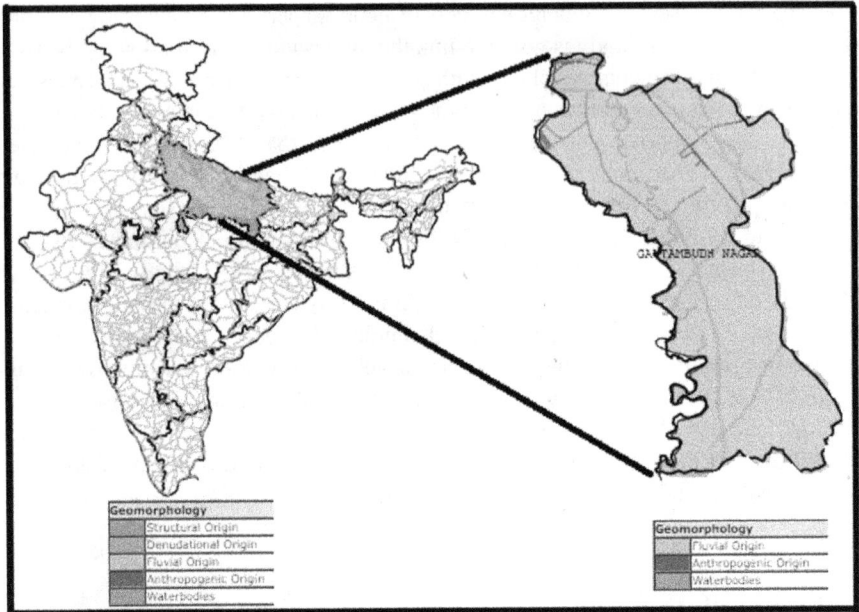

FIGURE 12.1 Study area map.

peak flood level. The development of dams along the rivers Yamuna and Hindon has reduced flood risks in the area [18]. The site's low level, on the other hand, is a barrier to enhanced water and sewage disposal. As a result, the area's hygienic condition is relatively poor. The soil is primarily silt that has been formed over a long period of time. Rainwater has a tough time percolating through the silty soil. This causes the voids to shrink, lowering the percolation rate.

12.2.3 LAND USE LAND COVER (LULC) CHANGE ANALYSIS AND BUILT-UP EXPANSION

Land cover is related to the perceived biophysical cover of the surface of the Earth and land use means how this biophysical cover is used [19,20]. LULC is influenced by socioeconomic and institutional factors as well as geological structure, altitude, and slope. The dynamics of LULC have changed dramatically in recent decades. Both human and environmental forces have contributed to this shift in the LULC pattern [21]. As a result, quantitative analysis of LULC change dynamics is an efficient method for controlling and comprehending landscape modification.

LULC vector data of 2011–12, multi-temporal satellite data of 2015–16 from Resourcesat-2 LISS III Ortho rectified Resourcesat-2 Data from LISS-III sensor of three seasons pertaining to 2015–16 (Kharif: August–October in Monsoon; Rabi: December–March in Post-Monsoon; Zaid: April–May in Pre-Monsoon) were downloaded from Bhuvan-NRSC, India.

12.2.4 VULNERABILITY, CAPACITY ANALYSIS, AND FLOODING RISK

The rivers Yamuna and Hindon, albeit essential ecological elements of Gautam Buddha Nagar's designated zones, are vulnerable to flooding (Figure 12.2) [22,23].

Noida is vulnerable because it was once a part of the Yamuna and Hindon rivers. By building an embankment, the land was reclaimed for development. It is a low-lying and sloping terrain. In the event that one of the embankments fails, the area is likely to be flooded on a huge scale – whether by the Yamuna or the Hindon [10].

The subsurface water is saline, with a high mineral content [24]. Existing flood prevention techniques are ineffective in protecting the area from severe flooding. Because the location is depressed and nearly flat, the casual flooding issue in Noida cannot be overlooked. Floods in the sewerage system of the Yamuna and Hindon rivers have caused multiple floods in the area in the past. Over the last 33 years, the Yamuna River has 25 times surpassed the danger level which is set at 204.83 m. The area has a monsoonal climate. The temperature fluctuates from 16°C to 45°C, and the annual rainfall is 700 mm, with almost 81% of this falling in the time of monsoon.

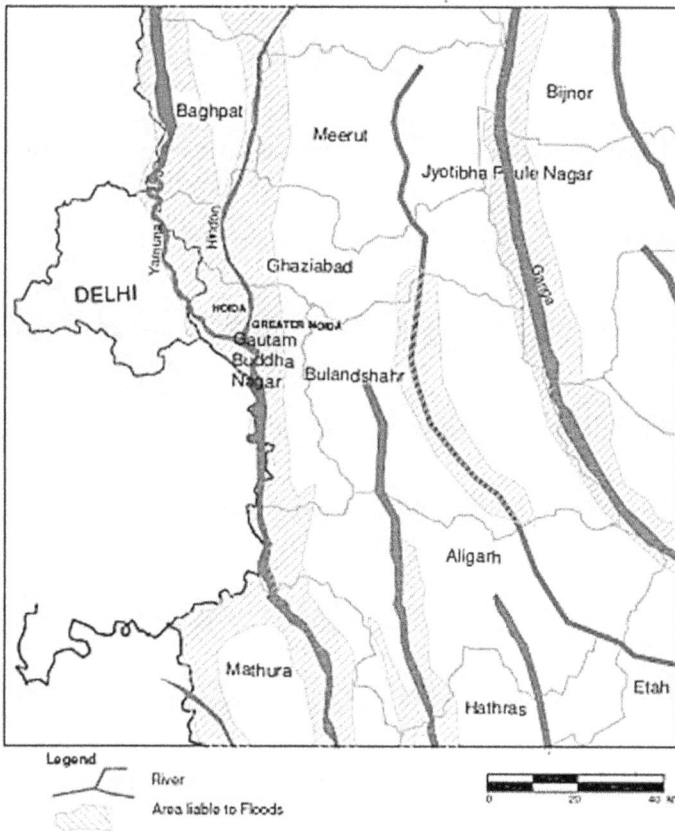

FIGURE 12.2   Flood risk map of Gautam Budh Nagar.

The wettest months of July and August receive roughly 59% of the monsoon season's rainfall. The catchment's runoff is roughly 1920 cusecs, with a runoff coefficient of 0.327. The city of Noida is served by five storm water drains. Drains No. 1 (which flows through sector 14) and 4 (which runs from sectors 34 and 51) meet drain No. 2 (which runs from sectors 37 and 50) and then the irrigation drain. Drain number 3 runs through sectors 8, 10, 11, 12, 21, 21A, 23, 24, 25, 32, and 33 before meeting the irrigation drain. Drain No. 5 connects to the irrigation drain after passing through the Noida Export Processing Zone (NEPZ).

12.2.5 Modelling of Urban Flash Floods: Analysis of Hydrologic Data and Rainfall Profusion Calculations

The surface sealing modelling raster file has data on the pervious and impervious portions of all grid cells. The spatially distributed excess rainfall estimates were calculated using this sub-grid variability [25]. To map impermeable surfaces, visual elucidation and field inventory and large-scale, ortho-rectified aerial images are studied. However, because such processes are time-consuming, data from Ikonos and Landsat high-resolution satellite sensors were used to create maps of the impervious-ness surface to compute the water transmission rate from the Bhuvan–NRSC datasets.

The technique develops local and severe flood situations using the following points (both geographical and mathematical) with computation and adjustments of drain flow and overflow in each land use type over 188 sectors [26–29]. As a result, the calculation of the amount of collected surface water in Noida on the last day of flooding is done using Equation 12.1 to compute the water transmission rate [30]:

$$K\left(\frac{m}{\text{sec}}\right) = \frac{1}{86400} \tag{12.1}$$

Manning's equation (V) and continuity equation (Q) (Manning 1890) were utilised to determine the capability of sewers to release flood waters and covering of uneven ground to limit outflow (Equation 12.2) [31]:

$$V = \frac{\left(\dfrac{1.486}{N}\right) * R^{0.67} * SO.50}{N} \tag{12.2}$$

where V is the flow velocity in m/s. The roughness coefficient (dimensionless) is denoted by the letter N (cross-sectional area, m^2/ wetted perimeter in metre), R denotes hydraulic radius in (m) and S stands for slope.

12.2.6 Observed Rainfall Variability and Changes

Precipitation data from the Indian Meteorological Department were supplied for examination of pattern, variance, and average weather systems from 1989 to 2018.

Monthly precipitation trends for every location were calculated using hourly rainfall, and quarterly region precipitation trends were created by algebraic averaging all location precipitation readings within the area [32,33]. The state's monthly rainfall series was calculated using the district's area weighted rainfall values.

12.3 RESULTS AND DISCUSSION

12.3.1 LAND USE AND LAND COVER ANALYSIS

The land use pattern of Noida, Gautam Buddha Nagar (Figure 12.3), has been continuously changing due to rapid and unregulated population increase, as well as industrial development and economic growth. Global warming and rising population, as well as rapid industrialisation and urban aggregation, urban expansion, and policy provisions have all been highlighted as potential causes of LULC shifts (Figure 12.4).

The extension of metropolitan regions because of urban slump and the development of public transportation systems allows a large portion of the labour force to travel daily, inducing changes in LULC in the suburbs. Due to this, the transformation of cultivation and other innate land covers into built-up areas and urban regions extends outwards. The most important factor is the transformation of innate land cover types into contrived land use types.

Construction, foliage, watercourses, and farming, along with other LULC classes, have been the results of the probabilistic classifier, a quantitative assessment in which images are assigned based on the majority class likelihood (Figure 12.5).

Figure 12.5 depicts the results of the LULC maps developed for the research region in 2006, 2012, and 2015. This figure shows the % change in LULC from 2005 to 2006, 2011 to 2012, and 2014 to 2015.

According to the above findings, there was a consistent shift in LULC from 2005 to 2006 as contrasted with 2011 to 2012. Between 2014 and 2015, the study region saw

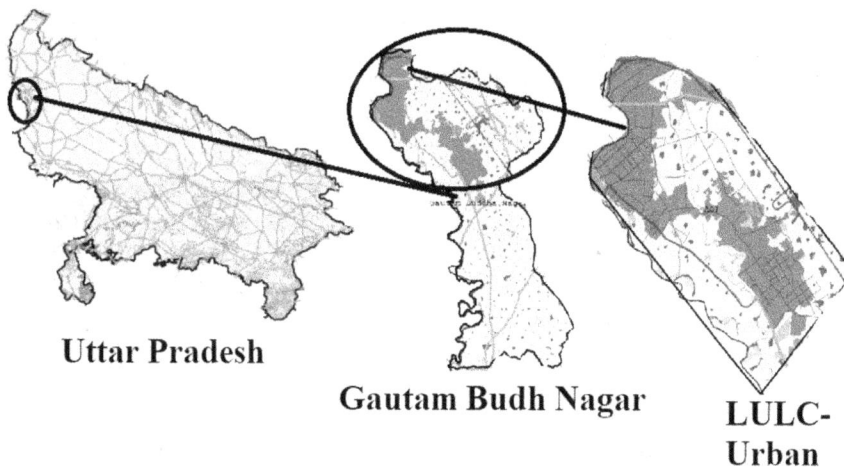

Uttar Pradesh

Gautam Budh Nagar

LULC-Urban

FIGURE 12.3   LULC pattern of Noida, Gautam Buddha Nagar.

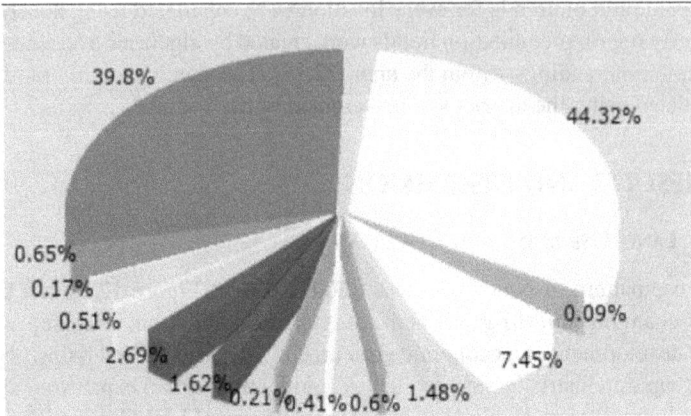

LULC Class	Area (Sq. Km.)
Cropland	261.1
Deciduous	0.51
Fallow land	43.88
Grasslands/ Grazing Lands	8.72
Inland Wetland	3.53
Plantation	2.44
Reservoirs/ Lakes/ Ponds	1.21
River/ Stream/ Canals	9.52
Rural	15.87
Sandy Area	3.03
Scrub Forest	0.99
Scrub land	3.84
Urban	234.48

You have selected 622.258 Sq.Km. Following LULC classes are available in the selected area. The approximate area distribution of each class (clipped area) is shown below:

FIGURE 12.4   LULC shifts observed in the Gautam Buddha Nagar district.

a significant impact on land use and, as a result, urban expansion. From 2005 to 2015, the percentage of built-up area in Noida rose rapidly. The entire built-up area in 2005 in Gautam Buddha Nagar was 210.72 km^2, and by 2015 it had expanded to 283.98 km^2. ove the period from 2005 to 2015, the purpose built area increased by 25.79% of the total area. Beneficial land use assessment was also highlighted by changing land cover class, with wastelands being reduced since they are substituted by vegetation regions, which increased in quantity within this period. In 2005, there were 1045.24 km^2 of vegetative land, which declined to 879.98 km^2 in 2015. As a result, the transformation of farmland to certain other land use categories, wherein fostered lands as well as other open spaces were supplanted by construction, roadways, sidewalks, as well as other infrastructural facilities, leads to a rise in urban vegetation, accounting for the vast bulk of the increment in the urban area.

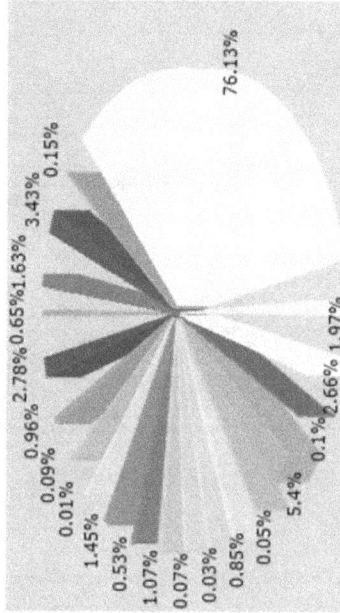

LULC Information (2005–06) for Uttar Pradesh
LULC Information (2011–12) for Uttar Pradesh
LULC Information (2014–15) for Uttar Pradesh

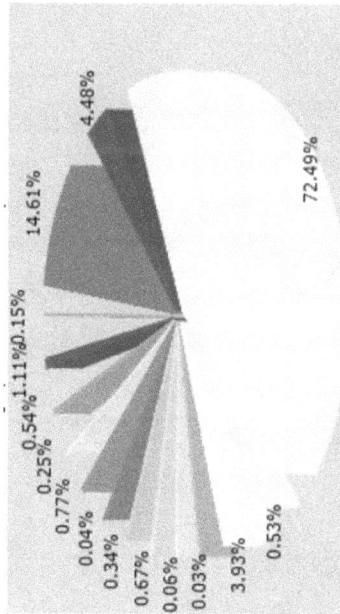

LULC Information (2005–06) for Gautam Buddha Nagar
LULC Information (2011–12) for Gautam Buddha Nagar
LULC Information (2014–15) for Gautam Buddha Nagar

FIGURE 12.5 LULC maps generated for the study area for 2006, 2012, and 2015.

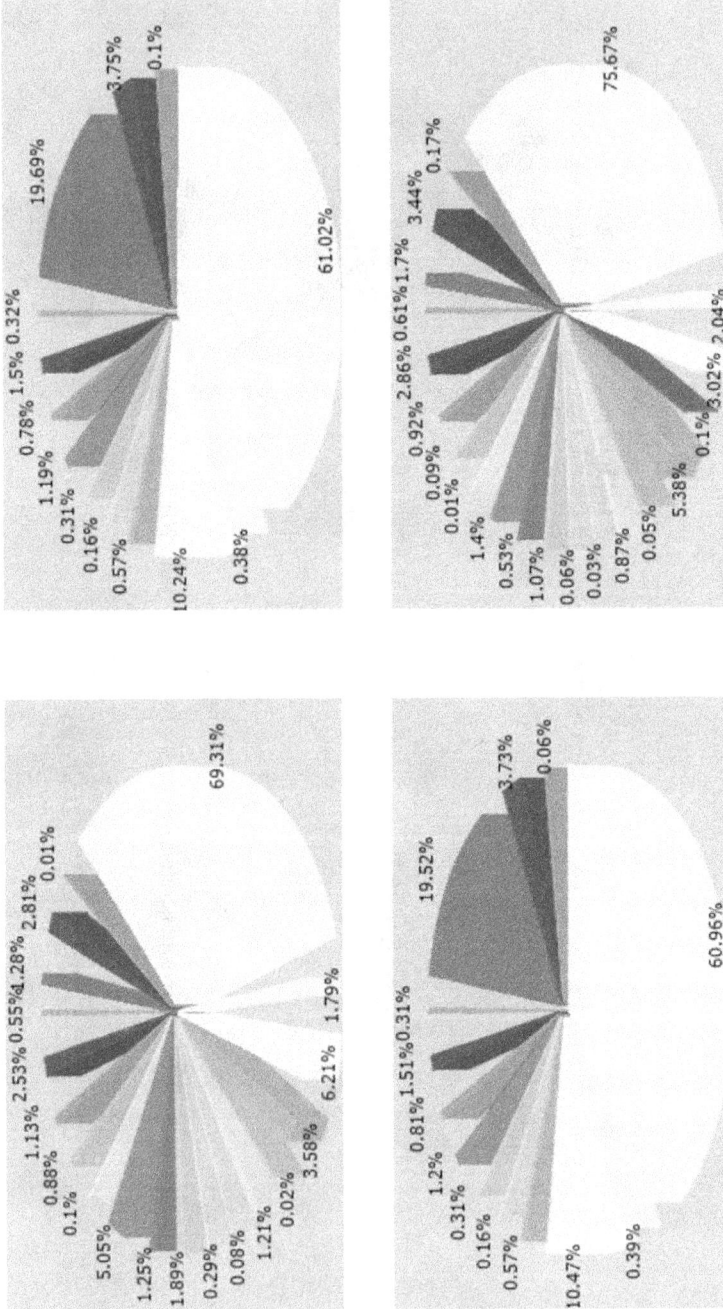

FIGURE 12.5 (Continued)

■ Builtup,Urban
■ Builtup,Rural
■ Builtup,Mining
　 Agriculture,Crop land
　 Agriculture,Plantation
　 Agriculture,Fallow
■ Forest,Deciduos
■ Forest,Forest Plantation
■ Forest,Scrub Forest
■ Forest,Swamp/Mangroves
■ Grass/Grazing
■ Barren/unculturable/Wastelands,Salt Affected Land
■ Barren/unculturable/Wastelands,Gullied/Ravinous Land
■ Barren/unculturable/Wastelands,Scrub land
■ Barren/unculturable/Wastelands,Sandy area
■ Barren/unculturable/Wastelands,Barren rocky
■ Wetlands/Water Bodies,Inland Wetland
■ Wetlands/Water Bodies,River/Stram/Canals
■ Wetlands/Water Bodies,Reservoir/Lakes/Ponds

FIGURE 12.5 (Continued)

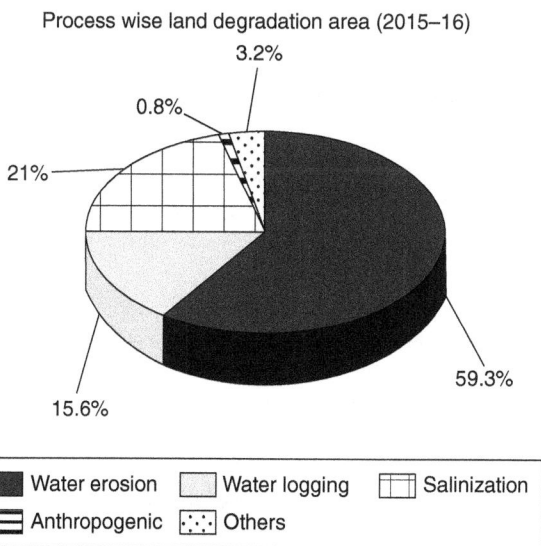

Process wise land degradation area (2015–16)

FIGURE 12.6 Estimation of gross erosion rates from the mainland of Uttar Pradesh.

TABLE 12.1
Imperviousness Degree for Various Land Use Classes, Estimated and Default Values from Satellite Imagery (Calculated by the Authors)

Land use	Imperviousness default degree	Calibrated value of Ikonos-Landsat
Low density	0.40	0.22
High density	0.60	0.71
City centre	0.80	0.48
Infrastructure	0.60	0.70
Roads/highways	0.60	0.41
Industry	0.80	0.96

TABLE 12.2
Runoff Coefficient for Land Use Types of Sectors in Reference to Marshall's Study

Land use	Code	Sectors	Runoff coefficient
Industrial	Id	9	0.90
Residential	Rs	12,26	0.60
Facilities	Fl	12, 22	0.70
Commercial	Cm	18, 27	0.80
Recreational	Rr	42, 25	0.40
Roads	Rd	64, 69	0.95
Mixed	Mx	49, 41, 50	0.60
Other	O	82, 46, 92	0.35

12.3.2 LAND DEGRADATION MAP AND DATA

Land deterioration is inextricably tied to soil erosion. Excessive soil loss because of inadequate management of land has major consequences for crop yield and crop security, necessitating the use of our soil resource in a sustainable manner. Soil erosion and sedimentation are important elements and indicators of the degradation of land [34,35].

Landscape development is primarily driven by erosion and sediment redistribution processes, and it also has an essential part in the development of soil. Similarly, the river's sediment loads have a significant impact on the system's operation. Because sedimentation and erosion have such a large impact on food supply, they also have a substantial socio-economic impact on the state.

The gross erosion rates from the mainland of Uttar Pradesh were estimated (Figure 12.6). The patterns of degradation are best studied at the size of watersheds, reservoirs, or river systems. Figure 12.6 illustrates the proportion of land affected by cumulative coastal erosion by type. A major section of the population of the total area

(59.33%) falls under water erosion categories and 21.02% of the area is affected by salinisation problems.

12.3.3 ANALYSIS OF HYDROLOGIC DATA AND INTEGRATION OF REMOTE SENSING PICTURES

After calculating the degree of imperviousness of each LULC using Ikonos and Landsat-7, interpolated results, the integration of both forms of study is required to obtain an accurate picture (Table 12.1). As a result, two flooded scenarios were created to extract surplus storm water output and peak rainfall intensity over time. Runoff = Precipitation + Infiltration + Interception + Evaporation is a common formula for calculating runoff.

The first step in creating flooding events for the year 2021 is to create a spreadsheet. The two prerequisites for producing flooding events of varying rainfall magnitude are the land use pattern and the degree of impervious area under each land use. The impervious area for each land use category was calculated using Ikonos and Landsat calibrated values, and the land use pattern and change were measured using changes in land use matrix and remote sensing data collected over decades. As a result, drainage capacity was required to evaluate the effects of different rainfall intensities [36,37]. Analysis of precipitation was done using Mayer's formula and Marshall et al.'s (1947) study for computing runoff for land use patterns changes in a typical model sector. The analysis was based strictly on Marshall et al.'s research by adding the results, multiplied by the size of every land-use types with its corresponding coefficient of runoff, and then dividing the total by the entire area of the relevant model area. Based on the compiled average values of various sectors, the calculated runoff findings for all types of model sector are shown in Table 12.2. The runoff from various land uses has been computed for various scenarios and combined to depict the critical flooding condition using the rational method provided by Marshall et al. (1947).

12.3.4 FLOOD HAZARD RISK IN GAUTAM BUDDHA NAGAR

Flooding has far-reaching consequences and poses a serious threat to public safety and regional economic sustainability. Flooding is a bigger issue in the GBN area and NGY since so many major commercial and administrative structures are close to the rivers' banks. The townships would suffer severe economic losses because of damage to farming lands, rural area herds, constructions, and urban buildings. Flooding poses a significant concern to the NGY townships in terms of both economic and environmental impacts.

12.3.4.1 Flood Threats from the Yamuna River

Yamuna's main stream begins from Yamnotri Glacier, which is located at a height of 6,400 metres above sea level. Yamuna river joins the Doon Valley after passing across the Himalayas. Many streams meet the river on its route to Haryana and the eastern and western Yamuna Canal Systems which supply Uttar Pradesh and Haryana have headworks. Approximately 163 kilometres north of Palla Village, the Yamuna enters Delhi's National Capital Region (NCR). It travels southeast for about 46 kilometres before exiting the NCT of Delhi at a place east of Jaitpur, below the Okhla Barrage.

The river Yamuna stretches can be split from its source to its terminus into the following reaches:

1. Tajewala upstream;
2. Tajewala to Wazirabad (Wazirabad to Tajewala);
3. Jaitpur to Wazirabad;
4. Etawah to Jaitpur;
5. Allahabad to Etawah.

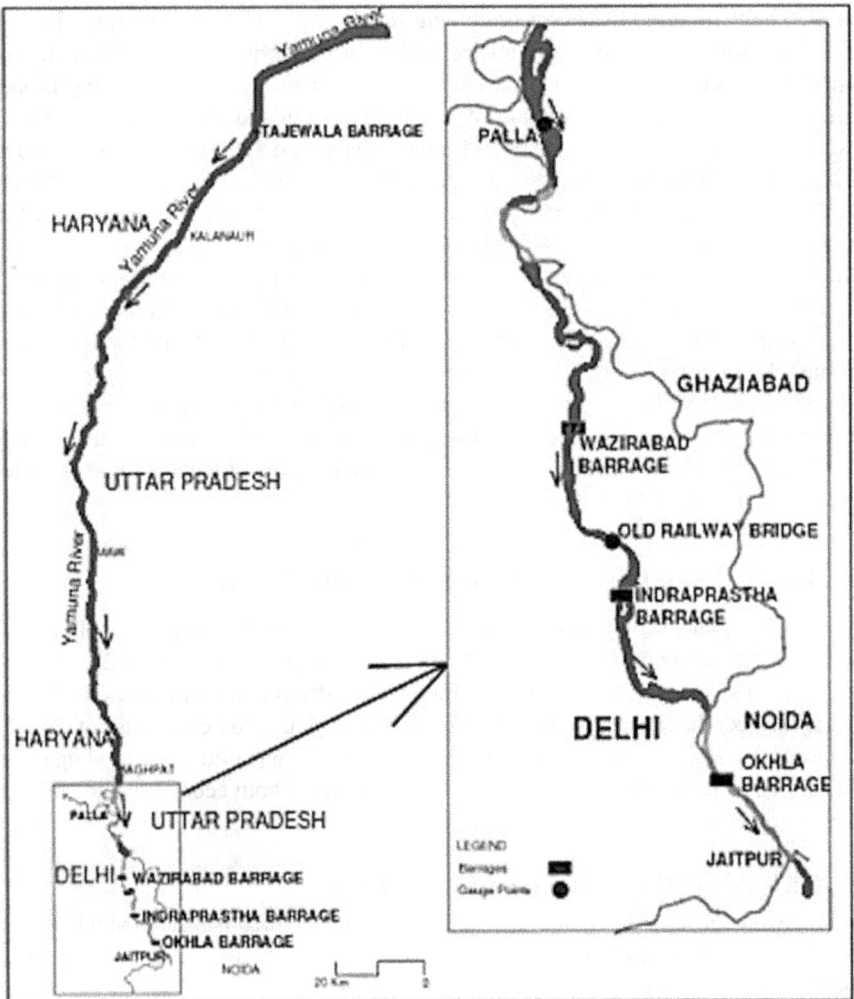

FIGURE 12.7   Yamuna River barrages.

The Yamuna's stretch from Tajewala to Jaitpur is critical for assessing the river's health. Due to floods, the townships of Noida and Greater Noida are vulnerable. There are three barrages on the River Yamuna: Wazirabad, Wazirabad, and Wazirabad. Indraprastha and Okhla are two cities in the area (Figure 12.7).

The gauge sites at Palla, Wazirabad, Station de l'ancienne, Indraprastha, and Okhla are found along the river's course from Palla to Jaitpur. Just downstream of the Noida toll bridge, the Hindon-cut canal meets the river Yamuna. The Shahdara and Noida drains eventually empty into the Yamuna River downstream of the Okhla Barrage. The Hindon eventually meets the Yamuna River at the barrage of Okhla in Haryana, where it finally joins with the Ganga River at Allahabad. The discharge from the drain in Noida and Shahdara drain with Hindon River contributes to the discharge in the river Yamuna during floods. Water is projected to rise in the river upstream of the barrage because of this.

12.3.4.2 The River Hindon Poses a Flood Risk

The Hindon River is a tributary of the Yamuna River. It emerges as a series of minor streams from the foothills of the Upper Shivalik (Lower Himalayas). It is a river that is entirely dependent on rainfall. The Indo-Gangetic Plain includes the river basin. The river has a total length of around 405 kilometres, of which roughly 275 kilometres (its lower stretch) are perennial, and 129 kilometres (its higher stretch) are non-perennial. Up the valley of Saharanpur town in northwest India, the river is dry during the quasi months, but it becomes a longstanding cascade of the area due to conflation with a few major rivers, with water disposal from the top of the Ganga Estuary and constant disposal of household and commercial effluents. Due to substantial diurnal temperature changes, the climate of the region is semi-arid. The yearly rainfall averages roughly 1190 mm.

The Hindon river flows through four of Uttar Pradesh's most industrialised and agriculturally advanced districts: Saharanpur, Ghaziabad, Muzaffar Nagar, and Meerut. At the southern point of Noida Township, the river enters the Yamuna.

12.3.4.3 Current Embankments

Currently, all the sides of the Yamuna in the capital city of Delhi have been embanked, between Palla inside the northwest to Jaitpur in the southeast. While the Flood Mitigation Department of the government is responsible for implementation, fixing, and preservation of flood mitigation works throughout the Yamuna's whole range in the NCR, the Irrigation Department of Uttar Pradesh is responsible for flood mitigation anywhere along the left bank up to the Noida Toll Bridge (Sector 14A), whereupon the Irrigation Department of Uttar Pradesh takes over.

A left afflux bund on the River Yamuna in Noida now runs for 5 kilometres upstream of the Kalindi Barrage. An 18-kilometre-long Yamuna Marginal Bund runs downstream of the Kalindi Barrage. The Uttar Pradesh (UP) Irrigation Department oversees the two bunds. Embankments have also been built on both sides of the Noida drain, the township's other important drainage route.

12.3.4.4 Major Flooding Causes in the Delhi–Noida and Greater Noida Areas

Discharges from the Tajewala headworks have a significant impact on flooding in Delhi and Noida. Excess water is released downstream in the event of heavy rain in locations upstream of Tajewala, producing floods in the downstream areas. However, even moderate rainfall has resulted in isolated floods in recent years. The high rate of runoff from metropolitan areas, which have been continuously developing at a very rapid rate, is a key cause of these local floods. The challenge of localised flooding is likely to intensify in the NCT of Delhi, Noida, and Greater Noida districts, since nearly the whole region is scheduled to be developed by 2025, with such little possibility for unstructured and gentle terrain areas which dampen runoffs and decrease the impact of flooding.

Another aspect that is anticipated to exacerbate the flood situation is that, by 2025, water demand is likely to skyrocket as the Noida and Greater Noida districts, reaching the maximum carrying capacity. Eighty percent of the water provided to such regions will be turned into waste liquid, which must be gathered and released, along with the increased surfaced flow caused by urbanisation of catchments, which must be drained by existing natural drains.

Since embankments are built on banks upstream of Tajewala, valley storage in the upper catchments has decreased in recent years. As a result, increased flooding is projected downstream of Tajewala in the near future for the same discharge as in previous years. This claim is supported by the fact that, although a discharge of 5.80 lakh cusecs of water from Tajewala created a flooding risk of 206.92 m at Delhi near the old railway bridge in 1989, a lower discharge of 5.40 lakh cusecs caused the same flood level in 1996. The scenario is likely to deteriorate as valley storage decreases, and Haryana might build major embankments to protect its farm lands from the effects of floods.

12.3.4.5 Flooding at Gautam Buddha Nagar's Noida Drain

Flood risk from the river Yamuna or the Noida drain could be dangerous as there is a considerable risk of backflow in the Noida drain, which might result in widespread flooding. Noida has a variety of industries. The situation is exacerbated by the fact that there is no regulator in place at the drain.

12.3.5 RAINFALL TREND, MEAN, AND VARIABILITY

Data from the Indian Meteorological Department provided daily precipitation data from 1990 to 2019 for trend analysis, variability, and average trends.

Monthly precipitation trends of every station are calculated from daily precipitation data followed by monthly district precipitation trends produced by taking the statistical average of all station precipitation values within the area (Figure 12.8). The state's monthly precipitation series was calculated using the district's area-weighted rainfall values (Table 12.3 and 12.4).

Figures 12.9 and 12.10 indicate the annual precipitation in mm for the months of June, July, August, and September, as well as the southwest monsoon season

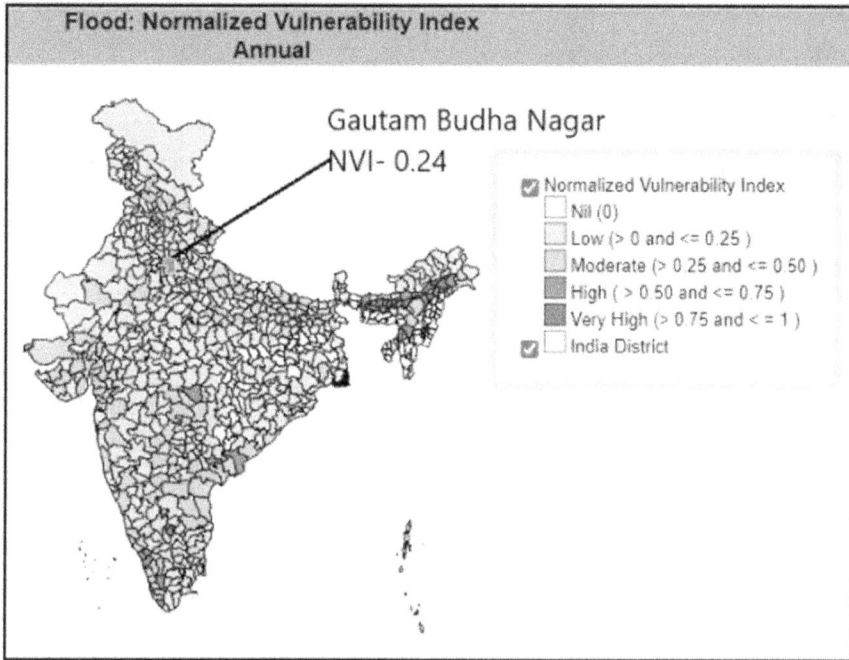

FIGURE 12.8 The state's monthly rainfall series.

TABLE 12.3
Uttar Pradesh Mean Precipitation (mm) and Variation Coefficient for Monsoon Months, the Southwest Monsoon Season, and Annual

	June	July	August	September	JJAS	Annual
Mean	96.2	239.5	220.2	143.1	699.1	769.5
CV	61.2	29.8	35.1	49.9	21.6	19.1

for the state and district. For each of the series, the trend lines are also displayed. Precipitation during the southwest monsoon season and annual precipitation are both on the decrease. The monthly precipitation of the southwest monsoon shows that September has a strong declining trend, but the other months show no notable trend. However, in June and August there is a negligible falling tendency, and in July, an insignificant increasing trend is observed.

During the southwest monsoon season, Uttar Pradesh receives roughly 89% of its annual rainfall. July has the highest precipitation (34% of the southwest monsoon rains), followed by August (31% of the southwest monsoon rainfall). Average precipitation and rains during the monsoon period are decreasing. During the research process, precipitation in September decreased significantly, although precipitation in other southwest monsoon months did not.

TABLE 12.4

Gautam Buddha Nagar Mean Coefficient (mm) and Variation Coefficient for Monsoon Months, the Southwest Monsoon Season, and Annual

	June	July	August	September	JJAS	Annual
Mean	55.3	176.0	206.5	139.6	577.2	652.4
CV	571	158	191	258	107	123

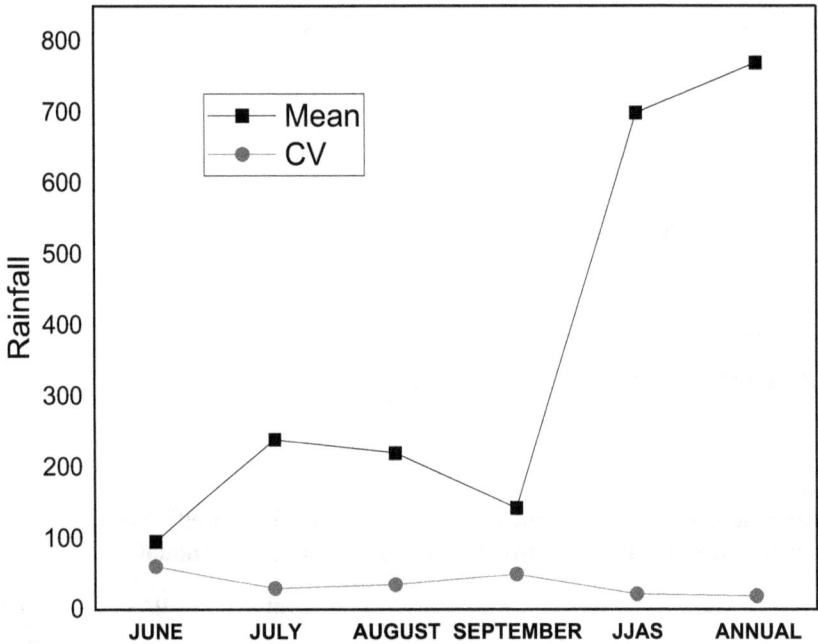

FIGURE 12.9 Uttar Pradesh mean precipitation (mm) and variation coefficient for monsoon months, the southwest monsoon season, and the annual.

12.5 CONCLUSIONS

The centre of Noida city is destined to spread onto the vital floodplains of the River Yamuna and River Hindon catchments in the future. This expanding centre, which is close to the present metropolitan city, will have to deal with a rising flooding danger, which will be caused mostly by increased activity on risk-prone lands or potentially by increased flood frequency. If the current flooding scenario and economic patterns continue, large land use dislocations in the city will be required to build a wider catchment for the Yamuna and Hindon Rivers, albeit the timing and scope of such dislocations are unknown. Maps showing flooded areas, when merged with current information about land use and flood depth, provide critical information for flooding destruction estimates in each micro urban unit. These findings demonstrate remote

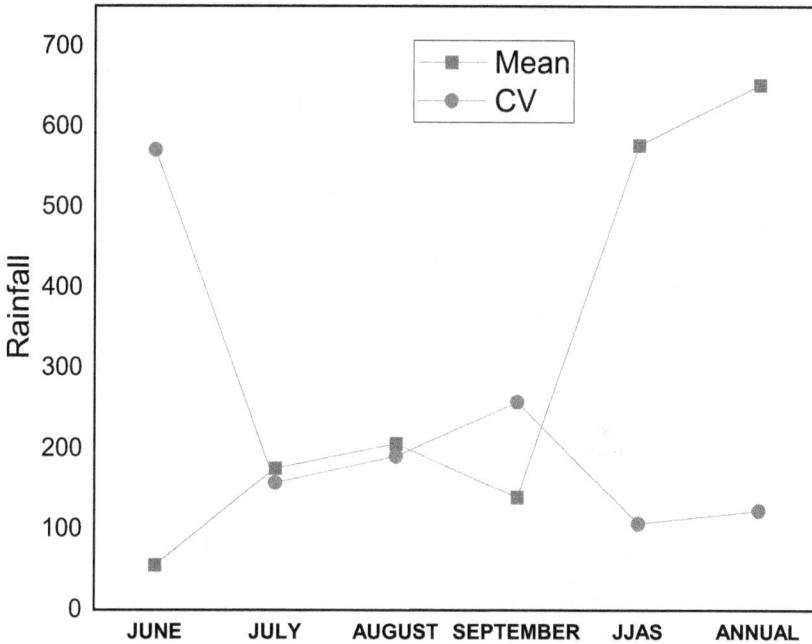

FIGURE 12.10   Uttar Pradesh mean coefficient of (mm) and variation coefficient for monsoon months, the southwest monsoon season, and the annual.

sensing's potential to produce multi-scale and multi-temporal outputs for micro-scale flood forecast and monitoring. The assessment of a possibly flood-prone zone on the Bhuvan-NRSC and the spatial trends of urbanisation are expanding with time, according to a synoptic overview. Because of its simplicity, precision, and long-term practical application, the integrated methodological framework, which integrates IMD and Bhuvan data analysis, remote sensing, and GIS with traditional methodologies like Manning's theory adopted for this study, has a significant advantage. The findings have immediate consequences for planning land use and urban development sustainable policies.

REFERENCES

[1]   Y. Chen, H. Zhou, H. Zhang, G. Du, and J. Zhou, Urban flood risk warning under rapid urbanization, *Environmental Research,* vol. 139, pp. 3–10, 2015/05/01/ 2015.

[2]   M. Misra, D. Kumar, and S. Shekhar, Assessing machine learning based supervised classifiers for built-up impervious surface area extraction from Sentinel-2 images, *Urban Forestry & Urban Greening,* vol. 53, p. 126714, 2020/08/01/ 2020.

[3]   D. Kumar, Mapping solar energy potential of southern India through geospatial technology, *Geocarto International,* vol. 34, no. 13, pp. 1477–1495, 2019/11/10 2019.

[4]   D. Birhanu, H. Kim, C. Jang, and S. Park, Flood risk and vulnerability of Addis Ababa city due to climate change and urbanization, *Procedia Engineering,* vol. 154, pp. 696–702, 2016/01/01/ 2016.

[5] H. Jain, V. Yadav, V. D. Rajput, T. Minkina, S. Agarwal, and M. C. Garg, An eco-sustainable green approach for biosorption of methylene blue dye from textile industry wastewater by sugarcane bagasse, peanut hull, and orange peel: a comparative study through response surface methodology, isotherms, kinetic, and thermodynamics, *Water, Air, & Soil Pollution,* vol. 233, no. 6, p. 187, 2022/05/19 2022.

[6] M. C. Garg and H. Jain, Membrane-based remediation of wastewater, in *Recent Trends in Wastewater Treatment*, S. Madhav, P. Singh, V. Mishra, S. Ahmed, and P. K. Mishra, Eds. Cham: Springer International Publishing, 2022, pp. 75–95.

[7] S. H. Mahmoud and T. Y. Gan, Urbanization and climate change implications in flood risk management: Developing an efficient decision support system for flood suscep-tibility mapping, *Science of The Total Environment,* vol. 636, pp. 152–167, 2018/09/15/ 2018.

[8] Ö. Ekmekcioğlu, K. Koc, and M. Özger, Towards flood risk mapping based on multi-tiered decision making in a densely urbanized metropolitan city of Istanbul, *Sustainable Cities and Society,* vol. 80, p. 103759, 2022/05/01/ 2022.

[9] R. S. Kookana, P. Drechsel, P. Jamwal, and J. Vanderzalm, Urbanisation and emerging economies: Issues and potential solutions for water and food security, *Science of The Total Environment,* vol. 732, p. 139057, 2020/08/25/ 2020.

[10] R. B. Singh and S. Singh, Rapid urbanization and induced flood risk in Noida, India, *Asian Geographer,* vol. 28, no. 2, pp. 147–169, 2011/12/01 2011.

[11] D. Kumar and S. Shekhar, Statistical analysis of land surface temperature–vegetation indexes relationship through thermal remote sensing, *Ecotoxicology and Environmental Safety,* vol. 121, pp. 39–44, 2015/11/01/ 2015.

[12] S. Rawat, A. Rawat, D. Kumar, and A. S. Sabitha, Application of machine learning and data visualization techniques for decision support in the insurance sector, *International Journal of Information Management Data Insights,* vol. 1, no. 2, p. 100012, 2021/11/01/ 2021.

[13] U. B. Prajapati, Chapter 9–Socio-economic perspective of river health: A case study of river Ami, Uttar Pradesh, India, in *Ecological Significance of River Ecosystems*, S. Madhav, S. Kanhaiya, A. Srivastav, V. Singh, and P. Singh, Eds.: Elsevier, 2022, pp. 167–186.

[14] Y. Zhang, D. Ryu, and D. Zheng, Using remote sensing techniques to improve hydro-logical predictions in a rapidly changing world, *Remote Sensing,* vol. 13, no. 19, 2021.

[15] A.T. N. Dang and L. Kumar, Application of remote sensing and GIS-based hydro-logical modelling for flood risk analysis: a case study of District 8, Ho Chi Minh city, Vietnam, *Geomatics, Natural Hazards and Risk,* vol. 8, no. 2, pp. 1792–1811, 2017/12/15 2017.

[16] O. Pabi, S. Egyir, and E. M. Attua, Flood hazard response to scenarios of rainfall dynamics and land use and land cover change in an urbanized river basin in Accra, Ghana, *City and Environment Interactions,* vol. 12, p. 100075, 2021/12/01/ 2021.

[17] N. Kikon, P. Singh, S. K. Singh, and A. Vyas, Assessment of urban heat islands (UHI) of Noida City, India using multi-temporal satellite data, *Sustainable Cities and Society,* vol. 22, pp. 19–28, 2016/04/01/ 2016.

[18] R. Sharma, L. Pradhan, M. Kumari, and P. Bhattacharya, Assessing urban heat islands and thermal comfort in Noida City using geospatial technology, *Urban Climate,* vol. 35, p. 100751, 2021/01/01/ 2021.

[19] Z. Hassan *et al.*, Dynamics of land use and land cover change (LULCC) using geospa-tial techniques: a case study of Islamabad Pakistan, *SpringerPlus,* vol. 5, no. 1, p. 812, 2016/06/21 2016.

[20] J. F. Gondwe, S. Lin, and R. M. Munthali, Analysis of land use and land cover changes in urban areas using remote sensing: case of Blantyre City, *Discrete Dynamics in Nature and Society,* vol. 2021, p. 8011565, 2021/12/23 2021.

[21] R. Sharma and P. K. Joshi, Mapping environmental impacts of rapid urbanization in the National Capital Region of India using remote sensing inputs, *Urban Climate,* vol. 15, pp. 70–82, 2016/03/01/ 2016.

[22] U. Saha and M. Sateesh, Rainfall extremes on the rise: Observations during 1951– 2020 and bias-corrected CMIP6 projections for near- and late 21st century over Indian landmass, *Journal of Hydrology,* vol. 608, p. 127682, 2022/05/01/ 2022.

[23] R. Ranjan, Payments for ecosystems services-based agroforestry and groundwater nitrate remediation: The case of Poplar deltoides in Uttar Pradesh, India, *Journal of Cleaner Production,* vol. 287, p. 125059, 2021/03/10/ 2021.

[24] H. Jain et al., Fabrication and characterization of high-performance forward-osmosis membrane by introducing manganese oxide incited graphene quantum dots, *Journal of Environmental Management,* vol. 305, p. 114335, 2022/03/01/ 2022.

[25] B. Feng, Y. Zhang, and R. Bourke, Urbanization impacts on flood risks based on urban growth data and coupled flood models, *Natural Hazards,* vol. 106, no. 1, pp. 613–627, 2021/03/01 2021.

[26] W. M. Elsadek, M. G. Ibrahim, and W. E. Mahmod, Runoff hazard analysis of Wadi Qena Watershed, Egypt based on GIS and remote sensing approach, *Alexandria Engineering Journal,* vol. 58, no. 1, pp. 377–385, 2019/03/01/ 2019.

[27] L. D. Ress, C.-L. J. Hung, and L. A. James, Impacts of urban drainage systems on stormwater hydrology: Rocky Branch Watershed, Columbia, South Carolina, *Journal of Flood Risk Management,* https://doi.org/10.1111/jfr3.12643 vol. 13, no. 3, p. e12643, 2020/09/01 2020.

[28] S. Ertan and R. N. Çelik, The assessment of urbanization effect and sustainable drainage solutions on flood hazard by GIS, *Sustainability,* vol. 13, no. 4, 2021.

[29] Mohamed Elmoustafa, Weighted normalized risk factor for floods risk assessment, *Ain Shams Engineering Journal,* vol. 3, no. 4, pp. 327–332, 2012/12/01/ 2012.

[30] S. Dazzi, I. Shustikova, A. Domeneghetti, A. Castellarin, and R. Vacondio, Comparison of two modelling strategies for 2D large-scale flood simulations, *Environmental Modelling & Software,* vol. 146, p. 105225, 2021/12/01/ 2021.

[31] B. Dong, J. Xia, M. Zhou, Q. Li, R. Ahmadian, and R. A. Falconer, Integrated modeling of 2D urban surface and 1D sewer hydrodynamic processes and flood risk assessment of people and vehicles, *Science of The Total Environment,* p. 154098, 2022/02/23/ 2022.

[32] R. Mahmood, S. Jia, and W. Zhu, Analysis of climate variability, trends, and prediction in the most active parts of the Lake Chad basin, Africa, *Scientific Reports,* vol. 9, no. 1, p. 6317, 2019/04/19 2019.

[33] A. Asfaw, B. Simane, A. Hassen, and A. Bantider, Variability and time series trend analysis of rainfall and temperature in northcentral Ethiopia: A case study in Woleka sub-basin, *Weather and Climate Extremes,* vol. 19, pp. 29–41, 2018/03/01/ 2018.

[34] M. J. de la Paix, L. Lanhai, C. Xi, S. Ahmed, and A. Varenyam, Soil degradation and altered flood risk as a consequence of reforestation, *Land Degradation & Development,* https://doi.org/10.1002/ldr.1147 vol. 24, no. 5, pp. 478–485, 2013/ 09/01 2013.

[35] Z. Li, X. Deng, F. Yin, and C. Yang, Analysis of climate and land use changes impacts on land degradation in the North China plain, *Advances in Meteorology,* vol. 2015, p. 976370, 2015/06/07 2015.

[36] A. Panda and N. Sahu, Trend analysis of seasonal rainfall and temperature pattern in Kalahandi, Bolangir and Koraput districts of Odisha, India, *Atmospheric Science Letters,* https://doi.org/10.1002/asl.932 vol. 20, no. 10, p. e932, 2019/10/01 2019.

[37] S. Gorai, D. Ratha, and A. Dhir, Adapting rainfall variability to flood risk: a case study of the Ghaggar River Basin, *Journal of the Geological Society of India,* vol. 97, no. 11, pp. 1347–1354, 2021/11/01 2021.

13 Changing Demographic Contours of Hyderabad City

Kalpana Markandey

CONTENTS

13.1 INTRODUCTION

Hyderabad was initially the capital of the feudal state of the Nizam of Hyderabad and was then the capital of Andhra Pradesh from 1956. When Andhra Pradesh was bifurcated into two states, Andhra Pradesh and Telangana in 2014, Hyderabad remained the capital city of Telangana. Hyderabad enjoyed a primate city status in the Nizam's dominions, in Andhra Pradesh and now does so in Telangana. Hyderabad witnessed a huge inflow of migrants from coastal Andhra in the 1980s, leading to profound cultural changes in the cityscape. Physical expansion associated with this migration and subsequent development of the IT sector in the era of globalisation was in the western part of the city, where the land values started spiralling upwards. Growth in the north was associated with educational institutions of international repute in that part of the city and in the northwest and northeast was associated with industrialisation. The expansion of the built-up area in the south was sheer accretion, especially before the 1980s. The area of Hyderabad increased from 175 sq. km in 1971 to 650 sq. km in 2011 and the population increased fourfold from 1.8 million to 6.14 million during the same period. It was been projected to have a population of 10 million in 2020.

The rapid growth of metropolitan suburbs is a characteristic feature of spatial change not only in Hyderabad but in most South Asian cities. Relatively lower land values initially attract people from the core city to relocate to these peri-urban areas from the core area. This, coupled with the migrants from adjacent villages, districts, or states, who establish hold over these areas, is responsible for the growth of these peri-urban areas. The transformation which follows in terms of altered land use, activity pattern, skyline, and social fabric of these areas is phenomenal.

Hyderabad city has a star-shaped layout and appears to grow along the main transportation arteries. The growth towards the northwest, north, and northeast is along the highways that connect the city to Mumbai, Delhi, and various district headquarters, respectively. These, incidentally, are also the directions in which there is major industrial development. Thus, urban–industrial–transportation development seems to go hand-in-hand in these areas and this is a significant post-independence phenomenon. These directions have, thus, been consistently pulsating with growth dynamics for the past almost 60 years. The growth on the southeastern periphery of the city is a case in contrast. It houses a large and growing residential area along the Vijayawada Highway. Being a residential area, it is a dormitory in nature compared to the other growing areas mentioned above that are industrial and hence economically active.

The later phase of development was witnessed on the western periphery where the Hitech City has grown at a very rapid pace. It is visualised as the area of the future, and this is reflected in a sharp increase in land values in this area over the past few years. Thus, growth and development seem to be taking place all around the city and south Hyderabad, largely comprising the Old City, is not to be written off in this context. Though an area with a high density of population and a relatively traditional society, it has also witnessed urban renewal, thanks to the efforts of the Quli Qutub Shah Urban Development Authority. The city seems to be growing and developing in all directions today compared to the scenario in the 1980s and early 1990s, where only the north and east were growing but the west and south were stagnant.

The skyline of the city has also witnessed a marked transformation since the 1980s. Vertical development has become the order of the day. Cement and concrete structures are replacing the high-roofed mansions of the bygone era. Demolitions of the latter are usually carried out without batting an eyelid or a thought about the environmental consequences of the works. Real estate promoters are having a heyday, as land and property values are soaring at an unprecedented pace. While the core areas of the city were the first to acquire vertical structures, starting in the 1970s; these are now found sprinkled all over the city extending up to the limits of the city and suburban areas, though their density and heights are greater in central city locations. These concrete jungles have raised the temperature of the city in the recent past and have been responsible for the emergence of a 'heat island' – a concept unknown to Hyderabad previously, which was known for its salubrious climate.

13.2   STUDY AREA

This study comprises the city of Hyderabad (Greater Hyderabad Municipal Corporation or GHMC as it is called), the core city of Hyderabad, which was earlier known as the Municipal Corporation of Hyderabad (MCH) as well as towns or municipalities which have been annexed by the ever-growing metropolis of Hyderabad in 2007 (Figure 13.1). These comprise Lalbahadur Nagar, Uppal Kalan, Malkajgiri, Kapra, Alwal, Qutubullapur, Kukatpally, Serilingampally, Rajendranagar, Gaddiannaram, Ramachandrapuram, and Patancheru.

STUDY AREA

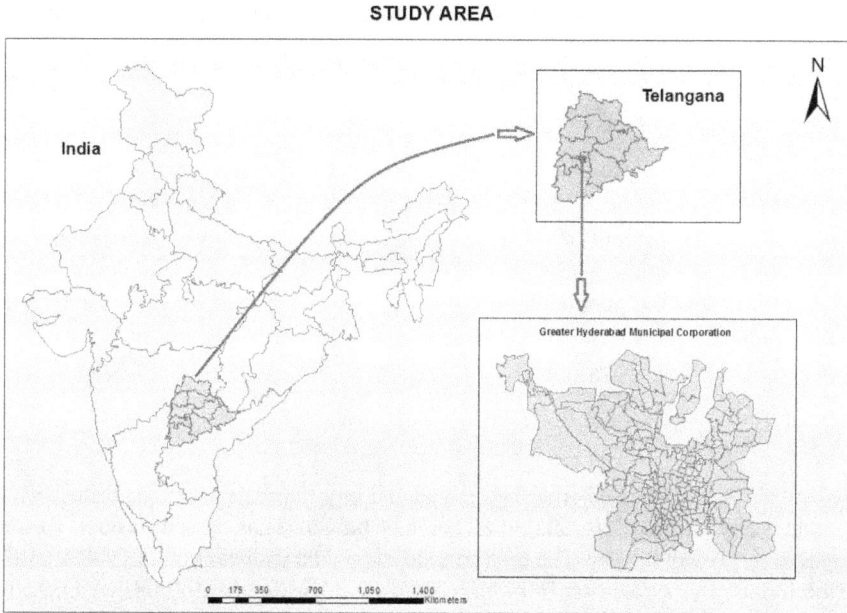

FIGURE 13.1   Study area.

Urban areas grow through different processes such as natural increases, migration, and annexation. Of these, the last mentioned has been found to play a vital role in the growth and expansion of Hyderabad in the recent past – both before the formation of Telangana and after the formation of the new state of Telangana of which Hyderabad is the capital. The rehashing of boundaries and repeated changes to the master plan for the development of this city have gone hand in hand with the development of the Outer Ring Road and the proposed Regional Ring Road. The spatial spread of the city is thus being propelled by some of these measures of the government.

Hyderabad is divided into five zones: the north zone, south zone, east zone, west zone, and central zone. The jurisdictions of these were revised after 2007 when the 12 municipalities – Lalbahadur Nagar, Uppal Kalan, Malkajgiri, Kapra, Alwal, Qutubullapur, Kukatpally, Serilingampally, Rajendranagar, Gaddiannaram, Ramachandrapuram and Patancheru – along with eight Gram Panchayats – Shamshabad, Satamarai, Jallapalli, Mamdipalli, Mankhal, Almasguda, Sardanagar, and Ravirala – were merged in Hyderabad. The 12 municipalities have subsequently acquired the status of Circles within the main city of Hyderabad. The growth in the peri-urban zone has been phenomenal during the recent past.

13.3   DATA SOURCES

Data from the Census of India as well as that from the internet have been used for this study. In addition, primary data have been used to gauge the ground reality. For this, a structured questionnaire was used to collect data from a select sample of households in the city.

13.3.1 SAMPLING FRAME

A sample of 300 households was taken. A multi-stage stratified sampling frame (Table 13.1) was adopted for collecting data using a structured questionnaire (Appendix I). Ten percent of wards in every circle of the city were chosen and 1% of households in sample wards were chosen. While choosing the samples, care was taken to represent the people along the entire socio-economic spectrum in a given sample unit. To that extent, the sample was also purposive. These data have been analysed to arrive at the demographic highlights of the city.

13.3.2 AN OVERVIEW OF CHANGE

Hyderabad has progressed from being a feudal city that was established in 1591 to a modern city with an information technology hub, and has passed through various stages in its growth history. The metropolitan stage had its inception in 1908 after the redevelopment processes set in motion after the great floods of 1908, which had a huge toll on the population and resources of the city. Hyderabad, as well as its twin, Secunderabad, were given a lot of revamping with massive restoration efforts after that. After facing political and economic vicissitudes, the city is now the sixth ranked in India in terms of its population (Figure 13.2) and the second ranked after Delhi in terms of area.

The profile of the population changes, as it also manifests economic and political change, from 1901 to 2021 is portrayed in Figure 13.3. In the early 20th century, Hyderabad had a very low population, which even declined in the 1911–21 decade owing to deaths on account of floods, famine, plague, and cholera. An additional factor that kept the population of Hyderabad rather low was that it had rather stringent rules about migration, which is a major factor in the increase of the population of urban areas. It would classify the people coming from outside the princely state as 'non-mulkis'. This was a severe deterrent to migrants who were not encouraged to take up jobs or even attend educational institutions there. The overall rate of increase in the population was thus low to start with in 1901 and even had a downside. This was so till 1961, when the vagaries of natural calamities, the partition of the country, and the migration of several families to the neighbouring country, etc. played their part in keeping the population relatively low. This is despite the lifting of restrictions on migration in the post-independence era, made for a quantum jump in the population as a result of in-migration at the 1951 census. Even post-1951, the migration flow from coastal Andhra and Rayalaseema was incessant. It may also be noted that opening large hospitals, extending the water supply lines, and laying a new sewerage system increased the life span of the people and hence the rate of natural increase of population also increased owing to a lower death rate [1]. After 1961, the spiralling growth of the city was a force to be reckoned with. The second 5-year plan brought industrialisation in its wake and with new industries being set up in Hyderabad in the

TABLE 13.1
Sampling Frame for Selecting Sample Households for the Survey

Sl. no.	Core/periphery	Zones	Circle	No. of wards	10% of sample wards	Name of sample wards	1% Sample population
1	Peripheral Municipalities	East Zone	Kapra Circle I	4	1	Cherlapally	20
2			Uppal Kalan Circle II	3	1	Habsiguda	20
3			LB Nagar/ Gaddi Annaram Circle III	10	1	Kothapet	20
4		West Zone	Serilingampally Circle XI and XII	4	1	Gachibowli	10
5			Ramachandrapuram Circle XIII	1	1	Ramachandrapuram	10
6			Patancheru Circle XIII	1	1	Patancheruvu	10
7			Kukatpally Circle XIV	8	1	Old Bowenpally	10
8		North Zone	Qutbullapur Circle XV	7	1	Gajula Ramaram	10
9			Alwal Circle XVI	3	1	Alwal	10
10			Malkajgiri Circle XVII	5	1	Moula Ali	20
11	Core City Municipalities	South Zone	Circle IV	27	3	Saidabad, Kanchanbagh, Pathargatti	20
12			Circle V	18	2	Falaknuma Begum Bazar	20
13			Circle VI Rajendra Nagar	4	1	Shivrampally	20
14		Central Zone	Circle VII	16	2	Karwan, Toli Chowki	20
15			Circle VIII	3	1	Sultan Bazar	20
16			Circle IX	16	2	Himayat Nagar, Vidya Nagar	20
17			Circle X	15	2	Ameerpet, Banjara Hills	20
18		Secunderabad Division	Circle XVIII	11	1	Marredpally	20

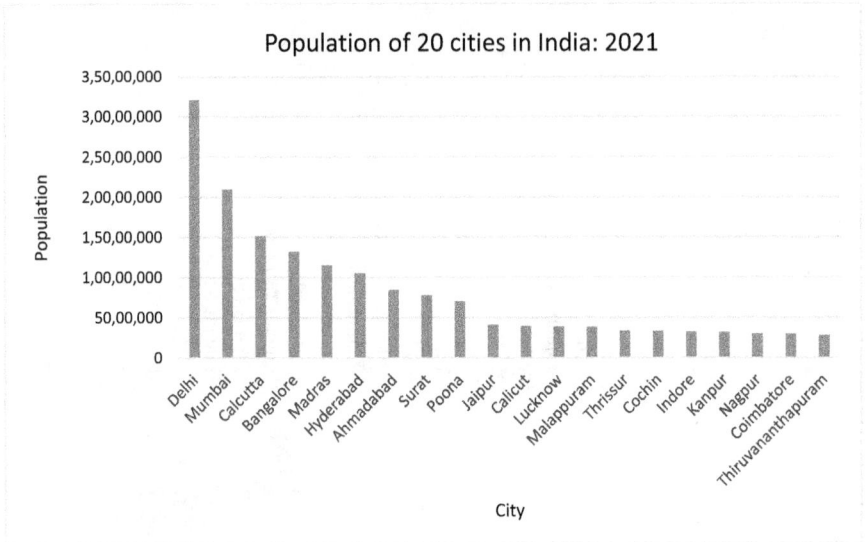

FIGURE 13.2   Populations of 20 cities in India.

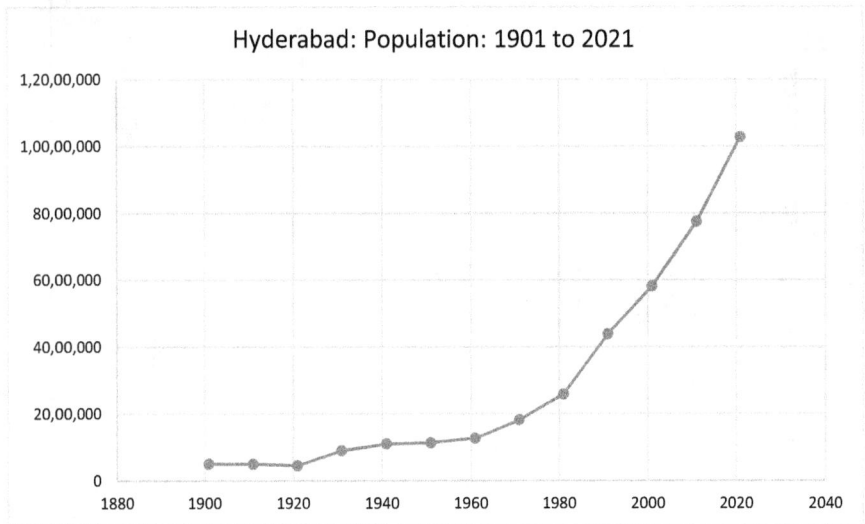

FIGURE 13.3   Population trends from 1901 to 2021.

Source: Census of India and www: HTTP Hyderabad India Metro Area Population 1950–2021/Macro Trends [2].

HYDERABAD CITY- POPULATION DISTRIBUTION - CENSUS 2011

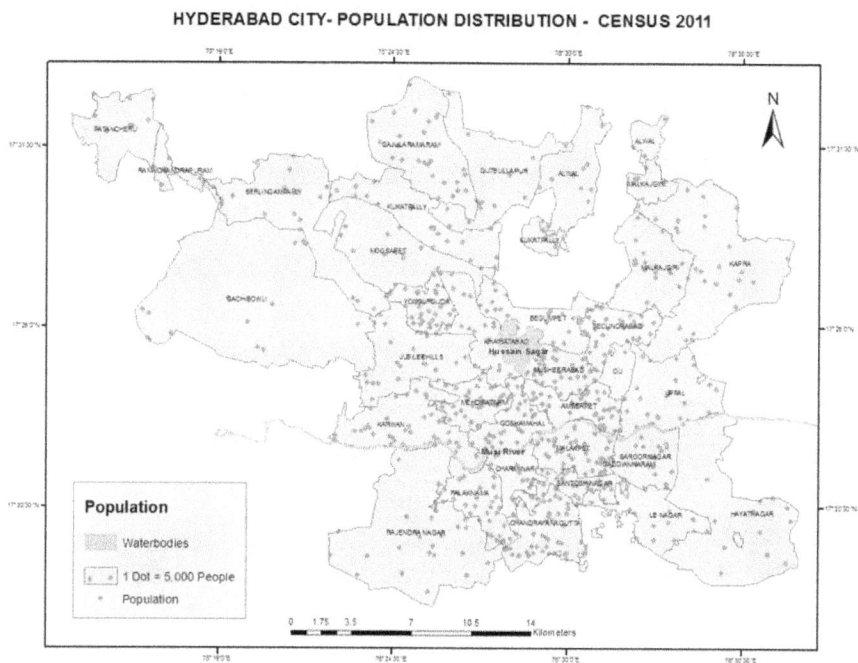

FIGURE 13.4 Population distribution trends.

northeastern and northwestern part of the city, in-migration gained impetus and the growth of the city has not looked back since then for one reason or the other. These include the migration of people from the then-Andhra region in the 1980s due to the winning of a particular party in the elections or the impact of globalisation and the setting up of IT units in the city after 1990, which acted as magnets for people from neighbouring states. The real estate sector has had a parallel run in this later period, where building activity and quite understandably migration, which is one of its concomitants, have witnessed an unabated increase during this period. Most of the spatial growth and also development in the recent past, as earlier, seems to have taken place astride the axial highways from the city which connect it to other state capitals and district headquarters

It is evident from Figure 13.4 that the population of the city is by and large concentrated in the central areas, mostly in areas that were part of the Municipal Corporation of Hyderabad (MCH) which was later designated as GHMC by appending the surrounding areas. The peripheral areas are more extensive in size and the population is not as densely packed as in the central parts of the city.

Although the Scheduled Caste population is spread all over the city, it is perceivable that the peripheral areas have it in a larger measure (Figure 13.5). Though this is deduced from secondary data, the master table about primary data throws some light on this factor where it is found that the Scheduled Caste population in the peripheral parts of the city is of recent origin and hence comprises of migrants settled

HYDERABAD CITY- SC POPULATION DISTRIBUTION - CENSUS 2011

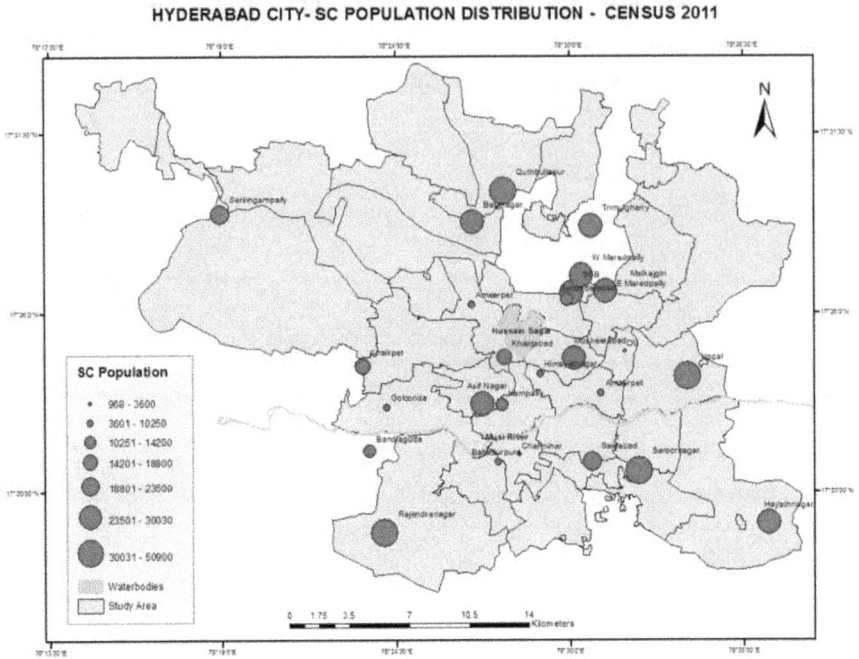

FIGURE 13.5   Population distribution trends – SC.

on the periphery of the city for want of additional accommodation in the central city locations and also close to their areas of origin in the rural areas.

Figure 13.6 depicts the Scheduled Caste (SC) and Scheduled Tribe (ST) population in Hyderabad, and it is evident from this figure that the SC and ST population is relatively more in the eastern part of the city compared to the other parts, being maximum in the Osmania University area. South Hyderabad and other old settled parts of Hyderabad have sometimes no or negligible representation of these groups. These are also incidentally the areas with a predominance of the Muslim population.

A glance at Figure 13.7 reveals that most of the people in Hyderabad are Telugu speaking and as one moves to the southern part of the city, Urdu-speaking people figure more on the scale. The central part of the city has some Hindi-speaking people in addition to Telugu-speaking people. This is on account of a large proportion of the business community of Marwadis who have settled in these parts. There is also a substantial Kayasth population and retired defence workers who live in these areas. It can also be observed from Figure 13.7 that the people in the peripheral areas are by and large Telugu speaking, as also those in Secunderabad.

Telugu and Urdu are the major languages spoken in Hyderabad, although Hindi, Tamil, Marathi, Kannada, Marwari, Malayali, Oriya, Gujarati, and Punjabi are also spoken by select groups in specific pockets of the city [3]. The Hyderabad version of Urdu in a vast majority of cases is the typical Dakhani or Hyderabadi Urdu, which is different from the Urdu spoken in other parts of the country or the world.

HYDERABAD CITY- SC ST POPULATION DISTRIBUTION- 2011

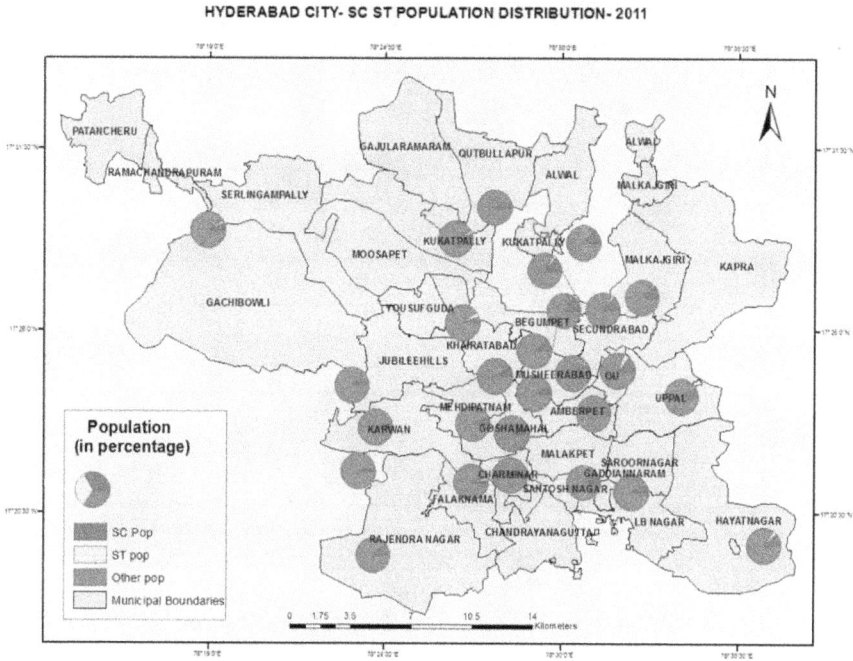

FIGURE 13.6   Population distribution trends – SC/ST.

On a similar note, Figure 13.8 conveys that most people follow the Hindu religion and those following Islam are relatively more in the southern and central parts of the city. Christianity is followed to a large extent in Secunderabad and some parts of north and south Hyderabad. As the Secunderabad area was a British Cantonment for a long time and has a majority of the Anglo-Indian population, the dominance of Christianity in this area is amply explained. The dominance of the Hindu population in the peripheral areas goes hand in hand with the dominance of the Telugu-speaking population in these areas who seem to have migrated to the city in the recent past and in the absence of adequate or suitable housing in the central part of the city took up residence in the peripheral parts. The other religions had a minuscule representation in a sample of this size and do not find a portrayal in the spatial data/map, although Hyderabad is a multi-cultural city with an eclectic medley of religions, languages, and castes.

This religio-linguistic scenario compares with a Muslim population of 52% in the city, a Hindu population of 46%, and Christians constituting 1.5% of the city population. Sikhs, Buddhists, Jains, Jews, Parsis, and others constituted the minuscule remainder of the population in 1951. Urdu was spoken by 52.3% of the population, Telugu by 34.3%, Hindi by 4.3%, Marathi by 3.1%, Kannada by 1.3%, Tamil by 1%, Gujarati by 0.8%, and English by 1.5% [1]. The dawn of the new era in Hyderabad saw a mingling of the cultures and also the emergence of Telugu as a predominant language, although the Hyderabadi Urdu language or a kind of Hindustani – the

HYDERABAD CITY- MAJOR LANGUAGES- 2020

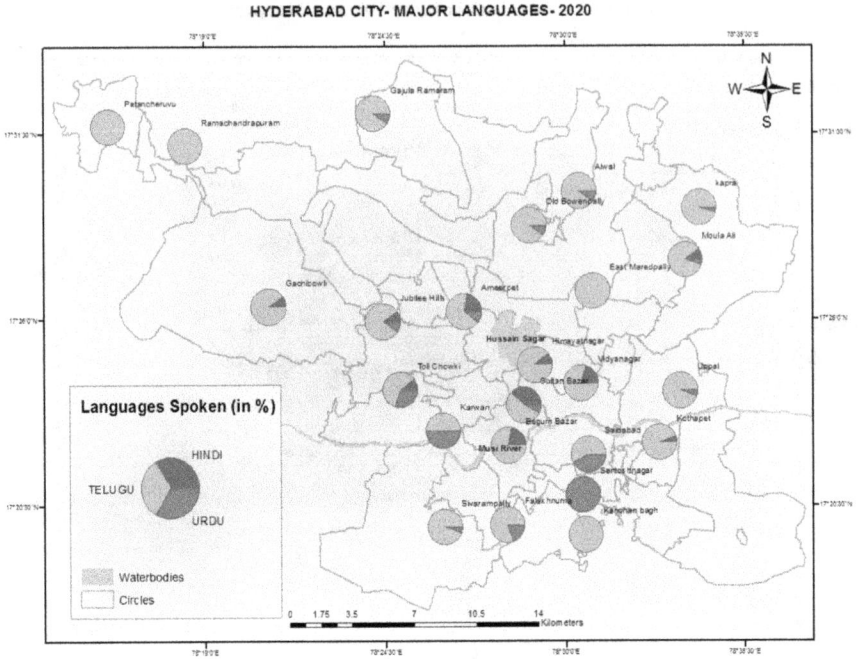

FIGURE 13.7   Major languages.

Source: Field Survey.

Dakhani language – is embraced by all, unlike in the other South Indian states and cities, where people do not converse in Hindi or Urdu to a great extent.

For reasons of simplification which was required for mapping on this scale, the castes have been broadly merged under the four broad categories of OC or the General Population, BC (Backward Caste) population, SC (Scheduled Caste), and ST (Scheduled Tribe). It is also majorly used for people who have not explicitly mentioned their caste but have stated the 'Official' category that they belong to.

While the OC and BC populations seem to dominate the caste structure of the city, central and some parts of north Hyderabad and Secunderabad have more of the OC population so far as the caste structure is concerned. Most other parts of the city are dominated by the BC population. The BC population is found in all the localities surveyed and makes for 50% or more than 50% of the population of nearly 50% of these localities. The SC and especially the ST population appears more on the relative periphery of the city (Figure 13.9). The SC category is found in a larger or smaller ratio in 91% of the localities surveyed, while the ST population is found in about 43% of these localities.

Most of the residents in about 50% of localities surveyed and under consideration here, belong to Hyderabad (Figure 13.10), though a majority in some of the areas on the eastern and western extremities of the city and some pockets in south Hyderabad have come from other areas. Most of those who live on the periphery of the city

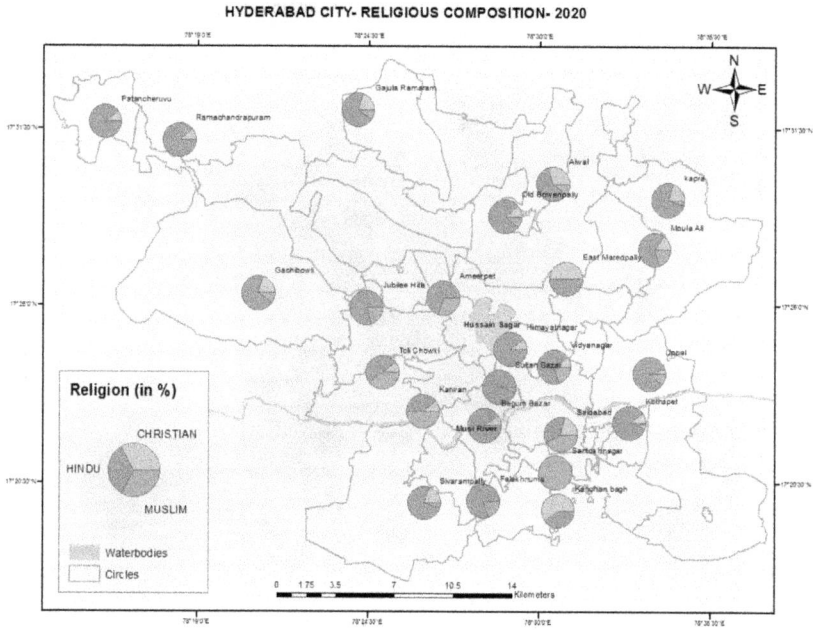

FIGURE 13.8 Religous composition.

Source: Field Survey.

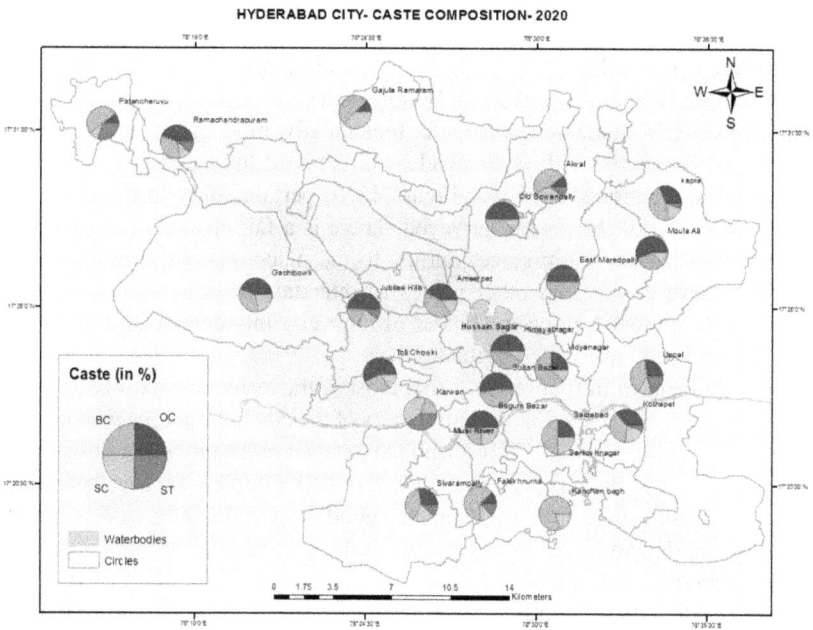

FIGURE 13.9 Caste composition.

Source: Field Survey.

HYDERABAD CITY- PLACE OF PREVIOUS RESIDENCE- 2020

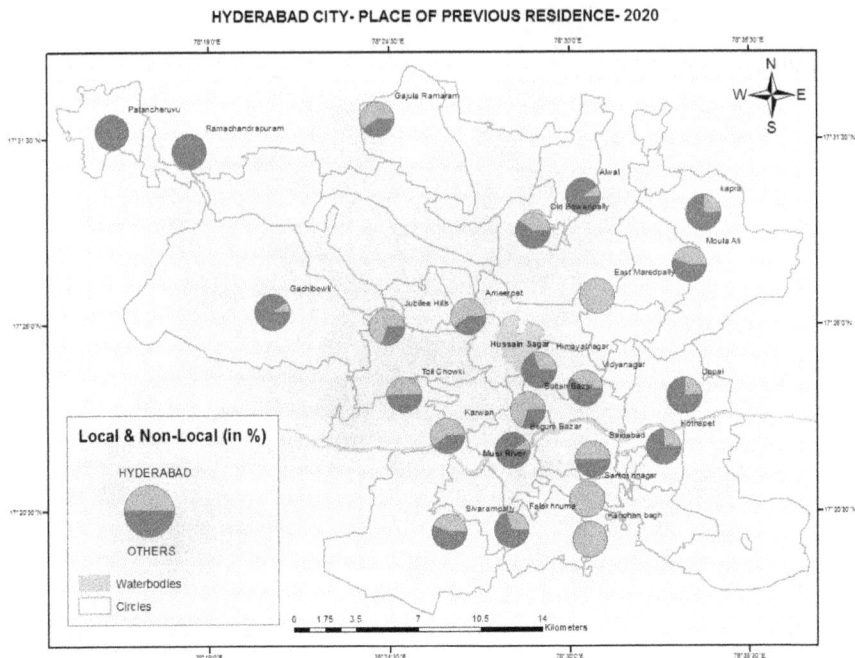

FIGURE 13.10   Places of previous residency.

Source: Field Survey.

have migrated from other places in Telangana, some parts of Andhra Pradesh like Vijayawada, Vizianagaram, Prakasam, Nellore, and Guntur, and also from other states and cities like Rajasthan, Bengaluru, etc. Incidentally, most of the cities mentioned here are well connected with Hyderabad by direct road links and also rail services. It may also be mentioned here that Figure 13.10 pertains only to the place of the previous residence of the people surveyed. There is a fair chance of other migrants making a two-stage or multi-stage journey to the place of their current residence. They could have come from other places as interstate or intra-state migrants and would then have moved along the ladder of intra-city migration within Hyderabad. They have not been represented in the figure.

It seems evident from Figure 13.11 that most of the residents #have been living in Hyderabad for 10–20 years. Many central parts of the city have people who have been residing there for 40–50 years. Those who have stayed in the city since birth are more in the eastern half of the city than in the western half. People who have been staying in the same locality since birth are found in a good proportion in the eastern circles of Moula Ali, Kapra, and Uppal. Many people in the southern circles have stayed there for more than 40 years. Remarkably, many of the people contacted in Falaknuma have been residing there for 50–80 years. However, those who have stayed in the city from birth constitute a small proportion of the overall population. A vast majority of the population being migrants is expected in a city of the size of Hyderabad. The

HYDERABAD CITY- YEARS OR DURATION OF STAY IN HYDERABAD- 2020

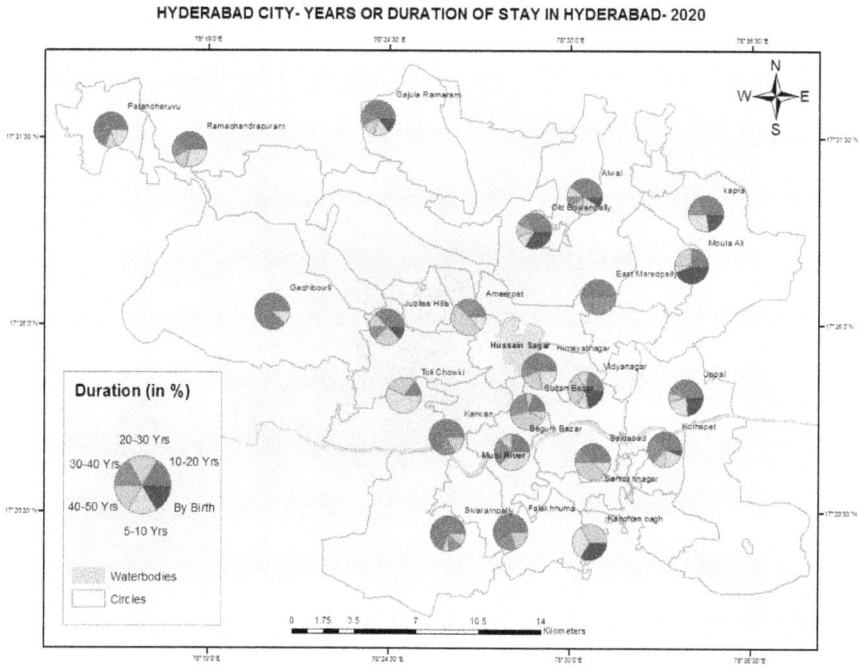

FIGURE 13.11   Duration of stay.

Source: Field Survey.

western half of the city is marked by those with a duration of stay of 10–20 years. That these are upcoming areas, amply explains this situation. In Ramachandrapuram and Patancheruvu, almost half of the respondents surveyed have been staying there for less than 10 years, most of them being migrants who had been lured by the industries and government institutions there. The central part of the city, very obviously enough, has people from all the periods on the chronological chart given in the index of the map as different generations have been inhabiting this area over time.

This is even though all localities have people from almost every category of the duration of stay in Hyderabad.

According to Figure 13.12, the northern, western, and central parts of the city have more of the graduate and post-graduate population. These are also the areas where people have more skilled jobs. Professional degree holders also figure prominently in these areas. The relative proportion of illiterates is higher in the southern part of the city. This aspect has to be addressed by the concerned authorities if this part of the city is to be uplifted.

A glimpse at Figure 13.13 reveals that so far as the eastern part of the city is concerned, people are in business, government jobs, or private jobs with a seeming absence of those who are self-employed and relative domination of those in private jobs. The dominance of those in private jobs is also found among those in some pockets of the southern, central, and northwestern parts of the city. Those

HYDERABAD CITY- EDUCATION LEVELS- 2020

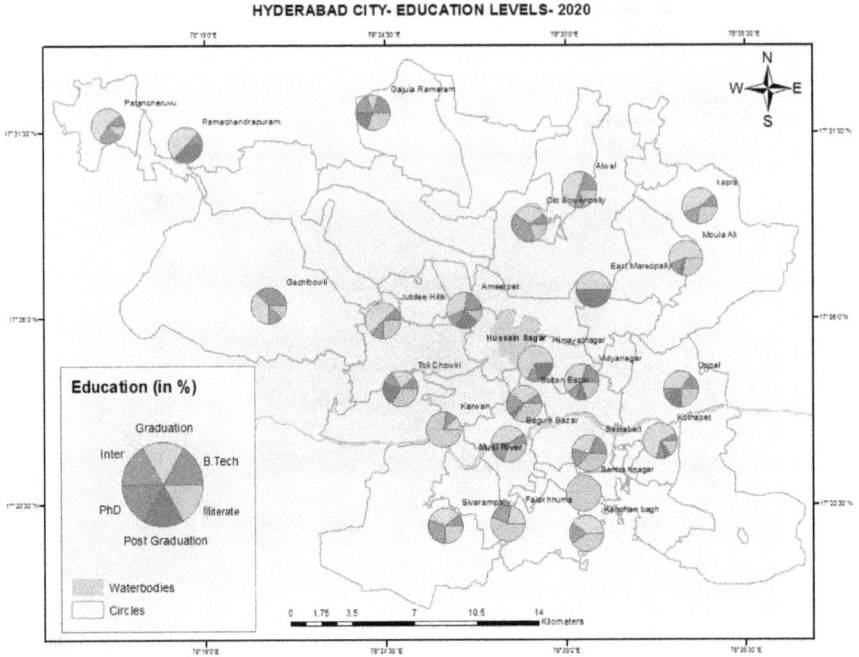

FIGURE 13.12   Education levels.

Source: Field Survey.

in government jobs are reported inordinately more in Moulali and are also high in Himayatnagar. They are also more in Patancheruvu and Ramachandrapuram, which have some of the esteemed national institutions such as Bharat Heavy Electrical Limited (BHEL), among others. Gachibowli also has a large number of government institutions and hence employment in them, these include, among others: the University of Hyderabad, Maulana Azad National Urdu University, Indian School of Business, International Institute of Information Technology, Hyderabad, National Institute of Tourism and Hospitality Management, Indian Immunologicals Limited, Kendriya Vidyalaya, and others. This is the case with Kanchan Bagh which has several defence labs and associated establishments.

This compares with a preponderance of people in 'services' at the 1951 census in Hyderabad, with a majority of the population being employed under the broad category of 'services' which can be broken up into transportation, manufacturing, finance, and administration in Secunderabad [1].

A glance at Figure 13.14 reveals that a majority of the people in every locality have a monthly income in the range of Rs. 20,000–50,000, followed by those in the range of 50,000–100,000. Very few people have an income of higher than Rs. 100,000; these are either in jobs like lecturers or in business or chartered accountancy. Some people in the eastern and central-western parts of the city have a monthly income of less than Rs. 20,000. They are engaged in private jobs, or occupations like tailoring,

HYDERABAD CITY- OCCUPATION - 2020

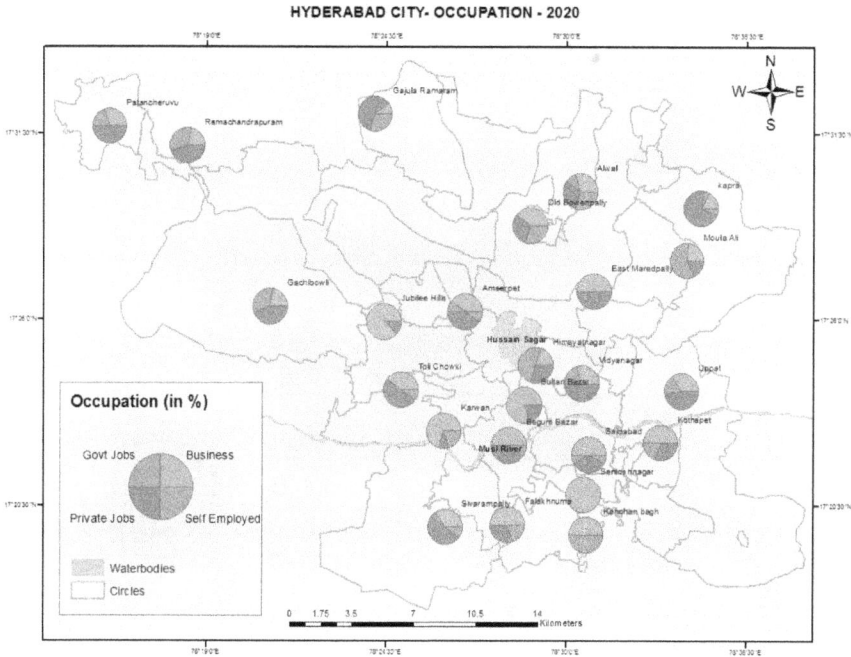

FIGURE 13.13 Occupations.

Source: Field Survey.

driving, security guards, or petty businesses like tea stalls, mutton shops, small grocery shops, bangle shops, and, surprisingly, even an ex-Member of Parliament. There can also be an element of under-reporting so far as incomes are concerned and one has to take the figures in a relative rather than an absolute sense.

13.4 SUMMARY AND CONCLUSIONS

Hyderabad witnessed a large number of changes in the 20th century. It had a low population to start with, with even a declining population at one stage, but has picked up inordinately so far as the rate of population increase is concerned. The composition of the population has also seen a sea change from a Muslim-dominated one to a predominantly Hindu population in most parts of the city. This also applies as far as the linguistic composition of the population is concerned, where Telugu is the dominant language spoken in most parts of the city while at the beginning of the 20th century it was Urdu.

The population of the city is largely concentrated in the core areas that were part of the Municipal Corporation of Hyderabad (MCH). The peripheral areas are more widespread in extent and the population is not as compactly packed as in the central parts of the city. Though the Scheduled Caste (SC) population is spread all over the city, it is perceivable that the peripheral areas have it in a larger measure. The SC

HYDERABAD CITY- INCOME LEVELS- 2020

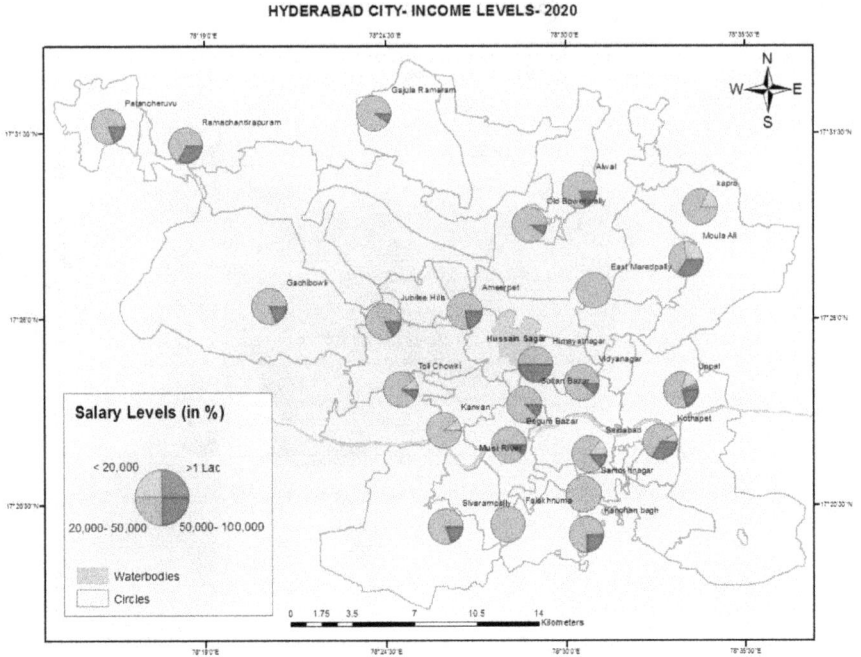

FIGURE 13.14 Income levels.

Source: Field Survey.

and ST population is relatively higher in the eastern part of the city compared to the other areas.

The city has expanded in the north-eastern and north-western parts due to the setting of industries there, it has grown in the west because of the growth of the information technology sector and in the north due to the location of educational institutions. It has extended areally along the roads joining it with major cities like Mumbai, Bengaluru, Vijayawada, and Warangal. The spatial growth of the city on both sides of the National Highways is very conspicuous. The building of the outer ring road (ORR) has encouraged real estate operations on the outskirts of the city. The ORR is also expected to help in the growth and development of satellite townships. This aspect can be tapped into by the government to set up these satellite towns as they can act as shock absorbers for the city of Hyderabad, where the prospective migrants to Hyderbad can settle in these towns instead of adding to the population of an already large city.

The demographic profile of the city can also help the authorities in their efforts to improve the lives of many by way of providing education in those areas where this is inadequate. Also, measures to optimise the skill sets of people in other areas where they are found to be highly qualified on certain counts can be taken. While this could serve as a preliminary study or a backdrop, a detailed enquiry could be carried out in specific pockets either to utilise the services of the people or to provide certain services to them to improve their quality of life.

ACKNOWLEDGEMENT

I thank the Indian Council of Social Science Research (ICSSR) for providing me with the National Fellowship (letter no. F. No. 1-06/18-19 NF dated 14.12.2018) and this work is partially taken from that project report.

REFERENCES

1. Alam, S.M, Hyderabad–Secunderabad (Twin Cities) A Study in Urban Geography, Allied Publishers, Bombay, 1965, Pp 80–81.
2. www: http Hyderabad India Metro Area Population 1950–2021/Macro Trends.
3. https://worldpopulationreview.com/world-cities/hyderabad-population/.

14 Assessing Sprawl Characteristics in the Peri-Urban Regions of Indian Metropolises through Geospatial Studies with a Special Study of Southern Chennai

A.R. Narayani and R. Nagalakshmi

CONTENTS

14.1 INTRODUCTION

Urban growth is the development of spatial pattens based on population, environmental, political, social, and economic changes. The typology of human activity and the movement of people from rural areas to urban centres fosters this growth process. The process is planned in some areas and is organic in others. The region of organic growth needs to be monitored constantly as they have a tendency to develop in a haphazard manner and result in habitats that are not suitable for sustenance. These

DOI: 10.1201/9781003331001-14

unsustainable developments manifest themselves more on the outer fringes of the city, and are identified as potential areas for future development. They result in the formation of squatter settlements and encroachments, stressing the existing infrastructure along with poor levels of sanitation and hygiene. Metropolitan cities are important for both urban growth as well as governance, as they attract a lot of investments and business ventures. They are seen as the economic drivers of the global and national market. The rapid influx of people to these cities also poses various threats to the growing economies that spill over in an unregulated and unplanned manner around the suburbs and fringe areas [1,2]. The Indian metropolises have turned into megacities in the recent past trying to fulfil the dreams of every citizen who wants to be a part of the urban realm. Planners have identified urban poverty as one of the main reasons for this unregulated growth which has also led to a high crime rate and increased social disparities [3,4]. The NURM Mission was framed to provide good urban infrastructure and governance, and basic services to the urban poor. The planned interventions can be attributed to the development of industries, IT parks, transport infrastructure, educational institutions, etc., that in turn fuel the settlement patterns around these regions. The promotion of mixed land uses and the availability of cheaper landbanks are also considered as the major reasons for rapid development in the peri-urban areas. This study aims to understand urban sprawl characteristics in the Indian metros of Delhi, Mumbai, and Kolkata with a special focus on the study of the peri-urban regions of southern Chennai. Multi-spectral satellite imagery from Landsat 5 and Landsat 8 has been used along with the spatial analyst tools in ArcGIS to identify the areas of sprawl. To study the changing conditions in southern regions of Chennai, change detection using NDVI and image differentiation techniques have been applied to understand the fringe transformation in detail. Secondary data sources include census reports and development plans from the respective planning departments for the assessment of sprawl conditions in the other metro cities. The results indicate that Chennai has expanded from the city along the transport corridors along the southern sides in a strip-type development. The growth is rapid after the dominance of the IT and auto sectors in the city [5,6]. In Delhi, the first level of expansion could be seen along the outer extent of the core city and this has been contained by promoting high-density developments. Mumbai could also be seen expanding around its edges with the formation of low-density informal settlements. The city of Kolkata is the industrial hub of the north-eastern hinterland. The central business district (CBD) is overly congested, and sprawling can be seen on the extents with the formation of new towns. In all these cities the central core retains its nature of being the nucleus of development. Planners need to look at two perspectives. One is to rehabilitate the central cores and free them from congestion, and the second is to plan for the newly developing areas with a vision of future requirements. Expansions of the cities are unavoidable due to the ever-growing needs of our population. However, self-sustained planning strategies will lead the way for sustainable cities for our future generations.

14.1.1 METHODOLOGY

Metropolitan areas are regions of high-density urban agglomerations, surrounded by industrial areas, commercial areas, and transit networks (Figure 14.1). The aim of this study is to assess the urban sprawl characteristics and compare the various drivers

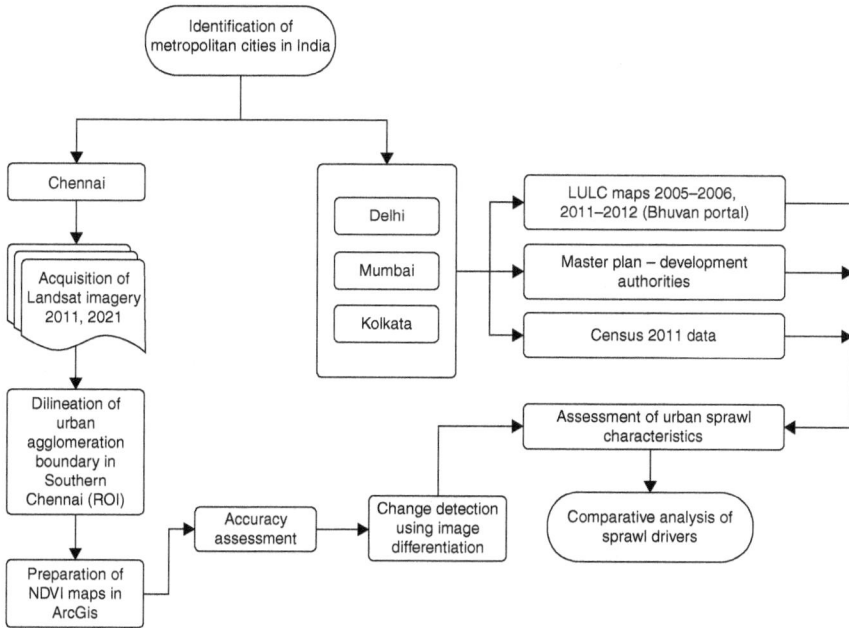

FIGURE 14.1   Methodology.

of sprawl in the different metropolises. To carry out a detailed study in the Chennai region, the methodology adopted includes GIS-based analysis. Satellite imagery of 30 m resolution has been acquired from Landsat 5 (TM) and Landsat 8 (OLI) for the study years 2011, and for 2021 from USGS Earth Explorer. The identified region of interest is delineated using polygon features and clipped for further studies. Image pre-processing is carried out for Landsat 5 image as given in the Landsat handbook and Landsat 8 images are directly used for the analysis. Normalised difference vegetation index (NDVI) is performed in ArcGIS, and land cover is classified into three classes, namely land, water, and vegetation. An accuracy assessment using the ground truthing method along with confusion matrix and multivariate kappa coefficient method is performed to assess the accuracy of the classification. Change detection is carried out using image differentiation techniques in ArcGIS to assess the characteristics of the sprawl conditions. For the other major metros, Delhi, Mumbai, and Kolkata, secondary data from Census 2011, land use landcover maps from NSRC (Bhuvan portal), and master plans from the development authorities have been analysed to assess the sprawl characteristics [6,7]. A comparative analysis of the urban characteristics of these megacities is summarised.

14.2   STUDY AREA

The identified metropolitan city of Delhi is the national capital of India, while the cities of Chennai, Mumbai, and Kolkata are the state capitals of Tamil Nādu, Maharashtra, and West Bengal, respectively (Figure 14.2). They are not just areas of dense urban settlements but the power houses of urban governance and engines of

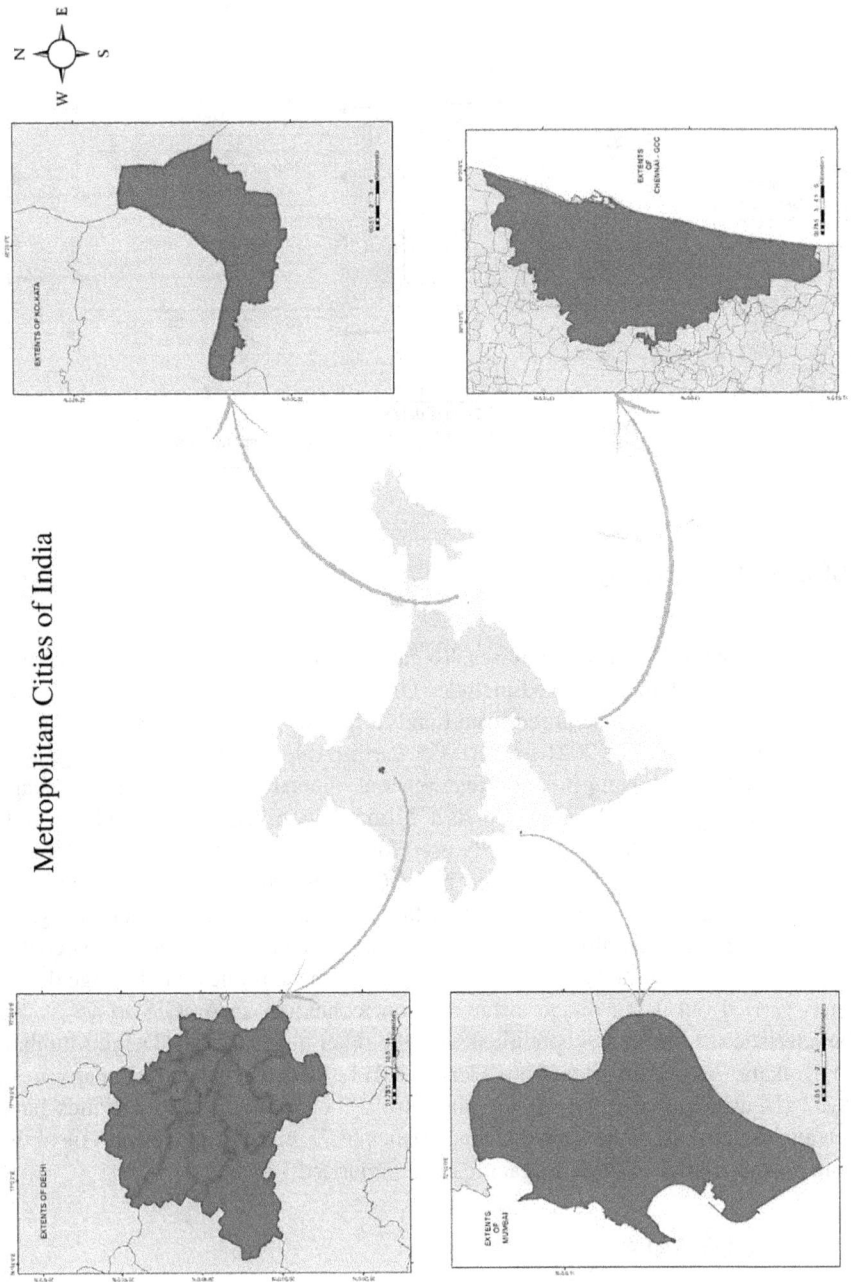

FIGURE 14.2   Metropolitan cities of India.

the national and global economy with significant contributions to the national GDP. With progress in development, they also contribute to environmental damage, urban poverty, and social polarisation.

14.2.1 PERIURBAN REGION IN THE SOUTH CHENNAI METROPOLITAN AREA

Chennai is the state capital of Tamil Nādu, lies at 13.0827°N, 80.2707°E, and has undergone unprecedented growth in the last two decades, especially after the advent of the IT and automobile industries. The old city areas have expanded 2.5 times to form Greater Chennai Corporation (GCC), and GCC has expanded 2.75 times to form the Chennai Metropolitan Area (CMA), as indicated in Figures 14.3 and 14.4. Future proposals indicate that the city has a capacity to grow eight-fold, covering the neighbouring districts of Chengalpattu, Kanchipuram, Vellore, and Tiruvallur [8]. Previous studies indicate that urban growth is higher towards the southern fringe areas. The area of interest for the purpose of this study is the urban–rural interface between Chennai and Chengalpattu regions comprising eight taluks, as indicated in Figure 14.5. The formation of the new district of Chengalpattu in 2019 opened new growth opportunities for the regions. Figures 14.6, 14.7, and 14.8 illustrate mapping of various population attributes from the 2001 census. It can be seen that the urban areas are accumulated along the transport corridors of Grand Southern Trunk Road (GST) which is the lifeline of industrial parks and satellite townships. The other transit corridors of Old Mahabalipuram Road (OMR) and East Coast Road (ECR) are also transitioning into urban areas with the setting up of IT parks. Dot density mapping (Figure 14.9) shows dense population dispersing along the transit networks towards the southern regions, most of which currently fall under the Chennai Metropolitan Area (CMA) [9].

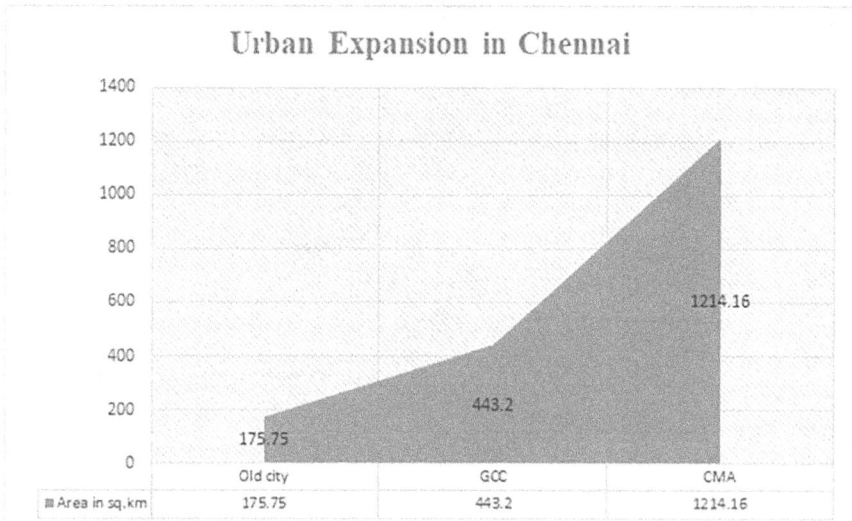

FIGURE 14.3   Graph showing rate of urban expansion.

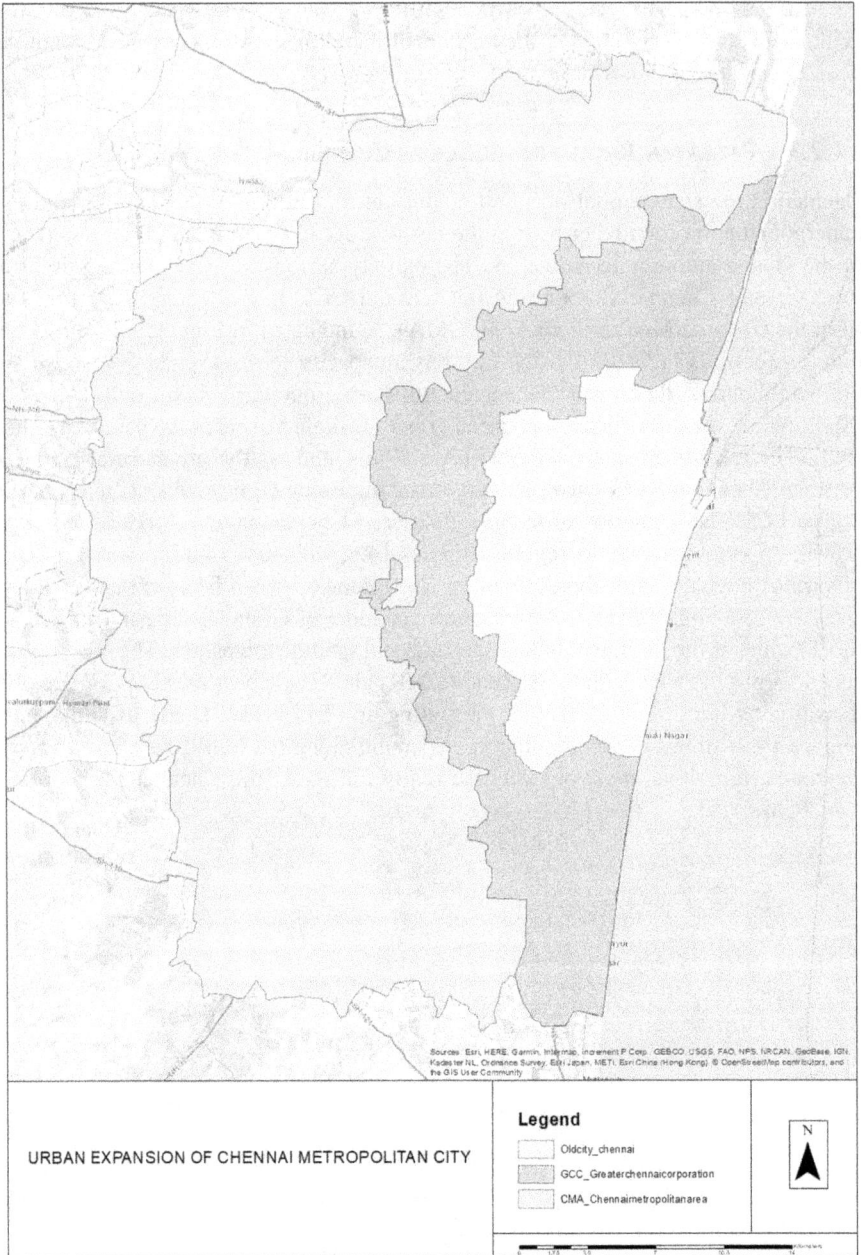

FIGURE 14.4   Expansion of Chennai city.

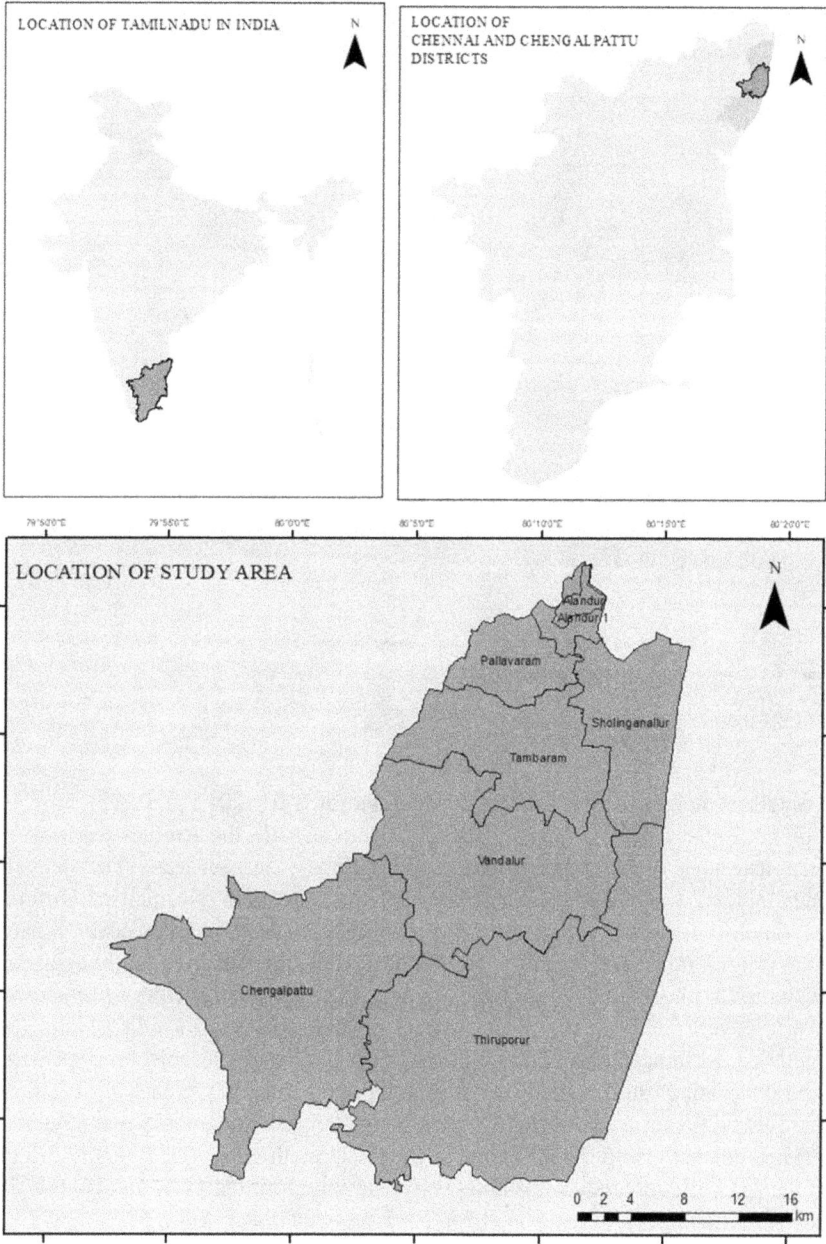

FIGURE 14.5   Interface between Chennai and Chengalpattu districts (ROI).

ANALYSIS OF SATELLITE IMAGERY

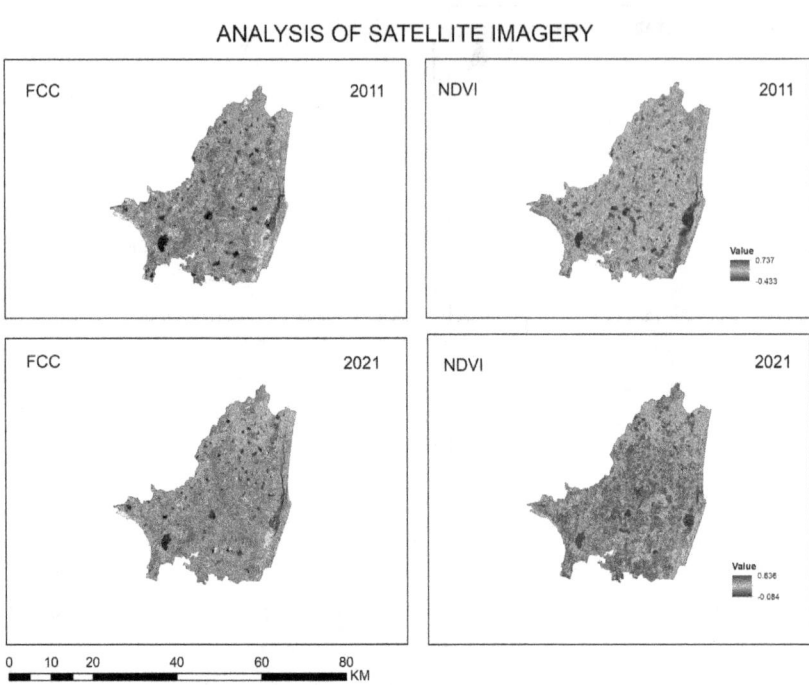

FIGURE 14.6   Satellite image analysis.

Satellite imagery from Landsat 5 and Landsat 8 for 2011 and 2021 have been acquired. False colour composite (FCC) is generated in the ArcGIS platform. The white and grey areas indicate built-up and impervious surfaces. The deep blue zones indicate water, and the red zones indicate vegetation. Normalised Difference Vegetation Index (NDVI) is a technique to analyse changes in green cover. Since the urban fringes are predominantly conversion of arable land, changes in these land uses are studied to understand the sprawl dynamics. The values range from –1 to +1, values close to –1 indicate no or very little vegetation. Values close to +1 indicate thick dense vegetation, as indicated in Figure 14.6 [10,11]. Accuracy assessment was performed using the ground truthing method with 90 sample points, 30 per class for each of the classified NDVI raster from the reference image. A confusion matrix was tabulated in ArcGIS. Further, using the multivariate statistical method of assessment to calculate the kappa coefficient value was computed using the formula given in equation 14.1. The overall accuracy of the NDVI for 2011 was calculated to be 85.56%, with a K value of 0.78, and the overall accuracy for 2021 was calculated to be 92.11% with a K value of 0.87 [12].

$$K = \frac{N \sum_{i=1}^{r} x_{ii} - \sum_{i=1}^{r} (x_{i+} * x_{+i})}{N^2 - \sum_{i=1}^{r} (x_{i+} * x_{+i})}$$

(14.1)

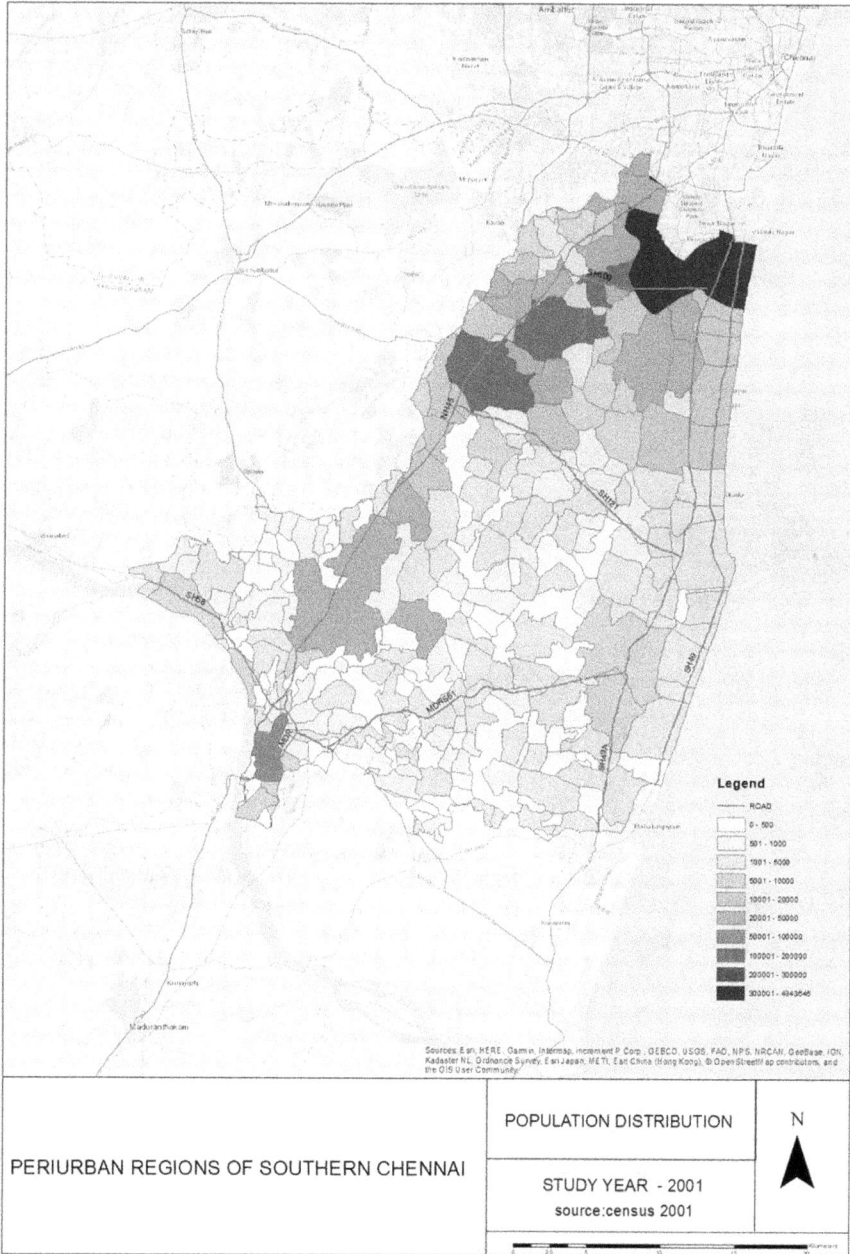

FIGURE 14.7   Map indicating distribution of population density in 2001.

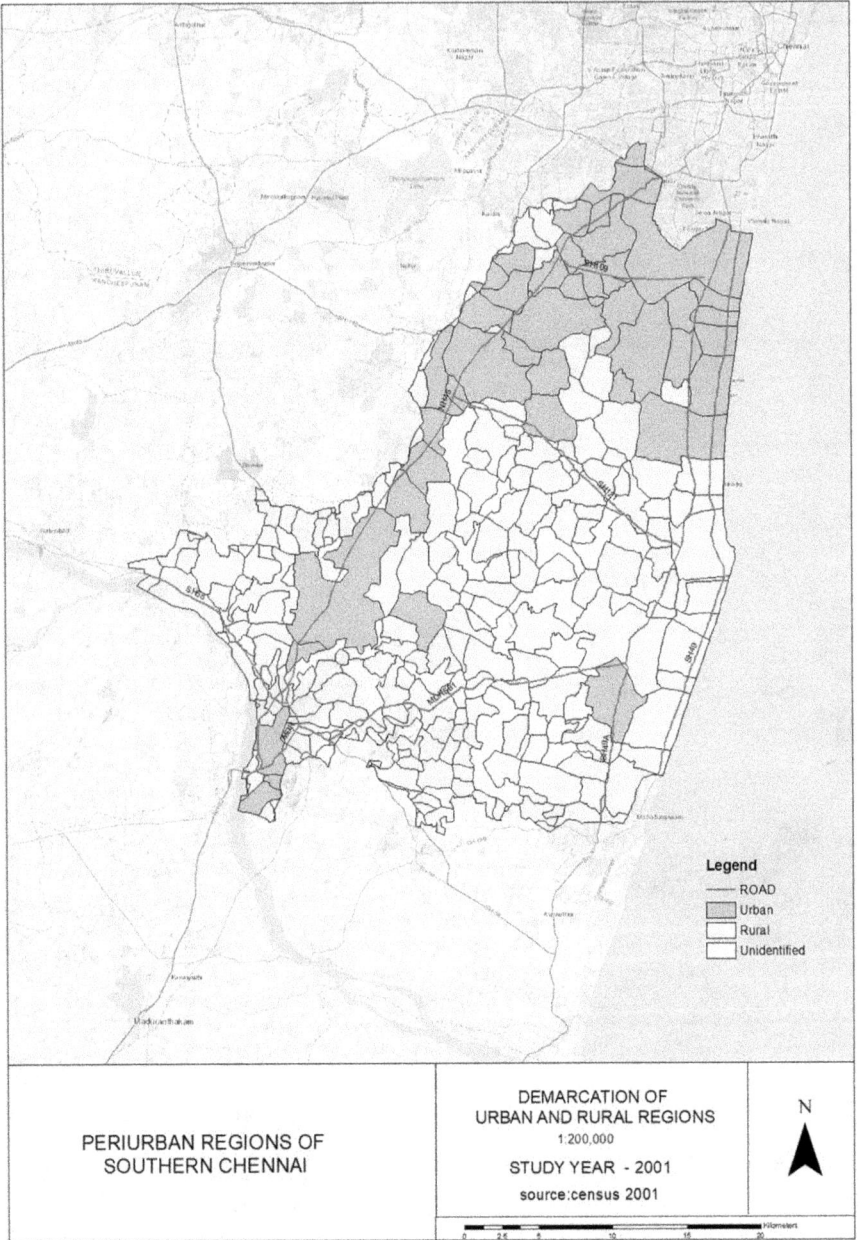

FIGURE 14.8　Map indicating urban and rural areas in 2001.

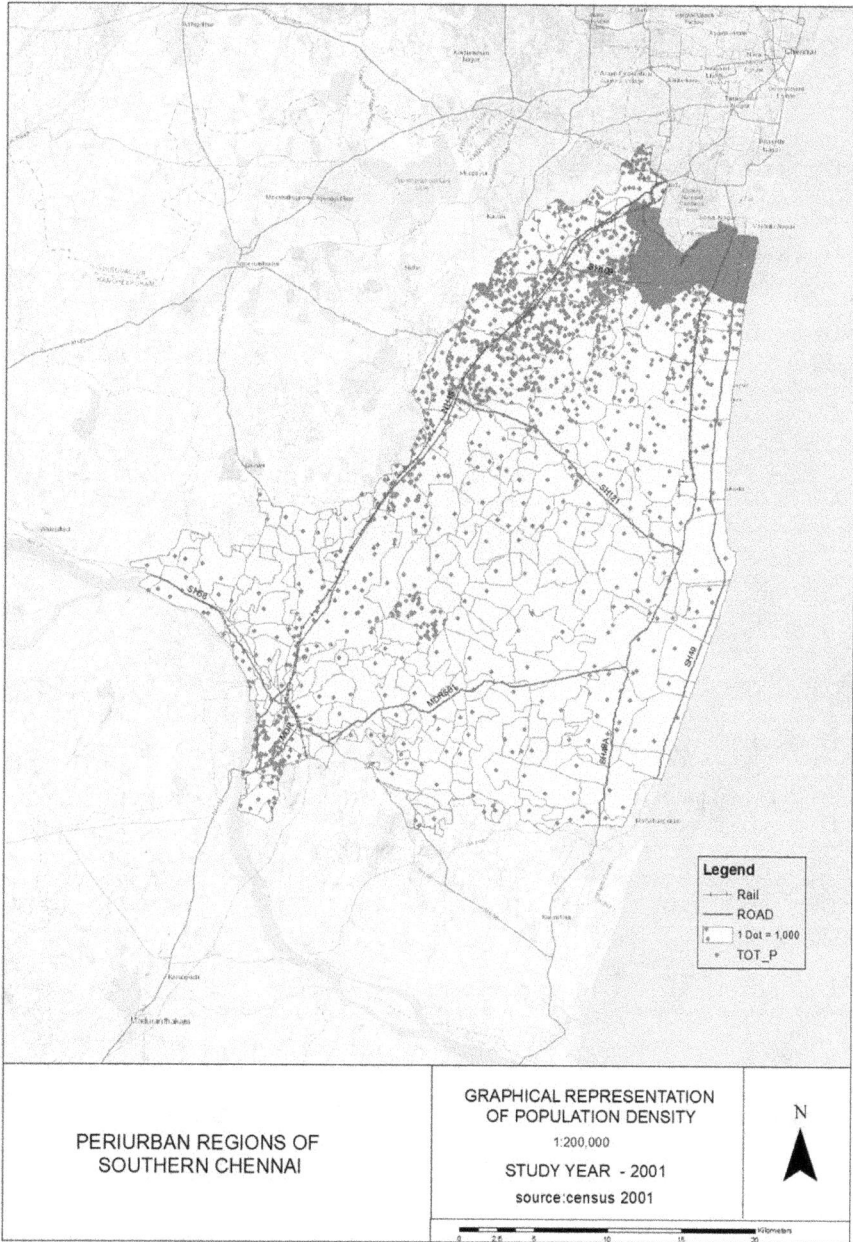

FIGURE 14.9   Map indicating dot density of population density in 2001.

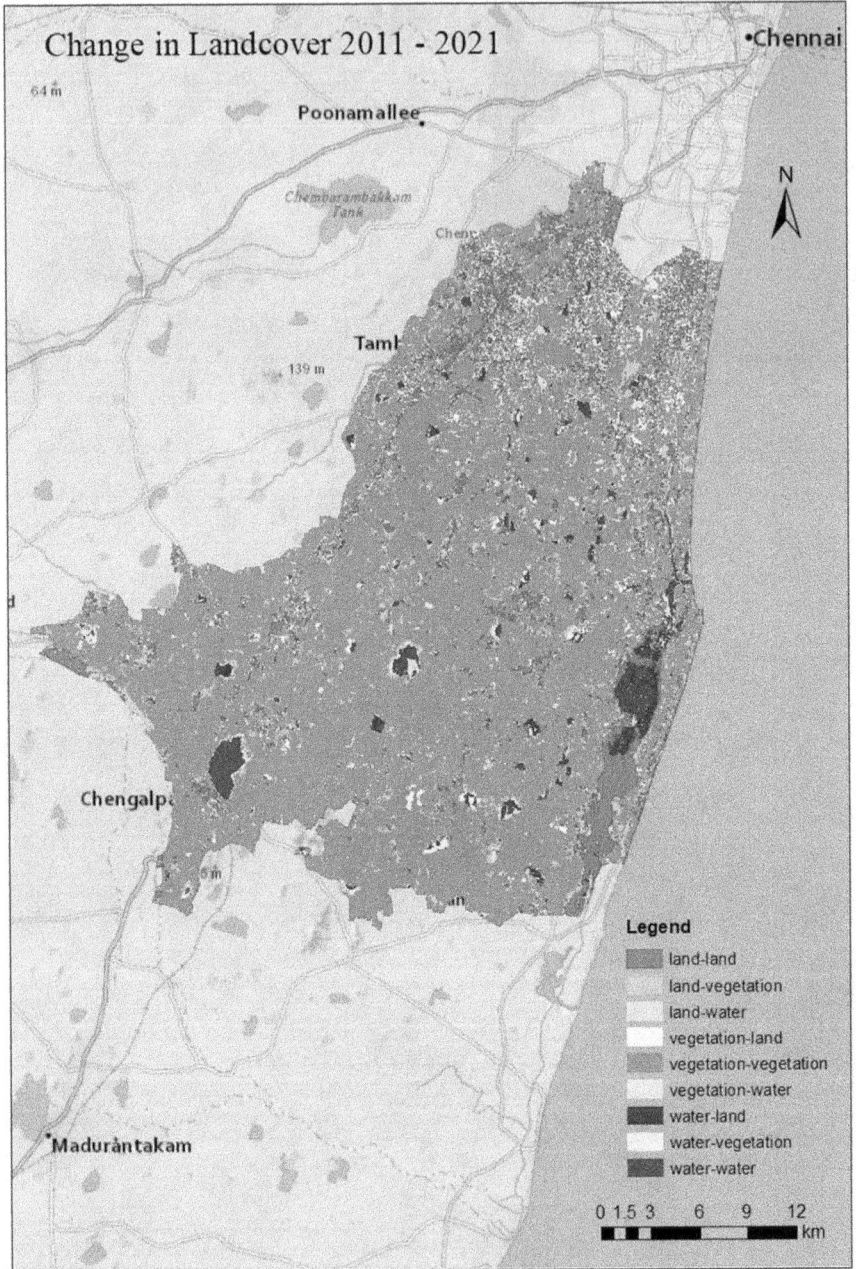

FIGURE 14.10 Change detection in landcover in the peri-urban regions of southern Chennai between 2011–2021.

TABLE 14.1
Change Detection in Landcover in the Peri-Urban Regions of Southern Chennai between 2011–2021

2011		2021		Change Detection	
class	Area (sq. km)	class	Area (sq. km)	change	Change in area (sq. km)
Water	39.46	Water	29.98	Water–water	21.22
Water	39.46	Land	148.19	Water–land	11.75
Water	39.46	Vegetation	826.65	Water–vegetation	6.47
Land	168.43	Water	29.98	Land–water	5.41
Land	168.43	Land	148.19	Land–land	84.77
Land	168.43	Vegetation	826.65	Land–vegetation	78.12
Vegetation	796.93	Water	29.98	Vegetation–water	3.34
Vegetation	796.93	Land	148.19	Vegetation–land	51.59
Vegetation	796.93	Vegetation	826.65	Vegetation–vegetation	741.85
				Total Area	**1004.52**

Changes in landcover classes were analysed using image differentiation techniques in ArcGIS. Figure 14.10 is a visual illustration of three landcover classifications, land, water, and vegetation, with the changes from 2011 to 2021. A detailed matrix showing the values of area changes has been tabulated in Table 14.1.

14.2.2 NATIONAL CAPITAL REGION – DELHI

The national capital region of Delhi lies between 28.418N, 76.848E by 28.888N, 77.358E, with hot dry summers and cold winters (Figure 14.11). The morphology of Delhi has a rich historical background. The settlements are denser at the core due to the radial planning and fortified developments. Previous studies indicate that there are expansions in the outer peripherals owing to the development of industries near Noida, Gurgaon, Ghaziabad, and Faridabad. Sprawling is majorly noticed towards the southwest districts due to economic reforms and policy changes. The population doubled between 1991–2001 due to migration. There was a boom in the construction of highways and metro systems to connect the different parts of the city seamlessly. Effective control of sprawl was established by shifting focus towards high-density vertical development. The increased population also has environmental consequences. Delhi has recently seen a spike in the air pollution levels and heat waves have become a common phenomenon, as most of the agricultural and pasture lands are being consumed to cater to the growing infrastructure needs. The land use landcover map of Delhi was acquired from the Bhuvan portal for the periods 2011–2012 and 2015–2016. The statistical data indicate six major landcover classifications which are further subdivided into 19 detailed classifications [13,14] (Figures 14.12–14.15; Table 14.2).

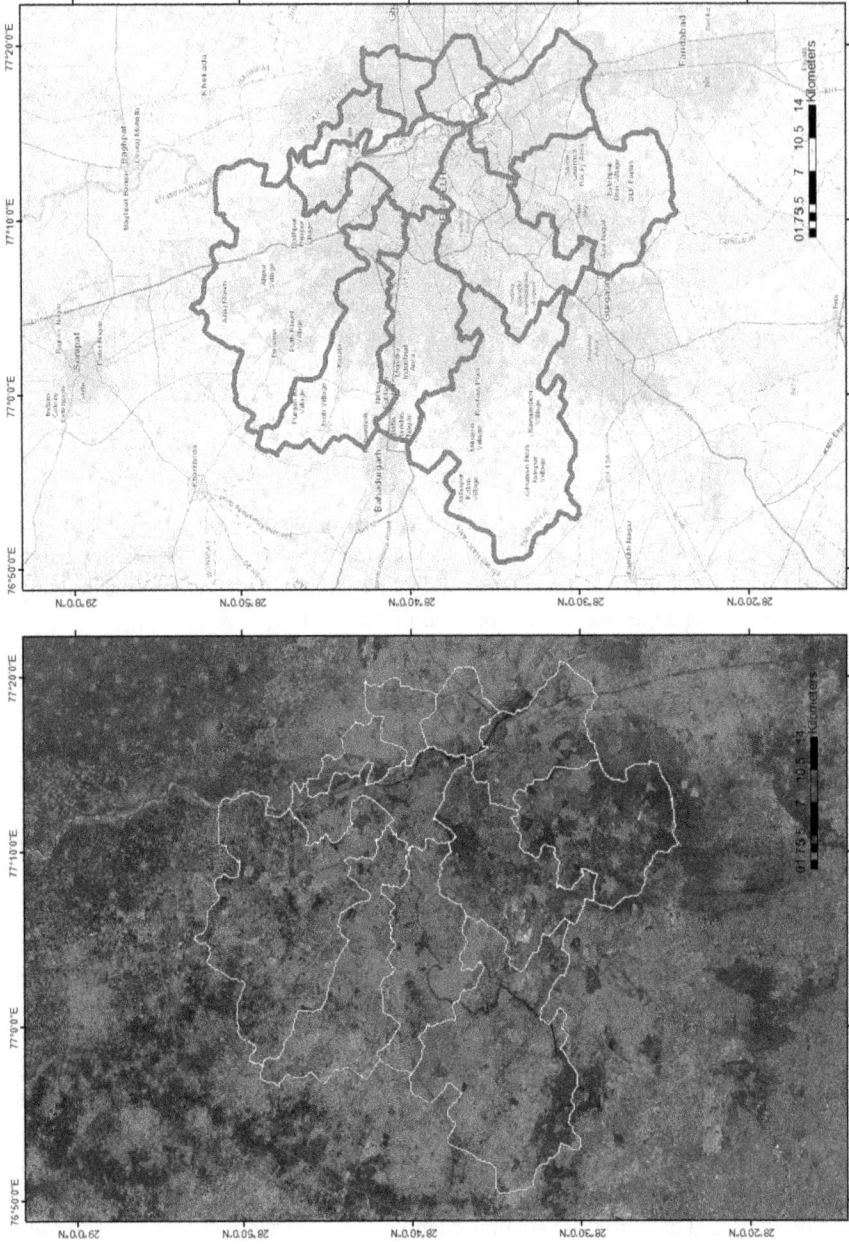

FIGURE 14.11   Boundaries of Delhi National Capital Territory.

FIGURE 14.12   Land use plan 2021. Source: Delhi Development Authority.

(a)

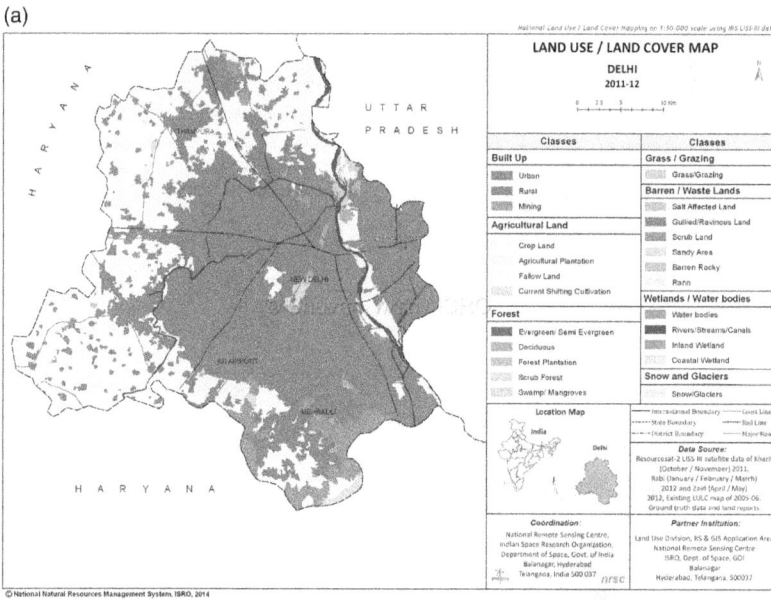

FIGURE 14.13a   Landuse/landcover map of Delhi (2011–2012).

Source: NRSC Bhuvan Portal.

(b)

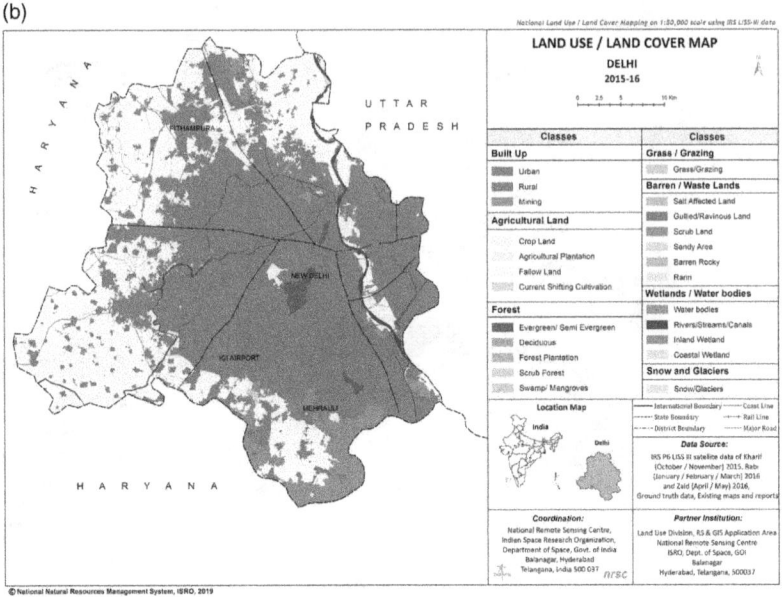

FIGURE 14.13b   Landuse/landcover map of Delhi (2015–2016).

Source: NRSC Bhuvan Portal.

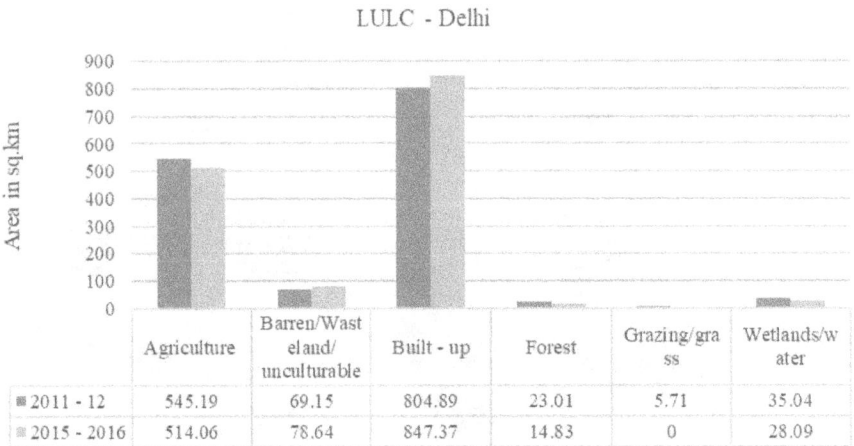

LULC - Delhi

	Agriculture	Barren/Wast eland/ unculturable	Built - up	Forest	Grazing/gra ss	Wetlands/w ater
2011 - 12	545.19	69.15	804.89	23.01	5.71	35.04
2015 - 2016	514.06	78.64	847.37	14.83	0	28.09

FIGURE 14.14   Comparison of LULC in Delhi.

Source: Adapted from NSRC statistical report.

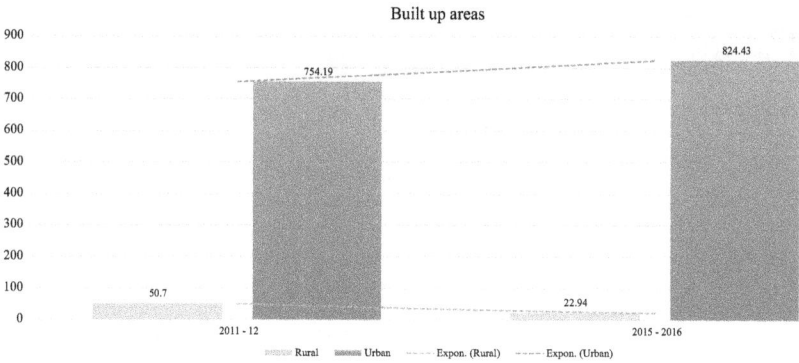

Built up areas

FIGURE 14.15 Graph indicating trends in urban and rural areas of Delhi.

TABLE 14.2
Land Use Landcover – Delhi

L1	L2	2011–12 (Area in sq.km)	2015–2016 (Area in sq.km)
Agriculture	Crop land	452.58	505.28
	Fallows	76.68	7.1
	Plantation	15.93	1.68
Barren/wasteland/ unculturable	Barren rocky	0.1	0
	Gullied/Ravinous land	6.05	0
	Sandy	0	0.62
	Salt affected land	0.15	0
	Scrub land	62.18	74.87
	Mining	0.67	3.15
Built–up	Rural	50.7	22.94
	Urban	754.19	824.43
Forest	Deciduous	10.58	1.18
	Evergreen	0	12.32
	Forest plantation	0.38	0
	Scrub forest	12.05	1.33
Grass/grazing	Grass/grazing	5.71	0
Wetlands/water	Inland wetland	4.1	3.77
	River/stream/canal	27.15	20.88
	Water bodies	3.79	3.44

Source: NSRC, Bhuvan.

14.2.3 Mumbai Metropolitan Region

Mumbai is a densely populated metropolitan city and is known as the financial capital of India. It consists of a group of islands 603.4 km in area located at 19.0760° N, 72.8777° E (Figures 14.16 and 14.17). The Mumbai urban agglomeration is one of the most populous UAs in India [15]. It consists of the six municipal corporations

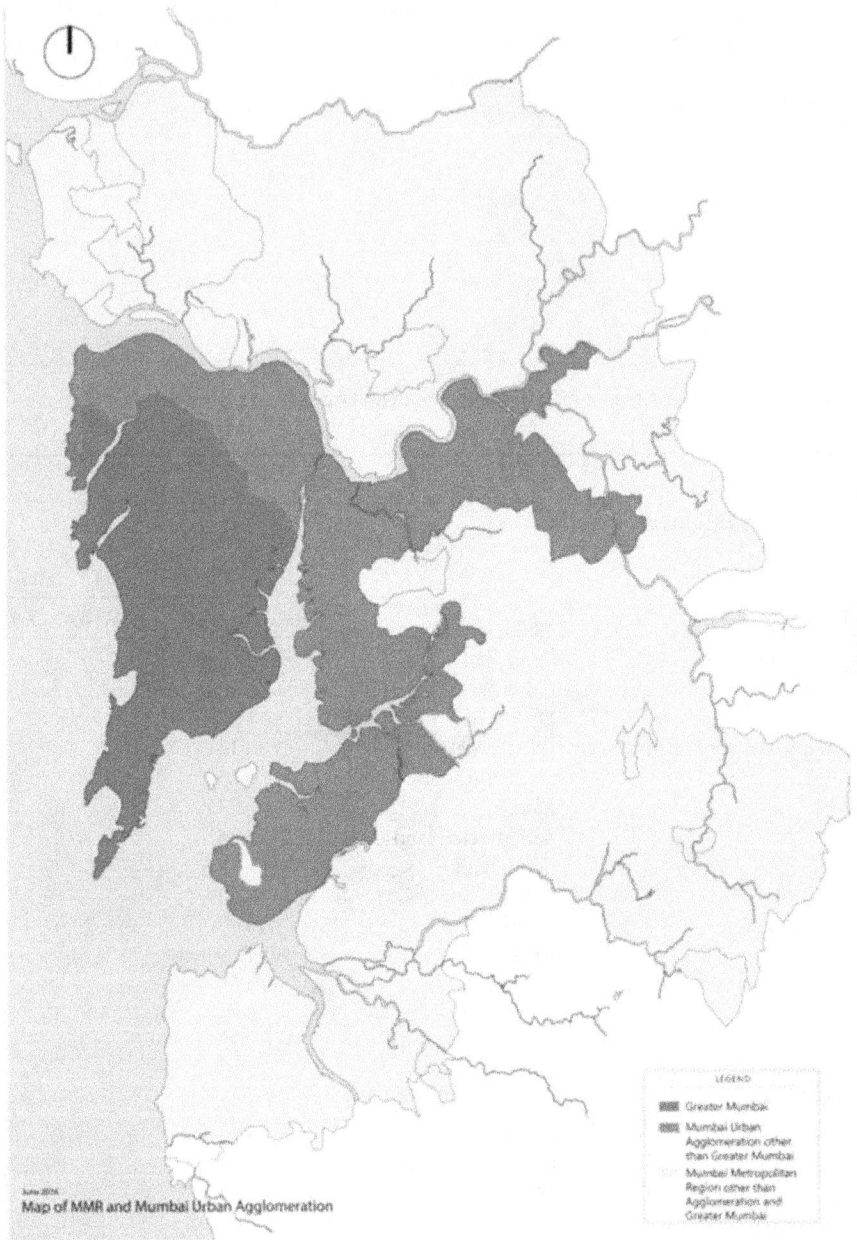

June 2006
Map of MMR and Mumbai Urban Agglomeration

LEGEND

Greater Mumbai

Mumbai Urban
Agglomeration other
than Greater Mumbai

Mumbai Metropolitan
Region other than
Agglomeration and
Greater Mumbai

FIGURE 14.16   Schematic map of Mumbai agglomeration.

Source: [15].

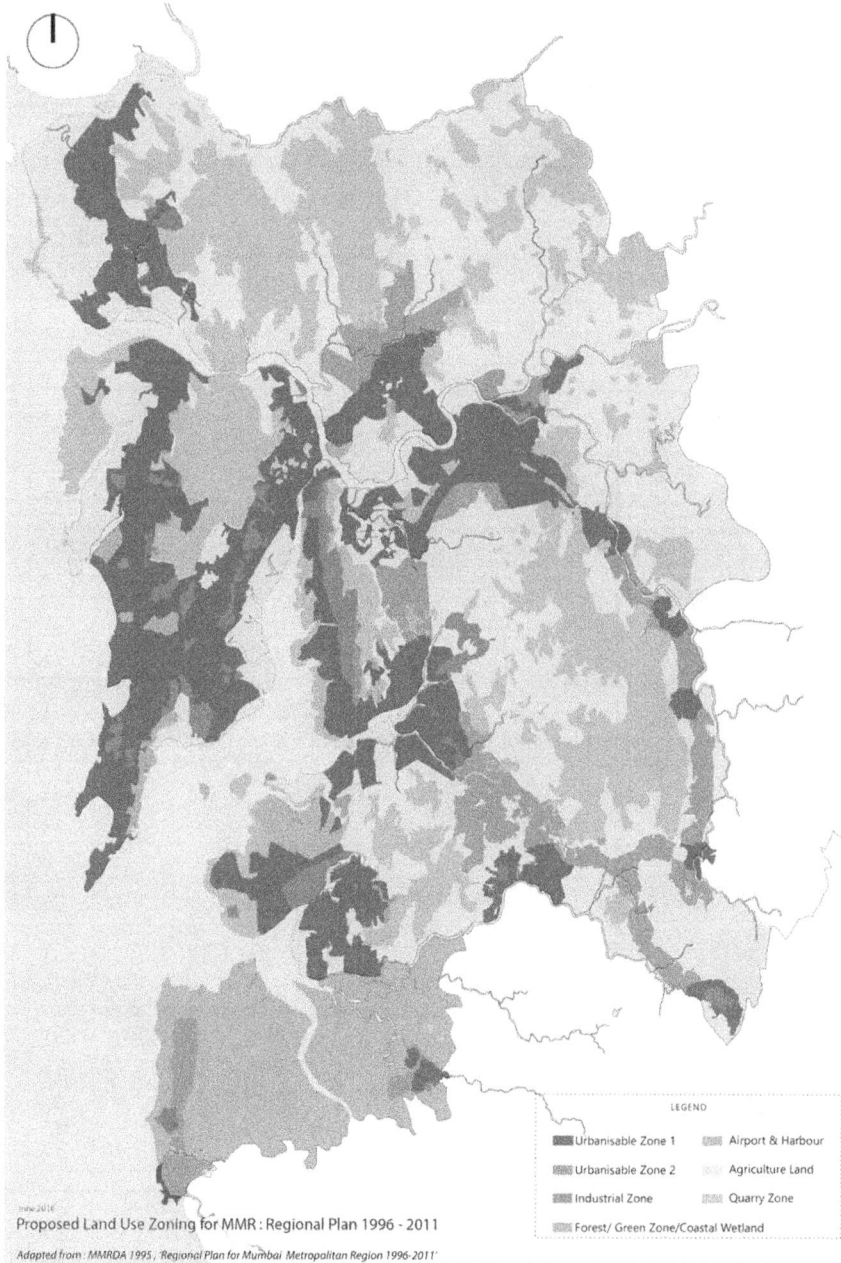

LEGEND

Urbanisable Zone 1   Airport & Harbour
Urbanisable Zone 2   Agriculture Land
Industrial Zone   Quarry Zone
Forest/ Green Zone/Coastal Wetland

Proposed Land Use Zoning for MMR : Regional Plan 1996 - 2011

Adapted from : MMRDA 1995 , 'Regional Plan for Mumbai Metropolitan Region 1996-2011'

FIGURE 14.17   Proposed land use zoning for MMR.

Source: [15].

TABLE 14.3
Land Use Landcover for Mumbai City and Suburban Mumbai

Landcover		2011–12		2015–16	
L1	L2	Mumbai city (area in sq. km)	Suburban–Mumbai (area in sq. km)	Mumbai city (area in sq. km)	Suburban–Mumbai (area in sq. km)
Agriculture	Crop land		4.77		3.04
	Fallows		0.58	0.08	0.61
	Plantation		1.45		3.15
Barren/wasteland/ unculturable	Barren rocky	0.08	0.32	0.46	0.44
	Gullied/ravinous land				
	Sandy area	0.04	0.05	0.16	1.84
	Salt affected land		0.08		1.07
	Scrub land		17.97		18.61
	Mining		0.47		0.6
Built–up	Rural		0.68		0.68
	Urban	65.64	216.6	65.64	216.24
Forest	Deciduous		48.13		47.15
	Evergreen		9.4		9.29
	Scrub forest		0.08		0.08
Wetland/waterbodies	Swamp and mangroves	2.6	54.64	2.6	54.64
	River/stream/canal		0.02		0.2
	Coastal wetland	88.53	81.78	87.95	80.6
	Waterbodies		0.11	0.11	7.74

Source: Statistical data NSRC Bhuvan portal.

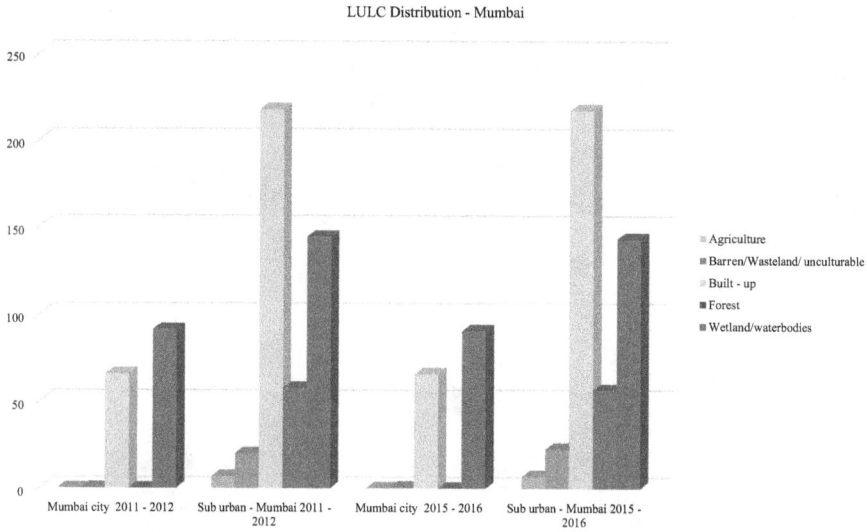

FIGURE 14.18   LULC – 2011–2012, 2015–2016.

Source: Adapted from Bhuvan statistical reports.

of Greater Mumbai, Mira-Bhayandar, Thane, Navi Mumbai, Kalyan-Dombivli, and Ulhasnagar and the two municipal councils of Badlapur and Ambernath. Lucrative employment opportunities are seen as the one of the major reasons of migration to the city. The central core of the city is seen as being much suited for economic development and is often found in combination with the informal settlements. Transport infrastructure projects like MRTS, MUTP, and an airport terminal have displaced many hutments in the recent past. The Jawaharlal Nehru National Urban Renewal Mission (JNNURM) was a national drive to eradicate slum dwellings and provide habitable spaces for the residents of informal settlements. The landcover has been classified into five classes, namely agriculture, barren/wasteland, built up, forest, and wetland areas. They are further classified into various subclasses as indicated in Table 14.3. The graphical illustration in Figure 14.18 allows us to compare the various trends in the various classes in the central core areas of Mumbai and the suburban regions of Mumbai [16, 17].

14.2.4   KOLKATA METROPOLITAN REGION

Kolkata is the capital city of West Bengal, located 22.5726° N, 88.3639° E. It is the industrial nucleus of the north-east and the hinterlands. The climate of the region is of a tropical wet and dry type. The core of Kolkata city spreads over around 200 sq. km, with a population of 4.5 million. Kolkata port is also one of the major trading hubs for the neighbouring countries of Nepal, Bhutan, and Bangladesh. The developments of world-class infrastructure and healthcare facilities are some of the reasons for the growing population. To counter the rapid urbanisation process there are proposals for new township developments such as Rajarhat New Town, Dankuni Township, and

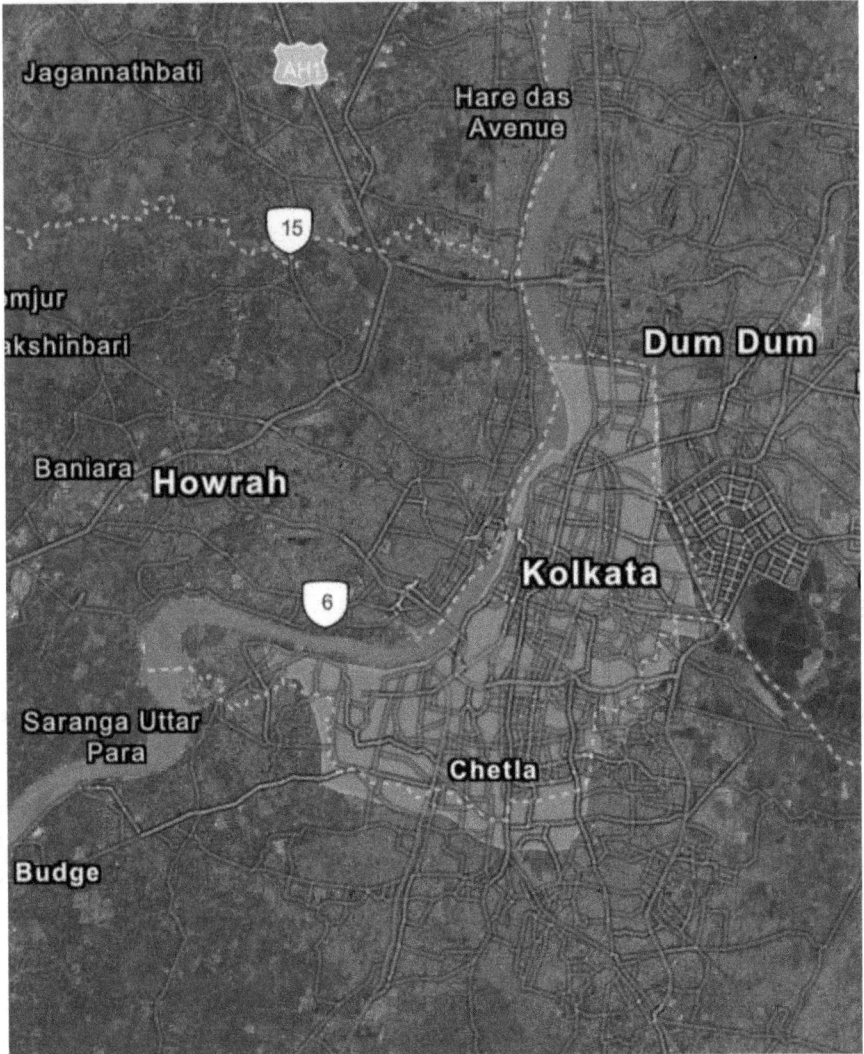

FIGURE 14.19   Boundaries of Kolkata Metropolitan Corporation.

Source: ESRI.

West Howrah Township to reduce the strain on the central business district. These are located just outside the urban extent and are deemed to be potential economic growth centres. These surrounding regions are predominantly rural and semi-rural economies. Haphazard urbanisation and improper water management are some of the key issues to be addressed due to the LULC changes. These environmental issues have a huge bearing on the development of the future new towns. A graphical illustration of the land use landcover distribution between 2010 and 2015 can be isualized in Figure 14.19. The percentage of change in the LULC distribution is illustrated in

Landuse Landcover distribution

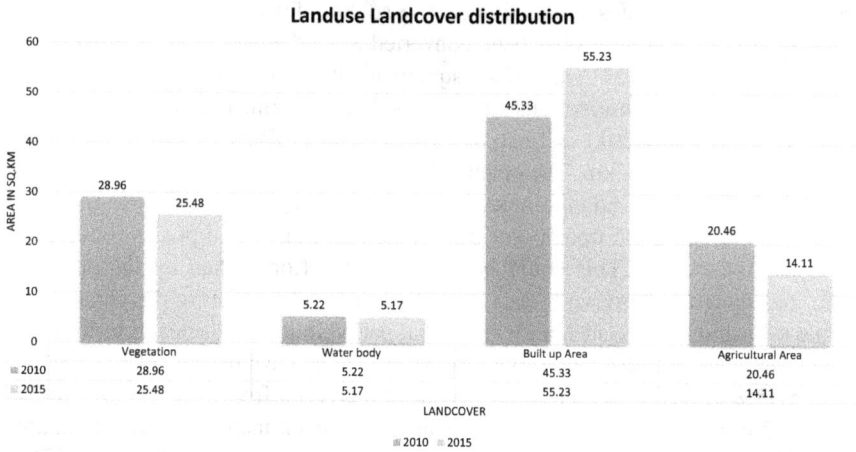

	Vegetation	Water body	Built up Area	Agricultural Area
2010	28.96	5.22	45.33	20.46
2015	25.48	5.17	55.23	14.11

LANDCOVER

2010   2015

FIGURE 14.20   Graph indicating the LULC distribution in 2010 and 2015 – Kolkata.

Trends in LULC distribution 2010–2015

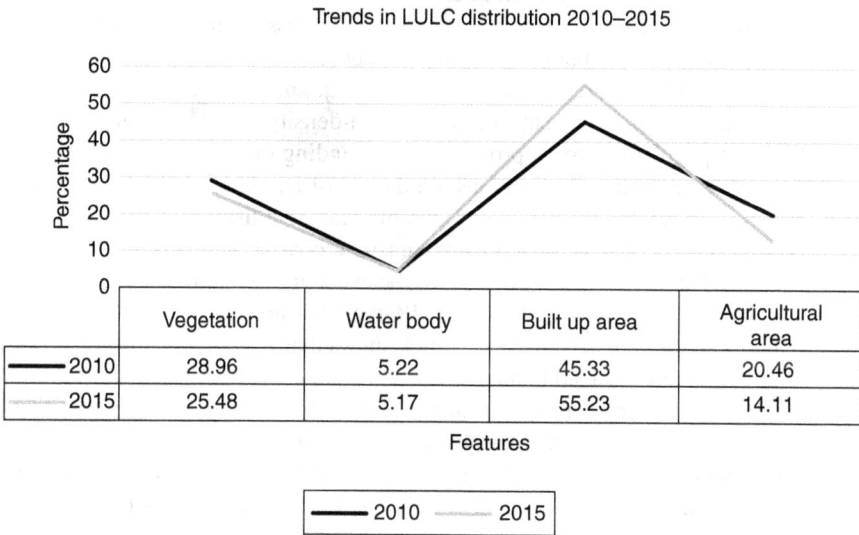

	Vegetation	Water body	Built up area	Agricultural area
2010	28.96	5.22	45.33	20.46
2015	25.48	5.17	55.23	14.11

Features

2010 ——— 2015

FIGURE 14.21   Trends in LULC distribution Kolkata.

Figure 14.20. The primary feature change indicates wetland and agricultural lands are predominantly occupied by built-up areas. Large-scale construction has taken up massive landbanks in the peri-urban fringe areas [18–20].

14.3   RESULTS AND DISCUSSION

The peri-urban regions of southern Chennai analysed in the study cover a land area of 1004 sq. km. The total area of waterbodies in 2011 was 39.46 sq. km and 29.98 sq.

km in 2021. This indicates a reduction of 9.48 sq. km of waterbodies. A maximum of 11.75 sq. km of waterbodies has been converted to land areas. The total area of land and barren surfaces in 2011 was 168.43 sq. km and it was 148.19 sq. km in 2021. This indicates a decrease in impervious surfaces by 20.24 sq. km. The areas of vegetation were 796.93 sq. km in 2011 and 826.65 sq. km in 2021. This indicates an increase in green cover of 29.72 sq. km. The change detection map (Figure 14.10) also indicates that the conversion to land and impervious surfaces is higher in the regions towards the city centre along the transit corridors. This also indicates a positive increase in the vegetation classes. This could be attributed to efforts taken by the governing authorities to protect reserve forests in the regions and promoting the plantation of avenue trees and green zones in the industrial parks. In the territories of Delhi region LULC, statistics indicate that agriculture land classes have decreased from 37% in 2011 to 35% in 2015. The built-up areas have increased from 54% in 2011 to 57% in 2015. Within the built-up areas there was a significant increase in the urban areas from 754.19 sq. km in 2011 to 824.43 sq. km in 2015, showing an increase of 70.24 sq. km. Subsequently, there has been a decline in the rural areas from 50.7 sq. km in 2011 to 22.94 sq. km, a decrease of 27.76 sq. km. The sprawl is predominantly around the industrial town on the fringe areas and is regulated within the city core due to the promotion of dense vertical developments. The loss of rural areas also pushes the shift from agriculture being the primary occupation to secondary and tertiary occupations. The Mumbai regions consists of 157 sq. km of city area and 445.99 sq. km of suburban regions. The central city is a high-density settlement with a popula- tion of 20,038 people/sq. km, approximately expanding on the peripherals into the suburban and fringe zones. The regions are devoid of any kind of rural settlements. They are surrounded by industrial and commercial establishments. It should also be noted that the zones have ecologically sensitive coastal wetlands and mangrove forests that need attention. Many efforts have been taken by the government to protect these ecological hotspots. As per Figure 14.18, there has not been much change in the land use between the years 2011–2015, which shows that growth might become static due to non-availability of land bank and the tendency to go for vertical high-density developments. However, there is a need for sustainable interventions to relieve the city's sprawling conditions in the peripheries. Kolkata Metropolitan is the centre core for the surrounding hinterlands. It has a unique feature of rural and semi-rural econ- omies close to the city centre. The landcover distribution, as given in Figure 14.21, indicates that between the periods 2010–2015 the vegetation classes have decreased by 3.48%, built-up areas have increased by 9.9%, and agricultural land has decreased by 6.35%. This also indicates that the population is shifting from a predominantly agriculture-based occupation to other secondary and tertiary occupations. The estab- lishment of the new towns at the fringes will act as satellite and counter magnet towns to reduce the stress on the existing core (Table 14.4).

TABLE 14.4
Comparative Analysis of Urban Characteristics

Characteristics	Delhi	Mumbai	Kolkata	Chennai
Status	National capital	State capital	State capital	State capital
Population In Metropolitan area as C-2011	11,034,555	12,442,373	4,496,694	4,646,732
Population In Urban Agglomerations as C-2011	16,349,831	18,394,912	14,035,959	8,653,521
Avg literacy rate	86.32%	89.78%	87.54%	90.23%
GDP	$293.6 billion	$310 billion	$150.1 billion	$78.6 billion
Predominant sprawl directions Direction of sprawl	Southwest	East	East	South
	Compact City Development	Sprawl Around the Periphery	Formation of New Towns	Sprawl Around the Transport Networks
Type of development	Vertical high-density settlements	Low-density peripheral settlements	Leap frog development	Linear development along the transit networks
Threats	Inadequate infrastructure and services	Unregulated growth of informal settlements	Decay of the central business district	Formation of unregulated settlements along the road networks
Recommendation	Self-sustaining targeted growth compact planning	Urban containment strategies including brown field development	Formation of planned satellite townships and rehabilitation of the central core	Transit-oriented development and planning preservation of corridors

14.4 CONCLUSIONS

The sprawling of the cities is seen as a natural phenomenon of expansion in all four metros that have been identified for the purpose of this study. The sprawl characteristics vary slightly based on some characteristics such as location, density, population primary occupation, development of infrastructure, etc. Conversion of arable land into industrial sites is predominantly on the rise. It is also noted that though the peri-urban fringes offer high potential zones for development, they need more infrastructure and services to provide for the rapidly converting rural areas. Some cities are trying to mitigate this by forming satellite townships that could divert the influx of people toward them. However, not all these towns offer the same living conditions and facilities. Mixed-use planning, transit-oriented development, and compact and high-density planning are some strategies that could be incorporated into the planning phases. Campaigns and awareness of the environmental damages of sprawl should be put forth to the population, and community participation should be insisted on in developing new cities. Sustainable water management systems, vertical farming, and waste management are some of the key concepts that need to be addressed in these new towns, which could lead to better living conditions and sustainable cities in the future.

REFERENCES

[1] M. Saxena and A. S. Sharma, Periurban Area: A Review of Problems and Resolutions. [Online]. Available: www.ijert.org

[2] H. S. Sudhira, T. v Ramachandra, and K. S. Jagadish, Map India 2003 Municipal GIS Urban Sprawl pattern recognition and modeling using GIS, 2003.

[3] V. Saini and R. K. Tiwari, A systematic review of urban sprawl studies in India: a geospatial data perspective, *Arabian Journal of Geosciences*, vol. 13, no. 17. Springer, Sep. 01, 2020. doi: 10.1007/s12517-020-05843-4

[4] V. Chettry and M. Surawar, Urban sprawl assessment in eight mid-sized Indian cities using RS and GIS, *Journal of the Indian Society of Remote Sensing*, vol. 49, no. 11, pp. 2721–2740, Nov. 2021, doi: 10.1007/s12524-021-01420-8

[5] K. Sundarakumar, M. Harika, S. K. Aspiya Begum, S. Yamini, and K. Balakrishna, "Land Use and Land Cover Change Detection and Urban Sprawl Analysis of Vijayawada City Using Multitemporal Landsat Data." Sundarakumar, K., Harika, M., Begum, S. A., Yamini, S., & Balakrishna, K. (2012). Land use and land cover change detection and urban sprawl analysis of Vijayawada city using multitemporal landsat data. *International Journal of Engineering Science and Technology*, vol. 4, no. 01, pp. 170–178.

[6] Dahiya, Peri-urban environments and community driven development: Chennai, Inda, *Cities*, vol. 20, no. 5, pp. 341–352, 2003, doi: 10.1016/S0264-2751(03)00051-9.

[7] A. Ohri and P. Yadav, Urban sprawl mapping and land use change detection using remote sensing and GIS thermal monitoring of Ganga river view project development of spatial decision support system for municipal solid waste management view project urban sprawl mapping and land use change detection using remote sensing and GIS, *International Journal of Remote Sensing and GIS*, vol. 1, no. 1, pp. 12–25, 2012, [Online]. Available: www.rpublishing.org

[8] "District Profile-2017 Chennai District." Accessed: Sep. 18, 2022. [Online]. Available: https://chennai.nic.in/about-district/district-profile/

[9] "Chennai District Executive Summary District Human Development Report Chennai District." Accessed: Sep. 18, 2022. [Online]. Available: https://spc.tn.gov.in/DHDR/Chennai.pdf

[10] H. Aithal and T. v. Ramachandra, Visualization of urban growth pattern in Chennai using geoinformatics and spatial metrics, *Journal of the Indian Society of Remote Sensing*, vol. 44, no. 4, pp. 617–633, Aug. 2016, doi: 10.1007/s12524-015-0482-0

[11] R. S. Defries and J. R. Townshend, NDVI-derived land cover classifications at a global scale, *Int J Remote Sens*, vol. 15, no. 17, pp. 3567–3586, 1994, doi: 10.1080/01431169408954345

[12] Congalton, R. G. (1991). A review of assessing the accuracy of classifications of remotely sensed data. *Remote Sensing of Environment*, *37*(1), 35–46.

[13] M. Jain, A. P. Dimri, and D. Niyogi, "Urban sprawl patterns and processes in Delhi from 1977 to 2014 based on remote sensing and spatial metrics approaches," *Earth Interactions*, vol. 20, 2016, doi: 10.1175/EI-D-15-0040.s1

[14] M. Jain, "Seamless Urbanisation and Knotted City Growth: Delhi Metropolitan Region Sridharan Namp School of Planning and Architecture," 2011. [Online]. Available: www.researchgate.net/publication/259895566

[15] H. Indorewala, S. Wagh, and U. Ramakrishnan, "City Résumé Mimbai," 2017. [Online]. Available: www.krvia.ac.in

[16] "The Chawls and Slums of Mumbai." Accessed: Sep. 18, 2022. [Online]. Available: https://deepblue.lib.umich.edu/bitstream/handle/2027.42/143823/A_12%20The%20Chawls%20and%20Slums%20of%20Mumbai.pdf

[17] Yedla, S. (2003). Urban environmental evolution: the case of Mumbai. *Dr. Indira Gandhi Institute of Development Research, Mumbai, India. Commissioned Report Prepared for Institute for Global Environmental Strategy (IGES), Japan.*

[18] A. Uttam and K. Roy, "Development of New Townships: A Catalyst in the growth of rural fringes of Kolkata Metropolitan Area (KMA)."

[19] S. Mukherjee, W. Bebermeier, and B. Schütt, An overview of the impacts of land use land cover changes (1980–2014) on urban water security of Kolkata, *Land*, vol. 7, no. 3. MDPI AG, Sep. 01, 2018. doi: 10.3390/land7030091

[20] U. Chatterjee, A. Biswas, J. Mukherjee, and S. Majumdar, *Advances in Urbanism, Smart Cities, and Sustainability*. CRC Press, 2022. doi: 10.1201/9781003126195

15 Geospatial Technologies for Groundwater Sensitive Urban Planning

Exploring the Status in Smart Cities of Rajasthan, India

Prerna Jasuja, Rina Surana, and Niruti Gupta

CONTENTS

DOI: 10.1201/9781003331001-15

15.1   INTRODUCTION

Sustainability calls for using resources in a manner that does not compromise quality and access to the resource for coming generations [1]. To achieve sustainability in human society, their needs should align with ecological integrity [2]. The concept was introduced through the Bruntland Commission in the 1990s. In the latter years of the 20th century, sustainability started getting linked to development or growth [2]. Striving for sustainability worldwide, the UN Sustainable Development Goals (SDGs) 2030 were introduced in 2015 [3]. Urban development must take place on principles of sustainability.

Water is an essential resource for human sustenance and ecology. With growing urban population and climate change, water faces a resource crunch, and it becomes essential to manage this resource, making water resource management (WRM) inevitable. WRM is necessary to ensure equitable access to and sustainable utilisation of this scarce resource. It is a key challenge in urban areas when the population increases and densifies at a rapid pace [4]. It is important to understand the nature and behaviour of water resources while dealing with WRM. Since the system of water, namely the hydrological cycle, is complex, and water changes various states spatially and temporally, WRM becomes difficult. Groundwater and surface water need to be managed for both qualitative and quantitative aspects. The management also needs to address the integrated nature of ground and surface water. It is important to consider both these resources in conjunction with sustainable physical development [5]. There are various concepts in which the integration of the management of land and water is discussed. Water-conscious land use planning, water-sensitive planning, wet growth, and integrated water resource management (IWRM) are some of these concepts [6–8]. Integrated planning for land and water has been recommended since the beginning of the 20th century [9] and urban land-use plans can allow this integration to be addressed. There are legal tools guiding urban development that can have a positive impact on both the quality and quantity of surface and groundwater resources and thus the quality of life in urban areas. This would require the collection of hydrological data, their processing and analysis, to finally arrive at informed decision-making within the urban planning framework.

Geoinformatics is the science and technology that deals with the acquisition, processing, analysis, and visualisation of spatial information. It is used in wide domains for resource management [10], environmental applications [11], disaster management [12], infrastructure development [13], and urban planning [14]. Satellite data information has increased over the years and has eased real-time information remotely, even in inaccessible and data-scarce areas. Rajasthan has semi-arid climatic conditions with scanty, erratic, and uneven distribution of rainfall. Thus, the cities of Rajasthan are highly dependent on groundwater resources. The utilisation of geospatial technologies for water resources and urban planning has been globally acknowledged but the integration of both has not been explored yet in the Indian context. This study explores the use of geoinformatics as a tool for decision-making in the management of groundwater resources for urban planning in the cities of Rajasthan, namely Ajmer, Jaipur, Kota, and Udaipur. The study provides insights into the current integration for groundwater-sensitive urban planning.

15.2 LITERATURE REVIEW ON GEOINFORMATICS TECHNOLOGY FOR GROUNDWATER RESOURCES AND URBAN PLANNING IN THE INDIAN CONTEXT

15.2.1 GEOINFORMATICS TECHNOLOGY AND ITS EMERGENCE IN INDIA

The tradition of cartography is not recent and dates back to Aryabhatta in the 6th century BC. During the British rule in India, surveying and map-making took a boost which also left an impact on the post-independent era. Traditional cartographic approaches shifted to aerial mapping with the launch of satellites [15]. India developed its Indian remote sensing (IRS) satellite program which was launched in 1988 [16]. Today, the Indian remote sensing satellite system has 53 satellites which include eight navigation satellites and 21 earth observation satellites [17].

Geoinformatics is generally used in analysing, modelling, managing, and storing diverse spatial information [15]. Since the beginning of the 21st century, its use has increased exponentially and it is seen as a tool to optimise the best use of resources and decisions [18]. It has a variety of components which include remote sensing (RS), photogrammetry, geographic information systems (GISs), and global positioning pystems (GPSs). GIS allows the layering of data on the same scale and eases the process of assessment by collating spatial data with attribute data. It takes both forms of data, namely raster and vector. Raster data are a form of grid data, whereas vector data are in the form of lines, points, and polygons. However, there are some limitations to geoinformatics and the data may require specialisation and expertise for correct interpretations. It also needs to be read in context to the time of data captured.

The National Centre of Geo-informatics (NCoG) under the Ministry of Electronics & Information Technology (Government of India) established a single-source GIS platform in 2015. Under this project, spatial data available from the central government such as land banks, mining, forests, water resources, industrial parks, etc. are collated with attribute-related data to provide information and support decision-making [19]. This centre caters to the requirements of central and state ministries and associated organisations across the country. The Indian Remote Sensing (IRS) satellite system has listed many applications of geoinformatics and promoted it in various national programs such as the GIS-Based Atal Mission for Rejuvenation and Urban Transformation scheme (AMRUT), Natural Resources data management system, Rajiv Gandhi National Drinking Water Mission (RGNDWM), Natural Resources Census (NRC), and National Spatial Data Infrastructure (NSDI). The Indian Space Research Organization (ISRO) also developed a geoportal Bhuvan in 2009 for nationwide data on various themes. It allows users to explore a set of thematic maps on various themes such as land use, groundwater, hydrology, water bodies, and basins to name a few. It allows 2D and 3D data visualisation and facilitates the creation of shapefiles for the user. Free and easy data dissemination to users through the Bhuvan portal can ensure better results in academia and organisations in the long run with improved e-governance. The development of such portals also supports the country's vision of 'Atmanirbhar Bharat' meaning a self-reliant India reducing its dependency on other countries for the development of GIS databases.

15.2.2 USE OF GEOINFORMATICS TECHNOLOGY FOR GROUNDWATER RESOURCES MANAGEMENT IN THE INDIAN CONTEXT

Groundwater is water beneath the Earth's surface. It is a replenishable but finite natural resource. Groundwater interacts with surface water, making integrated management of the resource necessary. It constitutes 30% of the world's freshwater and is an important resource. It supports a quarter of water withdrawal globally [20]. As per the United Nations World Water Development Report, India is the prime user of groundwater globally [21]. The country depends on groundwater for 85% of its rural domestic water demands, 50% of its urban water demands, and more than 50% of its irrigation requirements [22]. The quality and availability of this resource varies spatially mainly due to geological formations, physiographical, hydrological, climatic conditions, and human interventions through land use. The assessment of groundwater is necessary for achieving resource protection and sustainability. Both quantity and quality are required to be monitored for resource management. In the process of assessment, data management of varied domains is required that can be performed using GIS. After capturing data, various types of groundwater modeling based on quantitative and qualitative aspects can be performed [23,24].

For groundwater assessment, hydrogeology is an important component. Hydrogeology is assessed separately by groundwater departments through direct and indirect methods. Direct methods require field reconnaissance or drilling, while indirect methods are based on remote sensing. Digital Elevation Model (DEM) data captured through remote sensing give structure and lithology that are necessary for hydrogeological studies which help to locate groundwater [25]. Thus, remote sensing data provide baseline information on the presence and movement of groundwater. There is a need to study the hydrogeological situation in each area to assess the recharge capabilities of the underlying hydrogeological formations. Remote sensing along with GIS helps in locating lineaments which are the surface expression of fractures. These help in the identification of the groundwater potential zones and further locating wells for groundwater draft. It is suggested to validate data captured through remote sensing with ground-based data [26]. Thus, geoinformatics along with field investigation can help to save time and unnecessary drilling.

15.2.2.1 Geoinformatics as a Tool to Prospect and Disseminate Groundwater Data

One of the listed applications by the Indian Space Research Organization (ISRO) of geoinformatics in India is groundwater prospect maps. These maps give information on groundwater occurrence and suitable locations for constructing recharge structures [27]. The maps were prepared under the RGNDWM which is now handled under the Department of Drinking Water and Sanitation, Ministry of Drinking Water and Sanitation (MDWS), Government of India (GoI). The maps are available on a scale of 1:50,000 for the entire country. To comprehend these maps the expertise of a hydrogeologist is required. The scheme was launched for scientific source finding of drinking water for all habitation areas facing drinking water shortages. In this scheme, high-resolution satellite data along with ground data were utilised to prepare the maps. The data related to groundwater quality and water level of the wells monitored was collected from the Central Groundwater Board (CGWB) and State Groundwater Boards. Initially, under the RGNDWM, it was envisaged to produce these maps for rural areas only but the non-availability of these maps in various states resulted in a scientific database of groundwater prospects covering the whole nation [28]. These maps were prepared under the schemes by ISRO along with State Remote Sensing Centres (SRSC) [22].

These maps were prepared after an analysis of a variety of information such as lithology, geological structures, hydro-geomorphology, land use/land cover, etc. These maps give an insight into the geological parameters for groundwater exploration, namely lithology, geomorphology, geological structures, groundwater prospects, and probable depth and yield range of wells. Detailed hydrogeological and geophysical surveys can be taken up by hydrogeologists for aquifer mapping, enabling groundwater development, and planning recharge structures. The open-source Bhuvan portal also provides limited information to the user on the 1:50,000 scale separately on geomorphology, lineaments, lithology, etc., while authorised users such as hydrogeologists of groundwater departments can access the data with multiple layers. The data provided in the maps to the hydrogeologists include hydro-geomorphic units, their influence, and well inventory of yield, water table level, and quality of drinking water. The data can also assist in developing models. The accuracy and ease to disseminate data at the district or block level increases if the database is managed over GIS. Thus, it enables better availability of data, giving an opportunity to plan and manage the resource in better ways [29].

15.2.3   Use of Geoinformatics Technology in Urban Plans in the Indian Context

India is going through rapid urbanisation, with urban areas facing many challenges such as housing, infrastructure, urban services, water shortage, transportation system, and waterlogging. Urban plans or master plans guide the development of the city while considering the above challenges. These are legal tools that shape the decisions on the surface which have consequences for the coming decades. The process of urban planning alters the distribution of environmental processes and requires impact assessment. For this, geoinformatics can be used to study land use plans, social infrastructure, natural resource vulnerability, urban morphology, community planning, monitoring urban sprawl, monitoring of slums, transportation systems, catchment area analysis, green space, and the study of open or vacant land. It can highlight the chronological assessment of the above and help in making water-sensitive plans for the city.

Temporal data retrieved through satellites can help in examining earlier growth patterns to shape future decisions for better planning. In India, the sub-scheme for 500 cities was selected under AMRUT cities for the preparation of GIS-based master plans. Before GIS, AutoCAD software was used for making master plans. This scheme formally created a uniform spatial database using GIS for the selected cities. Under this centrally sponsored scheme, the plan-making process eased with the integration of both spatial and attribute data, which was a limitation of AutoCAD. This allows a thorough and more accurate assessment of urban growth, land use status, and infrastructure facilities for projected populations [30]. The format for the master plans prepared under the scheme covers urban utilities along with the land use plan. Thus, geoinformatics aid in improving plans through both the creation of a database and an analysis platform.

15.3 STUDY REGION: CITIES IN RAJASTHAN, INDIA

15.3.1 INTRODUCTION TO RAJASTHAN

Today, many temperate and tropical regions face the issue of water scarcity, while the situation is quite alarming in arid and semi-arid regions. Rajasthan state lies in the northwestern part of India, as shown in Figure 15.1. The state has a geographical area of 340,000 km², which is 10.4% of the country's total area. The population of the state is 68.5 million [31], which is 5.6% of the country's total population. Administratively, the state is divided into 33 districts. The state of urbanisation in Rajasthan according to the 2011 census is 24.8%. The state is expected to urbanise faster in future due to proposals of various schemes for the state by the Government of Rajasthan (GoR), and proposed national projects such as the Delhi–Mumbai industrial corridor, Dedicated Freight Corridor initiative, and the tourism sector [32].

15.3.2 CLIMATE

The major part of the state has an arid and semi-arid climate, while the rest in the southwest has a humid climate. It receives an average annual rainfall of 415 mm,

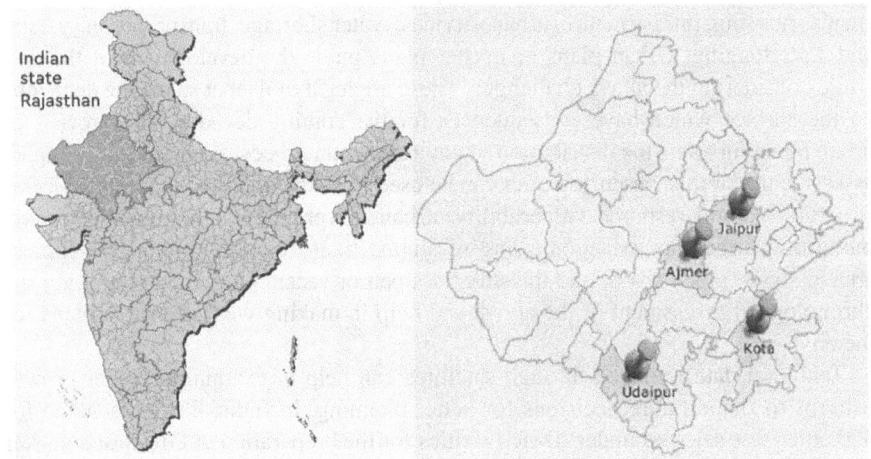

FIGURE 15.1 Location of Rajasthan state in India (left) and selected cities for the study (right).

which is lower than the national average rainfall of 1180 mm for the entire country [33]. The annual average variation of rainfall within the state ranges from 150 mm to 1500 mm. The rainfall is not evenly distributed and is concentrated from May to July. The average annual temperature is 33°C in the region, rising to 48°C in May and June.

15.3.3 PHYSIOGRAPHY AND BASINS

The Rajasthan plain slopes towards the west and south with a variation in elevation from 150 m to 900 m, as shown in Figure 15.2 [34]. Physiographically, Rajasthan is divided into four major regions: the western sandy plains, the Aravalli range with Vindhyan mountains in the middle portion, the eastern plains, and the south-eastern Hadoti Plateau.

The state is divided into 14 basins, as shown in Figure 15.3, of which the prominent six basins are Luni, Banas, Chambal, Mahi, Banganga, and Sabarmati.

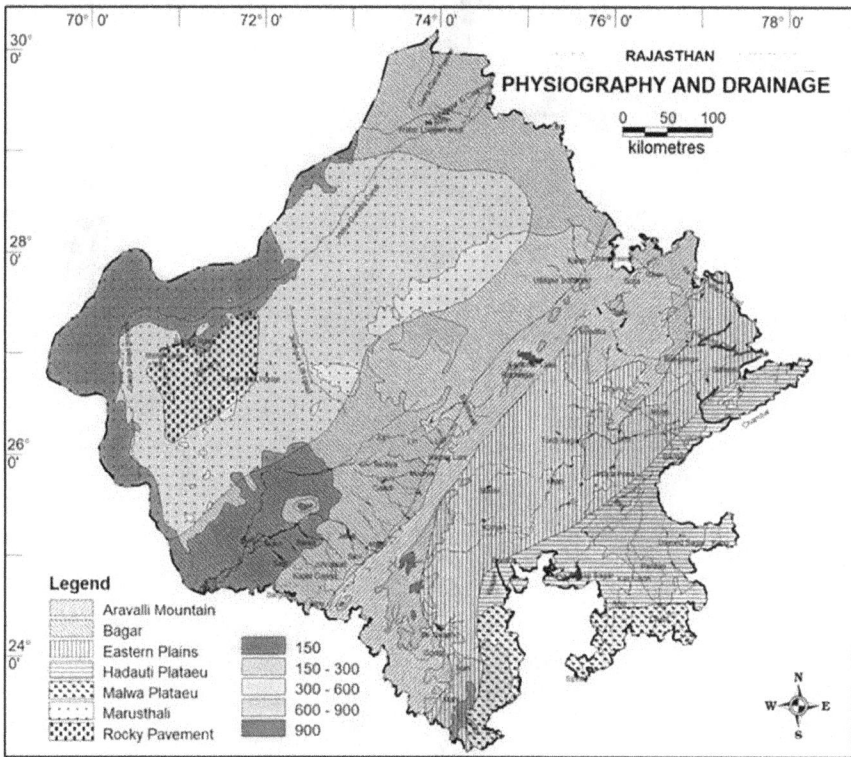

FIGURE 15.2 The physiography and drainage pap of Rajasthan State.

Source: [34].

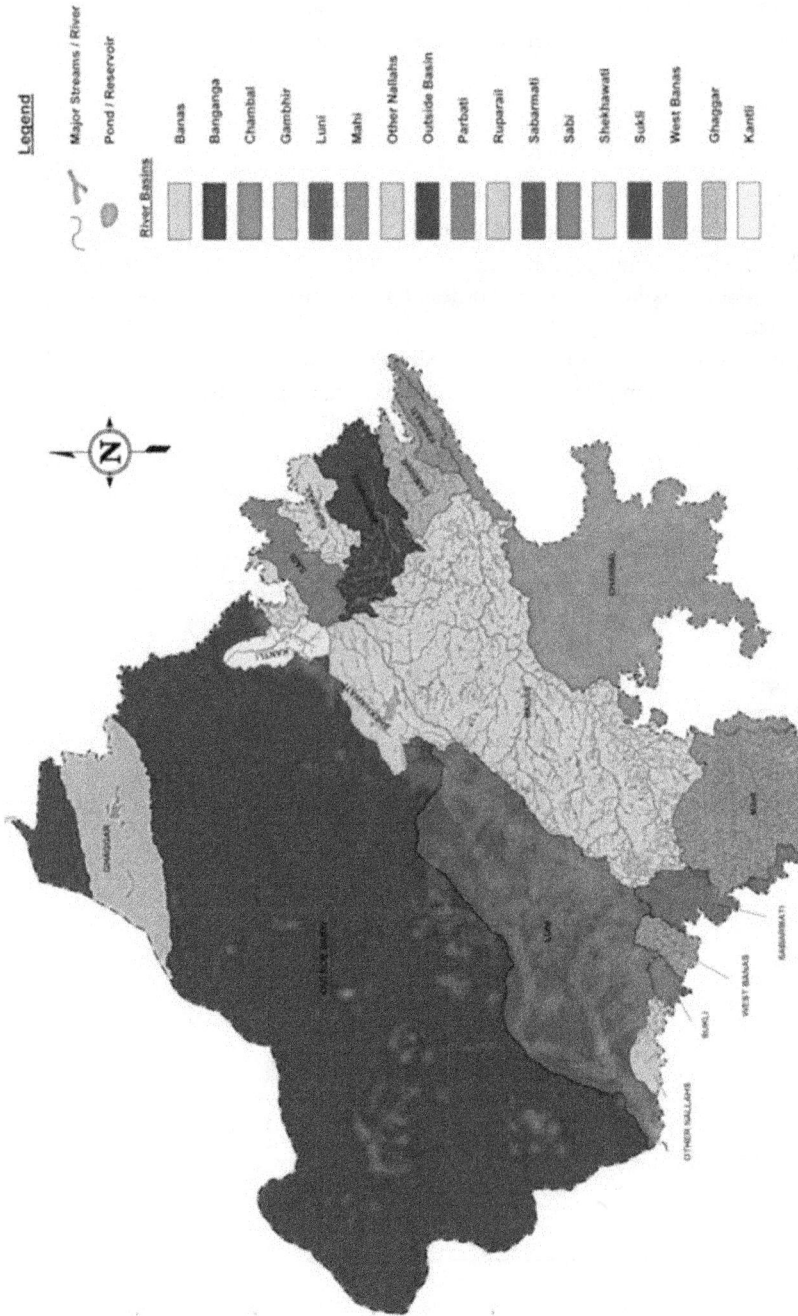

FIGURE 15.3   Drainage with river basins of Rajasthan state.

Source: [34].

15.3.4 Water Resources in Rajasthan

The arid climate with scanty, unevenly distributed, and erratic rainfall and high temperature results in high evaporative losses and ultimately substantial loss of scarce surface water resources. The state has limited surface water resources, with only one perennial river, Chambal, confined to the southeast part of the state. The Banas River supports the Hadoti region for its urban and rural water needs. The Indra Gandhi Canal supports the drinking and irrigation water needs of the state's northern districts. Traditionally, rainwater harvesting was extensively done at community levels. But in modern times, these structures have been neglected or destroyed, especially in urban areas. The Bisalpur dam on the Banas River within the central part of the state supplies water demands to the districts of Jaipur and Ajmer. Limited surface water resources have resulted in a high dependency and reliability on groundwater in the state [22]. The annual per capita water availability in the region is 640 cubic meters, which is low compared with the international standard requirement of 1000 cubic meters [35]. This availability is also continuously decreasing due to the rapid growth of the population, making water scarcity worse.

15.3.4.1 Groundwater Resources in Rajasthan

15.3.4.1.1 Hydrogeology of the State

The hydrogeology of a region determines the occurrence, distribution, and movement of groundwater. The state has heterogeneity in the hydrogeological formations, as shown in Figure 15.4. The state can be divided into three types of hydrogeological units, namely, unconsolidated sediments, semi-consolidated sediments, and consolidated rocks. These units have variations in groundwater potential. The unconsolidated sediments include older and younger alluvium hydrogeological formations and yield moderate to good quantities of groundwater. Semi-consolidated sediments include sandstone, shale, and limestone and these hydrogeological formations yield less water than unconsolidated sediments. Consolidated rocks such as slate, basalt, and granite, are nonporous and are natural barriers to groundwater recharge [36].

15.3.4.1.2 Availability and Dependency on Groundwater Resource

The state has 1.72% of the nation's groundwater, which translates to 11.36 billion cubic metres. According to the report published in 2014 by the state water resource planning department under the title 'State of planning of water resources of Rajasthan', 80% of drinking water and 60% of irrigation needs of the state are met through groundwater [22]. Because of the high reliance on groundwater, the gross annual water table depletion rate in the state is 1–3 meters [35]. According to the CGWB western region 2019, the stage of groundwater extraction (annual groundwater extraction/total annual extractable groundwater resource) in the state is 140% [34]. High dependency on groundwater has resulted in the overexploitation of groundwater. According to data from CGWB of 2017, for groundwater abstraction, Rajasthan has exploited groundwater resources heavily compared to other parts of the nation, as shown in Figure 15.5. Out of 295 assessed blocks of the state in 2017, 185 blocks have been

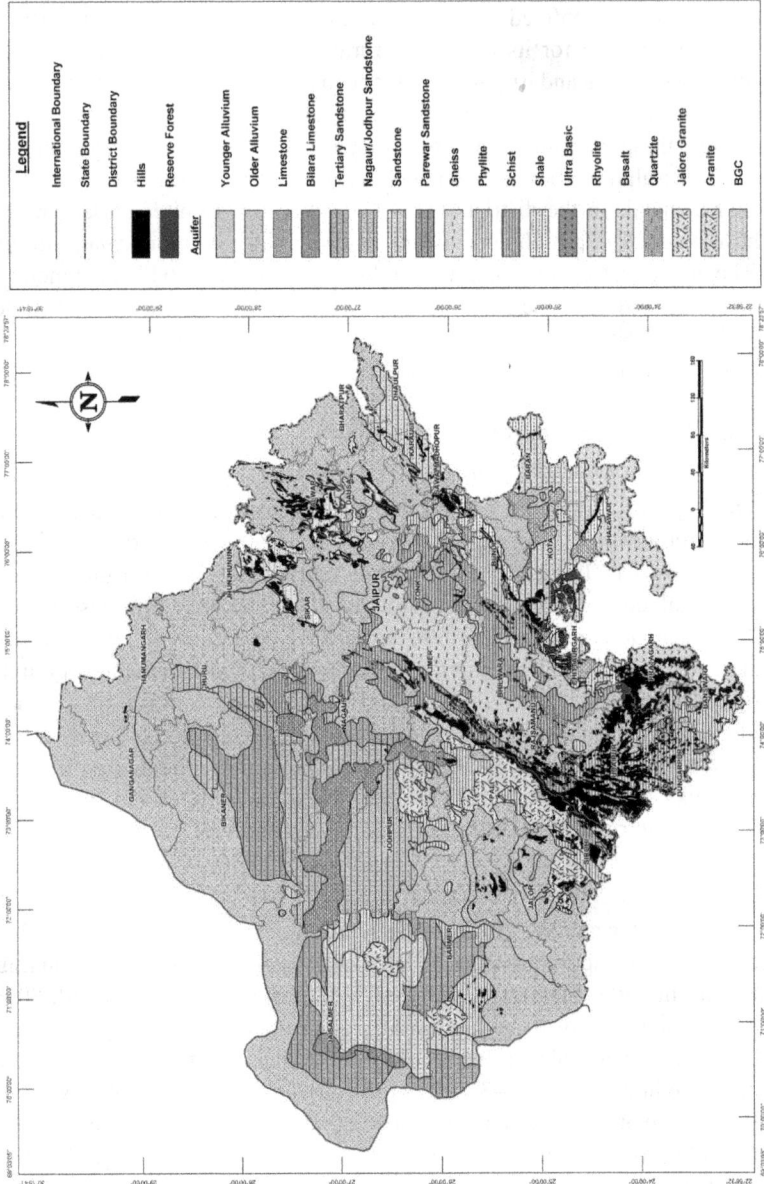

FIGURE 15.4 Types of aquifers in Rajasthan.

Source: [34].

CATEGORIZATION OF ASSESSMENT UNITS (AS IN MARCH 2017)

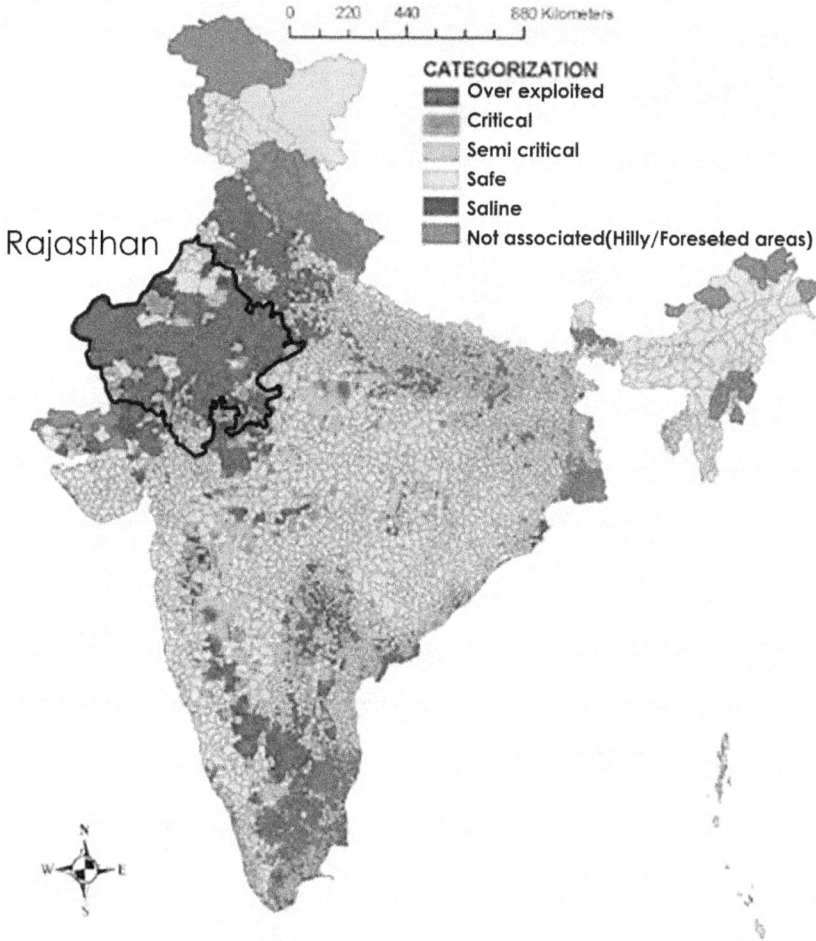

FIGURE 15.5 Category of abstraction of Rajasthan state in comparison to India.

Source: [34].

categorised as 'Over Exploited' while 33 fall in the 'Critical' and 29 in the 'Semi-Critical' category. The study shows that only 45 blocks are in the 'Safe' category and have the potential for development for future water demand [37]. In the northern blocks of the state, irrigated by the Indra Gandhi Canal system, the safe category is seen for abstraction as water from irrigation seeps into the recharge suitable strata.

Three blocks are not fit for usage because of salinity. The state faces quality issues of fluoride, nitrate, and salinity in the groundwater. The issue of fluoride contamination is found in Jaipur, Tonk, Ajmer, Bhilwara, Nagaur, and Sirohi districts. The presence of nitrate has been recorded in groundwater in Churu, Nagaur, and Jhunjunu districts. The presence of fluoride and nitrate in groundwater is mainly

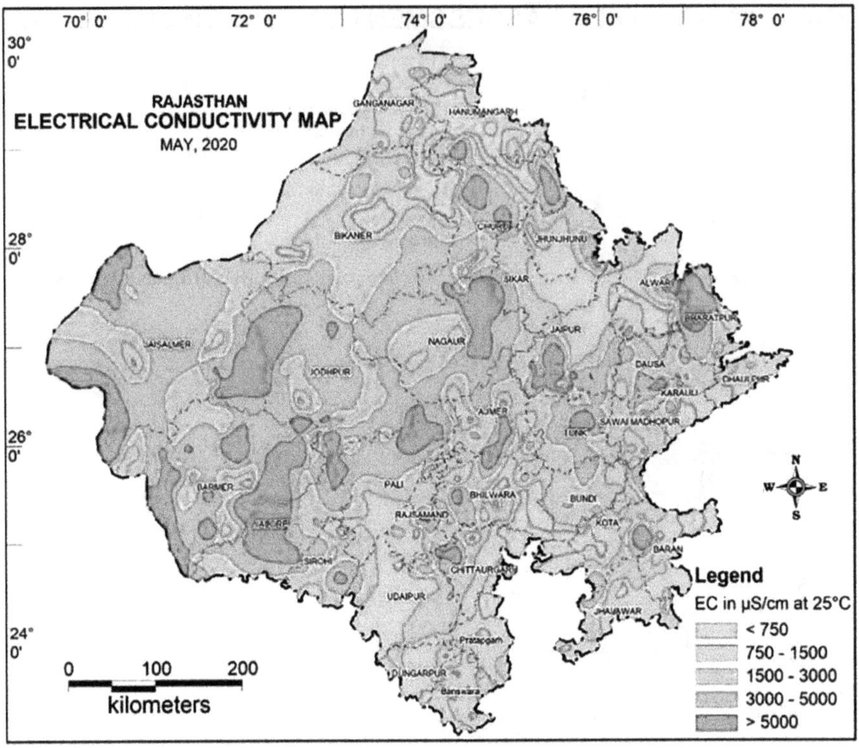

FIGURE 15.6   Electrical conductivity map for Rajasthan state.

Source: [34].

due to geogenic reasons [38]. Anthropogenic reasons for pollution have also been recorded in the state because of urbanisation and untreated sewage discharge into surface waterbodies and lakes as well as contamination as a result of pesticides and fertilisers in irrigated areas.

Electrical conductivity is associated with dissolved ions and is a prime indicator of salinity. It can be used for the assessment of the quality of groundwater. The electrical conductivity of potable water is between 0 and 1500 uS/cm. The value varies from 250 to 47,670 uS/cm within the state, as shown in Figure 15.6.

15.4   METHODOLOGY

A qualitative research approach was adopted for the study to understand under-lying opinions and reasons [39]. Exploratory research was conducted using semi-structured interviews through in-person meetings and telephonic communication [40]. The master plans, smart city proposals, and groundwater yearbooks issued by central and state groundwater boards at the district level of the four cities were studied. The authors also interviewed experts from the central groundwater board of the western region and the town planning department of Jaipur. This was done to gain an in-depth knowledge about ground reality of the utilisation of geospatial technologies for the integration of groundwater into urban planning. These cities were chosen by the authors due to the differences in spatial groundwater prospects and easy data availability. The information to integrate groundwater into urban planning were identified after an interview with two senior hydrogeologists from the central groundwater board in Jaipur. To gather information on the extent of integration of groundwater into urban planning, the retired chief town planner of Rajasthan and senior town planner were selected as experts. One official who works in the GIS department at the Jaipur Development Authority (JDA) was interviewed for gathering information on the utilisation of GIS and remote sensing data in urban planning at the state level.

15.5 FINDINGS

The selected four cities within Rajasthan, namely Ajmer, Kota, Jaipur, and Udaipur, lie in different hydrogeological, and climatic conditions and have different scenarios regarding groundwater quality and availability, as listed in Table 15.1. The senior hydrogeologists suggested use of hydrogeological information such as aquifer deposition, exploitation status, depth to groundwater table, and water quality conditions along with respecting the natural drainage channels and surface water bodies on the surface. The GIS database of suggested information is available within the department and can be accessed when required. The provision of wastewater infrastructure with space allocation for sewage treatment plants under a master plan for preserving groundwater quality was also suggested by the expert.

The experts indicated that the identification of possible water sources for the estimated population for the horizon year is done by the Public Health and Engineering Department (PHED) and the Groundwater Board for the preparation of the master plan. The allocation of land use and water source availability should be by the abstraction category of a block of groundwater. Primarily, this was explained by the expert through the instance of industrial land use allocation. The organisation Rajasthan State Industrial Development & Investment Corporation Ltd (RIICO) plays an important role in the industrial development within the state. It deals with site selection for industrial use, acquisition of land, financial assistance to the developer, government clearances, and concessions according to the state policy and department of industries [41]. The allocation of industries by RIICO in blocks that are overexploited was considered a setback for the already declining water tables of the state. The integration of development authorities and groundwater departments is required for the allocation of land use.

A suggestion to reduce the water standard of 135 LPCD was also provided by the expert to reduce the extraction of groundwater and check the depleting water table. The revision of the present policy of rooftop rainwater harvesting from a 300 sq. m plot to 162 sq. m is in the pipeline but the study/understanding of the hydrogeological conditions beneath has not been taken into account while formation of landuse plan. The recharge conditions are not uniformly available throughout the area, which is a key issue that requires the role of a hydrogeologist for preparation of the land use plan. The recharge conditions within the four different cities and within each city vary. Groundwater prospect maps could be utilised for the recharge conditions and suitability of recharge structures, but the information needs to be detailed at city level.

TABLE 15.1
Groundwater-Related Aspects of the Selected Cities for the Study

City	Rainfall average in mm (2011–2020)	Hydrogeology at the district level (refer to Figure 15.4)	Conditions for recharge are available through maps and the CGWB yearbooks Depth to the water table	Groundwater quality	Category of abstraction within the district (2017 data)
Ajmer	546	Younger alluvium, schist, gneiss, and hills	Suitable for recharge towards the north and south parts of the district with the younger alluvium zone	Salinity issues are present within the district	Over-exploited
Jaipur	566	Younger alluvium, older alluvium, gneiss, schist, quartzite, granite	Suitable for recharge	Salinity issues are present within the district	Over-exploited
Kota	861	Alluvium, limestone, sandstone, and shale	Suitable for recharge – younger alluvium in the northern part of the district	The majority of the district has water within the safe limit	Safe and semi-critical
Udaipur	735	Phyllite, hills	Not suitable for recharge	The majority of the district has water within the safe limit	Critical and safe

Sources: [34,37].

The hydrogeological studies at city level would further assist in robust data for decision making within urban plans for planning green over recharge areas. Broadly, the expert suggested that the sub-surface strata of Jaipur is suitable for recharge in comparison to the other cities Ajmer, Udaipur, and Kota. The allocation of green spaces or recreational land use over recharge areas can ensure the recharge of groundwater. When asked about the help that the geospatial technology has provided the expert answered that remote sensing and GIS have helped in creating an easily accessible and scientific database of the wells of the state and central groundwater authority over different geological conditions. Earlier, the maps and data were available in groundwater yearbooks but were not organised on geoportal and GIS platforms. The data of the central groundwater board and state groundwater board have also been put on one platform. Before this, the data were not readily available or organised.

The town planners from the Town Planning Department of Jaipur were selected to gather information on the utilisation of geospatial technologies for urban planning and for checking the integration of information suggested by experts from the groundwater board into master plans.

One official was interviewed for gathering information on the utilisation of GIS and remote sensing data in urban planning at the state level. The expert who works in the GIS department at Jaipur Development Authority (JDA) was interviewed for gathering information on the utilisation of GIS in the department. The official indicated that the consultancy-based work is carried out through contracts for the preparation of master plans on GIS for various cities of Rajasthan. Jaipur, Kota, and Udaipur were selected under the AMRUT scheme, but the master plan of Ajmer has also been developed on GIS. The utilisation of GIS is majorly done for mapping at present and not for analysis within the departments. The imagery utilised for mapping is obtained through drones or National Remote Sensing Centre (NRSC) for digitisation. A uniform database creation using GIS in the planning departments of Rajasthan is seen, but presently the analysis for data on various aspects of planning is not being undertaken. Full exploitation of the GIS with overlay analysis on various aspects such as land use, urban sprawl, slums, transportation systems, and green space, can ensure better master plans of the cities. Analysis can be done with GIS for checking land suitability wherein vulnerable natural resources such as water can be taken into account to decide or allocate land uses in urban areas.

To gather information on the extent of integration of groundwater into urban planning the retired chief town planner of Rajasthan and senior town planner were selected as experts. Both these experts had wide experience and thorough knowledge of the master plan process of the selected cities for the study, as shown in Table 15.2.

When asked for the parameters by the experts, the senior town planner and former chief town planner from the town planning department were suggested by the senior hydrogeologist of the central groundwater board, and the expert asserted that hydrogeological information of aquifer disposition parameters were not considered. During the meetings during master plan formulation, only broad discussions regarding water demand for the projected population took place. The departments involved in the discussion included officials from PHED, the Groundwater Board, and the Irrigation Department. The arrangements for water supply were checked for the projected population according to the horizon year. The expert at the position of senior town planner

TABLE 15.2
Various Aspects of the Selected Cities for the Study

City	Authority responsible for masterplan formulation	The latest master plan formulated by the authority on GIS
Ajmer	Ajmer Development Authority (ADA)	Ajmer master plan 2033
Jaipur	Jaipur Development Authority (JDA)	Jaipur master plan 2025
Kota	Kota Development Authority (KDA)	Kota master plan 2031
Udaipur	Udaipur Development Authority (UDA, formerly an urban improvement trust, Udaipur was looking into the master plan formulation till Feb 2022)	Udaipur master plan 2031

suggested the formation of separate water master plans and identification of projects within the master plan with the responsibility of departments that can ensure this integration. The experts suggested that industrial land use is required for the generation of employment in cities so to cater to the needs of the population and that water-intensive industries are not promoted in the region. The space allocation for sewage treatment plants and effluent treatment plants for polluting industries was checked through landuse under utilities. The wastewater infrastructure was checked by municipalities and land allocation was done by urban development authorities or urban improvement trusts. However, the monitoring aspect and implementation of schemes were not followed up and sometimes such projects were delayed because of a lack of funds and environmental considerations. Both experts admitted that the recharge policy of groundwater is not monitored. Broadly the rainwater harvesting policy of the state targets recharge or storage at the site for 300 sq. m. plot. But it is important to note that all the site are not suitable for recharge. The areas particularly suitable for recharge or storage has not been designated within the policy. The submission of a fee against the setup of rooftop rainwater harvesting allows escaping the arrangements for rooftop harvesting at the site.

While studing the master plan reports of the four cities it was found that only one city, Jaipur, had utilised GIS for land suitability analysis, and the other three had merely utilised GIS for mapping but did not perform any suitability analysis [39,40,42]. The map was developed based on the study of various geo-factors such as geology, hydrology, groundwater conditions, soil conditions, and environmental geohazards.

For the city of Jaipur, injection lines or recharge points have been identified to increase the water table by Jaipur Development Authority (JDA) with the help of the Geological Survey of India [43,44]. The major locations of these injection lines in the Jaipur region were mentioned at the regional level. While the suitability analysis has been found in the report for an existing profile, the master plan proposal report did not mention the planning of recreational land use based on land suitability analysis.

The ideal land use recommended for lineament zones within the master plan is the groundwater recharge sites and farmhouses, open recreation spaces, etc. [43]. The integration of the data for the suitability analysis is appreciable but it has not

TABLE 15.3
Various Aspects of the Selected Cities for the Study

City	Groundwater development stage	Historic source/ Groundwater reliance	Current source to meet water demands	Water resources identified for the future in the master plans	Recent interventions for groundwater recharge under the Smart City scheme
Ajmer	Over-exploited	Yes, from Bhaonta well field system (till 1990s)	Bisalpur dam and ground water	Isarda dam proposed along with dependency on existing resources	Installation of rain water harvesting system in all the public buildings and in open areas, recharge pits and bioswales in open spaces, parking areas, parks, and road median
Jaipur	Over-exploited	Groundwater and Ramgarh lake	Bisalpur dam and ground water	Isarda dam proposed along with dependency on existing resources	–
Kota	Critical	Surface water resource – Chambal from the 1960s	Kota barrage dam on Chambal River	Kota barrage dam on Chambal river	–
Udaipur	Critical	Lakes within the city	Lake Pichola	Anas river	–

Sources: [22, 42–46,48].

been mentioned in the proposal document. Thus, the available technology was used to generate knowledge but did not translate into urban planning decisions. The master plan reports of the three cities, namely Ajmer, Udaipur, and Kota, do not mention the hydrogeology, aquifer, or depth to water table aspects which were suggested by the hydrogeologists from CGWB [42–46]. The smart city scheme had several proposals, as mentioned in Table 15.3, for groundwater recharge in the case of Ajmer but the available hydrogeological information could have been better integrated in the scheme. The integration of data from the groundwater department has not been followed up [47]. The urban plans, policies, and schemes should have the integration of the data from various departments to guide decisions but this is not being undertaken.

15.6 DISCUSSION AND IMPLICATIONS

The utilisation of GIS and remote sensing for urban planning has not been fully exploited to assess urban issues within the area of the four cities and has been majorly limited to digitisation or mapping. The information of suitability analysis performed if utilised for the Jaipur master plan could have helped guide better decision-making.

Through this study, it is evident that the usage of geospatial technologies has been done in India for groundwater assessment and to prepare a database for the groundwater board departments. The attempts to capture groundwater information have increased although there has been less action on integrating the information into the land planning process. In this case, it was seen that the data were publicised on the Bhuvan portal but the importance of utilisation of the same has not been fully understood by urban and regional planners. The interview results show that the hydrogeological information needs better integration within the urban plans [28]. The other problem that needs to be addressed by planners is the complexity of understanding the information of hydrogeology. The involvement of hydrogeologists while making master plans can ensure greater sustainability of the resource.

While at the national level it has been directed in the report 'Reforms in urban planning capacity in India' that urban planning needs to address the natural drainage system, surface, and groundwater bodies with urban planning to ensure sustainable urban development, such measures were not found during implementation [49]. The national water framework bill also addresses the importance of scientific planning of land and water resources, considering catchments for holistic development [50]. Addressing this could also help urban areas overcome the issues of water shortage and water logging in one go.

Water is a state subject in India. Thus, water management, including its quality, is primarily the state's responsibility, but the central government has also taken steps to control water contamination in the country. For groundwater, there is a body at the national level, namely the Central Groundwater Board. State groundwater boards have also been set up that handle district-level groundwater boards under them. Groundwater governance is tough as the resource is not visible and is associated with the land parcel. Urban planning checks the land use on the parcels. However, after the 74th constitutional amendment, the responsibilities for urban planning lie in the hands of urban local bodies theoretically. In reality, however, urban planning falls in the domain of development authorities or town and country planning organisations

[32]. These institutions are required to align the decisions based on the hydrological process of water which demands the integration of the departments. The management of both the quality and quantity of groundwater resources together by departments would ensure resource sustainability. The State Water Policy of Rajasthan and The Model Groundwater (Sustainable Management) Bill, 2017, also suggest planning for checking water demand and supply balance and groundwater depletion and quality deterioration. The cities studied had also faced the issues of waterlogging which is a result of poor planning by not considering available geohydrology, topography, and recharge areas. There have been attempts to check the same in research in the USA, Denmark, and Canada through the guidelines and involvement of the planning and water departments [51,52]. There has been progress in some cities of India for groundwater sensitive urban planning. For instance in Pune collaboration of planning department with Advanced Center for Water Resources Development and Management (ACWADAM) has resulted in preservation of recharge areas in future expansion. The decisions can be made at the regional level and can be followed up at the urban scale.

There is an urgent need to look into the same as both depletion and contamination would render the water unfit for consumption and irrigation, and can damage the aquifer, ultimately affecting surface waterbodies, rivers, and wetlands and disrupting ecological dynamics. While the impact on the quality and quantity of surface waterbodies could be observed visually and through odor, the impact cannot be seen for both aspects in the case of groundwater. However, in the long run, the water dependency on polluted water would be harmful to ecological integrity and human needs.

The perspective of integration needs reform globally. An assessment of the present SDG 6 from a groundwater perspective shows that the present indicators under SDG 6 talk about water resources, but groundwater resource sustainability is not addressed. A region has three options to meet its water demands: dependency on existing resources, importing water from water surplus regions, and developing new resources. The development of resources through recharge and quality protection can assure water security within the region. This, in the long run, can also help cut the transportation of water from a distant source.

15.7   CONCLUSIONS

Geospatial technologies have helped in data collection and management for urban planning and the groundwater domain. This study has found that the collection of data required in both domains has been eased with the introduction of technologies, but the integration of the existing and extensive data of groundwater in land use planning is sorely missing. Improved governance and meetings of both departments during the plan formulation process can help in realising the sustainability of groundwater as a resource. The transition to recognise groundwater as a key component in the urban planning process is necessary to ensure water security for the future. The state policy for groundwater recharge in Rajasthan should be modified on the same grounds considering the spatial variation in the hydrogeological formations. Geoinformatics is a tool to monitor and manage resources judiciously. Although deploying technology does not ensure the concepts of sustainability, as seen in this case, technology can be seen as a tool to realise the concept become a reality. It should be seen as an aid to realise the vision within the urban plans to address the intertwined nature through improved governance. The policies and decisions between various departments should be in one direction through organised discussions and meetings. There should be a clear division of responsibilities for interdisciplinary cooperation that would help in groundwater-sensitive urban planning to achieve sustainable development. The guidelines or GIS-based tools could be explored for the decision-making process to ease this integration.

REFERENCES

[1] C. A. Ruggerio, Sustainability and sustainable development: A review of principles and definitions, *Sci. Total Environ.*, vol. 786, p. 147481, 2021, doi: 10.1016/j.scitotenv.2021.147481

[2] A. Heymans, J. Breadsell, G. M. Morrison, J. J. Byrne, and C. Eon, Ecological urban planning and design: A systematic literature review, *Sustainability (Switzerland)*, vol. 11, no. 13. MDPI AG, Jul. 01, 2019, doi: 10.3390/su11133723

[3] UN, "The Sustainable Development Agenda," 2015. www.un.org/sustainabledevelopment/development-agenda-retired/.

[4] T. Scholten, T. Hartmann, and T. Spit, The spatial component of integrative water resources management : differentiating integration of land and water governance, *Int. J. Water Resour. Dev.*, vol. 36, no. 5, pp. 800–817, 2020, doi: 10.1080/07900627.2019.1566055

[5] S. Foster, Is UN Sustainable Development Goal 15 relevant to governing the intimate land-use/groundwater linkage?, *Hydrogeol. J.*, vol. 26, no. 4, pp. 979–982, Jun. 2018, doi: 10.1007/s10040-018-1782-6

[6] E. Li, S. Li, and J. Endter-Wada, Water-smart growth planning: linking water and land in the arid urbanizing American West, *J. Environ. Plan. Manag.*, vol. 60, no. 6, pp. 1056–1072, 2017, doi: 10.1080/09640568.2016.1197106

[7] C. T. Arnold, Is wet growth smarter than smart growth?: The fragmentation and integration of land use and water, *Environ. Law Report.*, vol. 35, no. 3, pp. 10152–10178, 2005.

[8] V. Pusalkar, V. Swamy, and A. Shivapur, Future city – challenges and opportunities for water-sensitive sustainable cities, in India, in *E3S Web of Conferences*, May 2020, vol. 170, doi: 10.1051/e3sconf/202017006017

[9] M. R. España-Villanueva and L. M. Valenzuela-Montes, Criteria for assessing the level of land–water integration in planning instruments in Andalusia, Spain, *Water Int.*, vol. 41, no. 5, pp. 716–737, Jul. 2016, doi: 10.1080/02508060.2016.1167477

[10] J. Sheffield et al., Satellite remote sensing for water resources management: Potential for supporting sustainable development in data-poor regions, *Water Resour. Res.*, vol. 54, no. 12, pp. 9724–9758, 2018, doi: 10.1029/2017WR022437

[11] K. Gharehbaghi and C. Scott-Young, GIS as a vital tool for environmental impact assessment and mitigation, *IOP Conf. Ser. Earth Environ. Sci.*, vol. 127, no. 1, 2018, doi: 10.1088/1755-1315/127/1/012009

[12] R. M. Teeuw, M. Leidig, C. Saunders, and N. Morris, Free or low-cost geoinformatics for disaster management: Uses and availability issues, *Environ. Hazards*, vol. 12, no. 2, pp. 112–131, 2013, doi: 10.1080/17477891.2012.706214

[13] P. Persai and S. K. Katiyar, Development of information evaluation system for smart city planning using geoinformatics techniques, *J. Indian Soc. Remote Sens.*, vol. 46, no. 11, pp. 1881–1891, Nov. 2018, doi: 10.1007/s12524-018-0844-5

[14] P. Kumar Rai and V. K. Kumra, Role of geoinformatics in urban planning, *J. Sci. Res.*, vol. 55, pp. 11–24, 2011.

[15] S. Gupta, H. Karnataka, and P. L. N. Raju, Geo-informatics in India: Major milestones and present scenario, *Int. Arch. Photogramm. Remote Sens. Spat. Inf. Sci.–ISPRS Arch.*, vol. 41, no. July, pp. 111–121, 2016, doi: 10.5194/isprsarchives-XLI-B6-111-2016

[16] Department of Space, "The Saga of Indian Remote Sensing Satellite System," 2014. www.isro.gov.in/saga-of-indian-remote-sensing-satellite-system.

[17] Department of Space, "Union Minister Dr. Jitendra Singh says, ISRO has launched a total of 129 satellites of Indian Origin and 342 foreign satellites belonging to 36 countries since 1975," 2022. https://pib.gov.in/PressReleasePage.aspx?PRID=1797 196 (accessed Apr. 22, 2022).

[18] S. Burch, "Accelerating a Just Transition to Smart , Sustainable Cities," no. 2021, 2022.

[19] NCoG, "National Center of Geoinformatics," 2015. https://ncog.gov.in/ (accessed Apr. 22, 2022).

[20] T. Shah, A. D. Roy, A. S. Qureshi, and J. Wang, Sustaining Asia's groundwater boom: An overview of issues and evidence, *Nat. Resour. Forum*, vol. 27, no. 2, pp. 130–141, 2003, doi: 10.1111/1477-8947.00048

[21] UN, *The United Nations World Water Development Report 2022-Making the invisible visible*, vol. 19, no. 3. Paris, 2022.

[22] Water Resource Department, "Final Report No . 4 . 4 Ground Water Study by Agroclimatic Zones Main Report State of Planning of water resources," no. 4, 2014.

[23] C. Rajanayaka, J. Weir, T. Kerr, and J. Thomas, Sustainable water resource management using surface-groundwater modelling: Motueka-Riwaka Plains, New Zealand, *Watershed Ecol. Environ.*, vol. 3, pp. 38–56, 2021, doi: 10.1016/j.wsee.2021.08.001

[24] W. Hussainzada and H. S. Lee, Hydrological modelling for water resource management in a semi-arid mountainous region using the soil and water assessment tool: A case study in northern Afghanistan, *Hydrology*, vol. 8, no. 1, pp. 1–21, Mar. 2021, doi: 10.3390/hydrology8010016

[25] D. K. Butler, Groundwater resource assessments, *Int. J. Sustain. Dev. World Ecol.*, vol. 7, no. 3, pp. 173–188, 2000, doi: 10.1080/13504500009470039

[26] Hoffmann and P. Sander, Remote sensing and GIS in hydrogeology, *Hydrogeol. J.*, vol. 15, no. 1, pp. 1–3, 2007, doi: 10.1007/s10040-006-0140-2

[27] N. R. S. C. (NRSC) I. space R. O. ISRO, *Concept of Ground Water Prospects Maps preparation using Remote Sensing and Geographic Information System for Rajiv Gandhi National Drinking Water Mission Project*. 2015.

[28] Bhuvan, "Ministry of Drinking Water and Sanitation Bhuvan–Bhujal (Ground Water Prospects and Quality Information System) National Rural Drinking Water Program," 2017. https://bhuvan-app1.nrsc.gov.in/gwis/ (accessed Apr. 22, 2022).

[29] Bhuvan, "Scope and limitations of Ground Water Prospects Maps," 2019. https://bhu van-app1.nrsc.gov.in/gwis/index2.php.

[30] MoUD, *Formulation of GIS Based Master Plan for AMRUT Cities*. 2015.

[31] Census, *Size, Growth Rate and Distribution of Population*. 2011.

[32] GoR, *Rajasthan: Urban Development Policy*. 2017.

[33] India Meteorological Department, "Monsoon Report-2020 (Rajasthan) Meteorological Centre Jaipur," vol. 2020, 2020.

[34] CGWB Western Region, "Ground Water Year Book Rajasthan," no. November, 2019, [Online]. Available: http://cgwb.gov.in/Regions/WR/Reports/Year Book 2019-20_English corrected 3-11-20 sent.pdf.

[35] S. M. Hooda, "India–Rajasthan water assessment and potential for private sector interventions," 2017.

[36] C. R. K. R. Rana, "Govt. of India Ministry of Water Resources Central Ground Water Board Assessment of Ground Water Resources A Review of International Practices," pp. 1–101, 2014, [Online]. Available: www.cgwb.gov.in.

[37] Central Ground Water Board, *National Compilation on Dynamic Ground Water Resources of India*, no. July. 2019.

[38] Central Ground Water Board, "Groundwater Scenario with special reference to Rajasthan."

[39] D. Silverman, Doing qualitative research, *Sage Publ.*, vol. 1999, no. December, pp. 1–6, 2006.

[40] Purvis and A. Dinar, Are intra- and inter-basin water transfers a sustainable policy intervention for addressing water scarcity?, *Water Secur.*, vol. 9, Apr. 2020, doi: 10.1016/j.wasec.2019.100058

[41] Government of Rajasthan and India, "About RIICO," 2018. https://industries.rajasthan.gov.in/riico/.

[42] Kota Development Authority, "Kota Master Plan 2012–2031 draft," Kota.

[43] Jaipur Development Authority, *Master Development Plan-2025 Jaipur Region (Existing Profile)*, vol. 1. 2011.

[44] Jaipur Development Authority, *Development Plan-2025*. 2013.

[45] Udaipur Development Authority, "Udaipur master plan 2013–31," 2015.

[46] Ajmer Development Authority, *Master Development Plan Ajmer 2013–2033*. 2013.

[47] CGWB, "Ground Water Scenario Udaipur District Rajasthan," 2013.

[48] Government of India, "Smart Cities," 2022. https://smartcities.gov.in/.

[49] NITI Aayog, "Reforms in urban planning capacity in india," *Gov. India,* no. September, p. 144, 2021, [Online]. Available: www.niti.gov.in/sites/default/files/2021-09/UrbanPlanningCapacity-in-India-16092021.pdf.

[50] "Draft National Water Framework Bill, 2016," 2018.

[51] R. Lavoie, A. Lebel, F. Joerin, and M. J. Rodriguez, Integration of groundwater information into decision making for regional planning: A portrait for North America, *J. Environ. Manage.*, vol. 114, pp. 496–504, 2013, doi: 10.1016/j.jenvman.2012.10.056

[52] R. Thomsen, V. H. Søndergaard, and K. I. Sørensen, Hydrogeological mapping as a basis for establishing site-specific groundwater protection zones in Denmark, *Hydrogeol. J.*, vol. 12, no. 5, pp. 550–562, 2004, doi: 10.1007/s10040-004-0345-1

16 Computing the Accessibility of the Settlements Using the Network Analysis of Indore City Region
A Regional Approach

Aman Singh Rajput and Chetna Singh

CONTENTS

16.1 INTRODUCTION

A city region is an area that demonstrates the relationship between a city and its surroundings, and is used as an appropriate unit for specific business and administrative purposes. It is also a region that is connected to other areas in terms of a particular set of related circumstances, whether they relate to the land or the people, such as industry, farming, population distribution, or the general sphere of influence of a city. A region's growth is based on its ideal expansion and aims to raise the standard and quality of life there. In addition to the typical policy implementation within the boundary, regional development includes socioeconomic processes structured within distinct political and cultural environments [1,2]. Many academics have defined a city region based on the nature of reliance, hierarchy, and administrative divisions as settlements have developed over time. In today's world of inescapable urbanisation, a city (administrative or financial) emerges inside a broader region that dominates the other counterparts and supports several infrastructures and amenities on which the hinterland population is dependent. There are now more urban regions than ever before, whether because of population growth, boundary extension, rural-to-urban migration, or the ever-expanding urbanisation process. As a result, there is an increase in the need for land for infrastructure and housing [3]. The rise of places immediately beyond the city limits is caused by the scarcity and high value of land in urban areas. Urban sprawl is another name for this process. The existence of railway lines or corridors serves as fuel for this process because they improve the connectivity of the regions.

The expansion of the city region is the result of people moving to the most desirable sites outside the city limits [4]. The locational preference is primarily determined by the growth influencing the land uses in a region and the development of the transportation network. The ability of a site to be accessed by, or to reach other locations, is measured by its accessibility. According to the principle of natural mobility, a location's attractions have an impact on how people move through it, and vice versa. Regional road networks, however, are what define the distribution of attractions and movement at a specific location because they provide a means for access to specific towns. According to the principles of natural economy, people choose their desired locations from a variety of options, and accessibility is a key factor in determining how easy it is to take advantage of opportunities in a particular region. As a result, accessibility and appeal are related; the more accessible a location is, the more people it will draw in.

The spatial structures in a region – hamlets, villages, towns, and cities – are mutually dependent on each other and together form a unified system of settlements. It is these dependencies which help to provide an understanding of the settlement spacing along with its functional nature and the interaction between them supported by transportation lines. Their regional spatial form has three elements: (1) location of specialised industries; (2) distribution of services and infrastructure and (3) transportation nodes and linkages. Transportation nodes and linkages are associated with spatial connectivity in the regional structure. On a regional scale, some roads carry a higher volume of traffic whereas others cater to low vehicular traffic volume [5]. Thus, what are those parameters that we as planners use, to decide which road will act as an arterial route and affects the development along it? Henceforth, this research focuses on what information can be extracted from the road network, which can explain the arrangement of the settlements at a regional scale.

16.2   ACCESSIBILITY AND REGIONAL DEVELOPMENT

Accessibility is its most common factor and refers to the ease of reaching opportunities. These opportunities may be services, amenities, goods, activities, or destinations. It is also defined as a sum of opportunities that a person acquires to reach and enjoy the activity at a given location. Mathematically, it is the function of the ability of an individual to access a particular location. These abilities may be time, cost, safety, security, etc. [6]. Thus, it can be concluded that it is concerned with the opportunities provided by transport and related infrastructure to travel. The accessibility of the location guides the attraction potential of the space, which influences the developers' decisions [7]. There are two types of accessibility when it comes to its operational form – relative accessibility and integral accessibility [8]. Relative accessibility, as the work suggests, is relative to some other point and is defined as the degree to which two places are connected on the same surface, whereas integral accessibility is the degree of interconnection with all other points [9]. Integral accessibility is used in the regional planning context as it takes into consideration all the settlements within [10]. The inclusion of accessibility in regional planning can be two-folds – firstly, planning an efficient road network system and secondly in 'development zone' formation through the integration of accessibility and socio-economic development.

A variety of perspectives have been considered while defining accessibility and therefore it holds a multiplicity of meanings. It can be defined in terms of physical distance, cost impedance and locational opportunities provided through various land uses and geographic characteristics, in Ref. [11] it is defined as 'potential of opportunities for interaction'. Similarly, in Ref. [12] it is defined in terms of transport mode efficiency as 'the ease with which any activity can be reached from a location using the particular transport system'. The human perspective is defined by Ref. [13] as 'the freedom of individuals to decide whether or not to participate in different activities at a given spatial location'. Finally, connecting the concept of land use and transport, Ref. [14] defines accessibility as 'the extent to which the land use transport system enables individuals or goods to reach activities or destinations using transport modes'. Thus, the term accessibility has been defined from various perspectives and accordingly given different components to analyse and parameters that govern it.

The concept of accessibility consists of and depends on four major components (refer to Figure 16.1) defined as per Ref. [14] and [25], which are:

1. **Transportation component**: The transportation component deals with the time taken to travel, the cost of the vehicle and the condition of the transport infrastructure [15].
2. **Land use component**: Related to the land use characteristics, that is, how the activities are spatially distributed in the region and the nature of destinations in each zone which attracts people along with the demand and supply of amenities in the residential areas [16].
3. **Temporal component**: This component deals with the time-related measures of accessibility, in general from the time taken by the individual to travel to their destination and the temporal availability of the services [17].
4. **Individual component**: This component deals with the socio-economic and demographic profile of the users. These indicate the ability to pay, affordability and opportunities required by the people of that zone. This component is analysed in integration with other components across the region [18].

The research works focus on the network component and try to draw the methodology for accessing the accessibility of the settlements at a regional scale.

Network measures are the most important measure and are computed using the street network or any other network based on the graph theory [17]. The network is

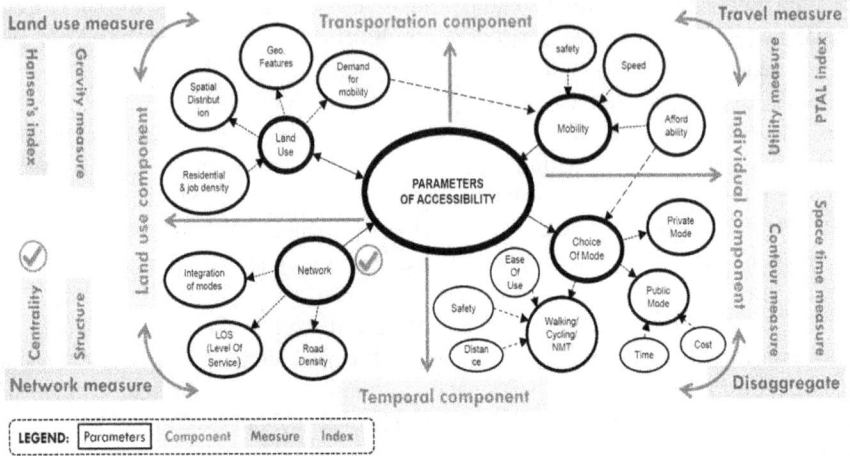

FIGURE 16.1   Components, parameters, and measures of accessibility.

described as being composed of nodes and links, in which two nodes are connected via a link. Network measures provide an authoritative methodology to quantify the regional distribution and are the basic dataset for measuring accessibility using network measures [19]. Thus, the computed accessibility has a direct impact on the distribution of activities in the region.

16.2.1   CASE STUDY – INDORE

Indore tier II city is also described as the commercial capital of Madhya Pradesh. This city has been selected based on the population size, fewer geographical constraints and availability of secondary data. The city is on a relatively plain terrain and major constraints for spatial development are the airport present in the west and the environmentally sensitive zones Pipaliyapala Tank on the south-eastern side and Sirpur Tank in the southwest. In terms of regional outlook, there are two levels of the regional landscape defined for Indore. The statutory one is the Indore Agro-Industrial region which consists of seven districts Indore, Dewas, Ujjain, Ratlam, Dhar, Mandsaur and Jhabua. This region has been notified as per Madhya Pradesh Nagar Tatha Gramin Nivesh Adhiniyam, 1983. Another metropolitan region has been identified by the Town and Country Planning Organisation, Madhya Pradesh, during the preparation of the Indore Master Plan–2021. An influence zone of 4798 sq. km had been identified around the city as the Indore influence region, consisting of Indore, Dewas, Ujjain, Ratlam, Mandsaur, Dhar and Jhabua. Relatively, the region is more urbanised compared to other regions identified in the master plans of Madhya Pradesh cities. Out of the seven, Dewas, Ujjain and Indore are present in triangular form and are more urbanised. Dewas is 35 km and Ujjain is 37 km from Indore city. The city also falls under the influence of the DMIC (Delhi Mumbai Industrial Corridor) influence area. From a regional perspective, the city influences the urban fringe with economic activities such as those proposed in the peri-urban area by the state government in the form of a super corridor in the north direction and the Transport Nagar adjacent to the municipal boundary.

The region considered for the research is the buffer area of 8 km from the Indore municipal boundary (refer to Figure 16.2), as per URDPFI guidelines [20]. The municipal boundary during the 2011 census had been considered due to consistent data availability across the settlements. The total number of settlements around IMC in the region is 138. The region supports a total population of 2,351,718 and an area of 813 km². The region is a part of the Indore metropolitan region identified in the Indore master plan 2021. At a regional scale, the city enjoys a locational advantage within the nation as well as the state. The national highways passing through the region, NH 3 and NH 59, connect the city to major centres in India, such as Delhi, Mumbai, Gujarat and Bhopal. The city lies along the religious and industrial route via SH 27, connecting the city to Ujjain in the north and Mhow in the south. There are two ring roads around the IMC, the internal ring road and the outer ring road (commonly known as the AB road).

16.3 METHODOLOGY

The regional accessibility of the settlements in the region is calculated using graph theory. In a graph theory approach, the complete road network of the region is viewed as a graph in two dimensions. The coordinate system of x and y is viewed as the nodes and links in the network. In the network measure approach '.... . expressed the accessibility of a point in a system as a function of its location in space for all other points in the system'. Graph theory was majorly used in sociological studies to analyse the interconnectedness of individuals in society. The regional road network has four classifications – national highway, state highway, district roads and village roads. A node is the terminal point, or the intersection point, generally the representation of the location such as a city, administrative division, or transport terminal. A link is the connection between nodes and is an abstraction of the transport infrastructure present. For the research, the nodes are represented as the settlements and the links as the regional road network. There are two ways of analysing network accessibility using graph theory – space syntax and multiple centrality assessment. Multiple centrality assessment uses the real-time travel distance between two points. As discussed in the previous section, the network centrality takes up the fact that in a region better location results from its centrality [21]. There are three indices used for the research based on centrality:

a) *Closeness:* The closeness centrality of a node is defined as the inverse of the sum of all the distances required to travel to all other nodes or settlements in the region, as shown in Figure 16.3 [22], along the shortest path. The higher the value of closeness, the higher the accessibility. This value indicates that the higher the closeness of a settlement the closer it is to the surrounding settlements.

b) *Betweenness*: This is defined as the total number of shortest paths passing through a particular node (settlement). This is the most important centrality index as it gives an individual knowledge of the traffic attracting potential

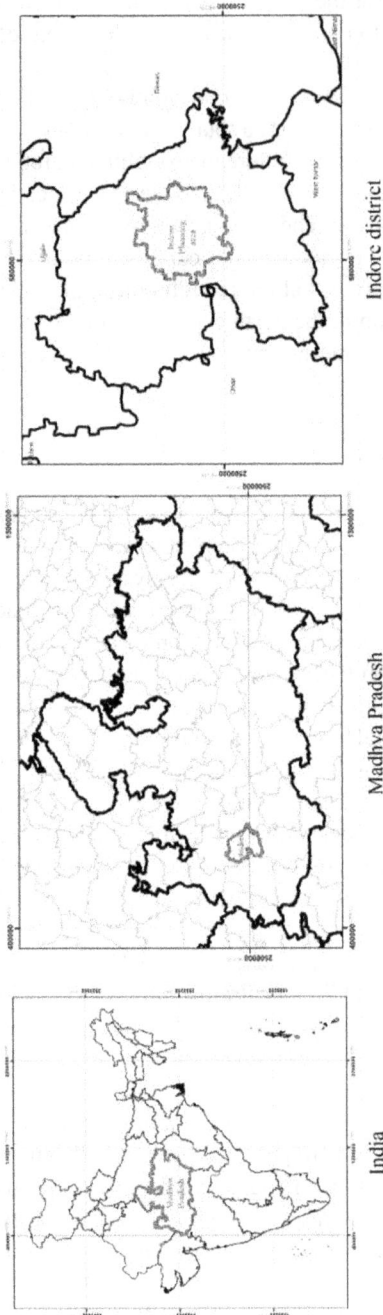

FIGURE 16.2   Location map of Indore City Region.

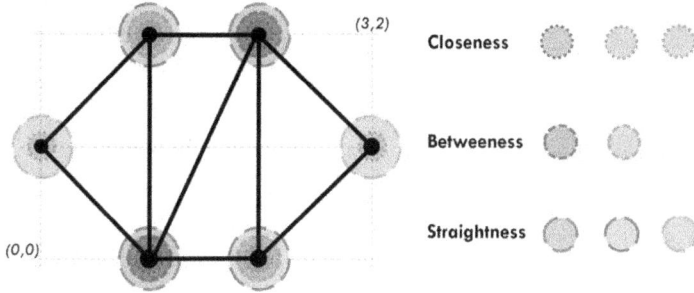

FIGURE 16.3 Schematic representation of the road network over the graph with centrality indexes. An infographic which provides information for the same number of linkages which nodes have higher centrality measures.

[23]. Since it depicts the frequency of passing the settlement for reaching the other settlements, these are major nodes which attract land uses such as industrial, commercial, public semi-public, etc. The higher the betweenness value, the higher is the accessibility of the settlement.

c) ***Straightness***: This index is like the commonly used detour index in regional analysis of road network connectivity [24]. It represents the efficiency of the road network system which is being developed. It is defined as the sum of the ratio of Euclidean distance (displacement) to the actual distance to reach from the node in consideration to all other nodes. The more straight the settlement is in joining with the other nodes, the higher is its straightness value and, in turn, the higher is its accessibility.

Graph theory has been applied to the regional road network of the study area. The settlement's centre point was considered as the node, and the road network of the region was converted to links. The network analysis toolbox was used to calculate the shortest path of travel between different settlements (Figure 16.4). The OD matrix for the same was processed to calculate the closeness centrality of the settlement through the inverse sum of the shortest path to reach all other settlements. The base layer required for calculating the centrality of the region is the road network comprising the major roads and the settlements which would act as nodes. For the road network, the data available from the open street maps were downloaded for the region, however, the file obtained was an outdated one. The file was updated in ArcGIS using the editor tool and extended and trimmed. The settlement level map was acquired from the census of India website for the district of Indore. This layer was then cropped for the entire region, thereby using the feature-to-point tool of ArcGIS the settlements were converted to points. Merging the two layers in ArcGIS gave the output file for conducting the network analysis. For the calculation of the centrality indices, the network analyst toolbox in ArcGIS was used. Two computations were made: the closest facility analysis and the OD matrix analysis. The closest facility analysis gave the real distance from one node to all other nodes along the shortest paths.

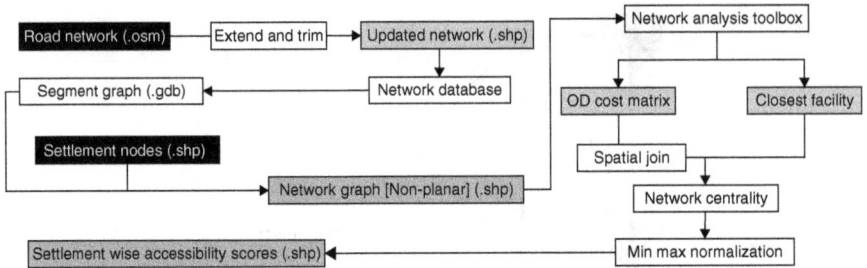

FIGURE 16.4 Flowchart depicting the tools and methods used for calculating the centrality on GIS.

From the attribute table, the data were copied to Excel, where an OD matrix was generated for the real distances. Similarly, the OD matrix analysis gave the Euclidean distance to travel from one node to all other nodes and was converted to an OD matrix displacement file in Excel. In Excel, for every settlement the sum of all the distances required to travel along the shortest path was added and the inverse of it was calculated to compute the closeness values of the settlements. Next, to analyse the straightness index, the ratio of Euclidean distance to real distance was calculated and added. To calculate the betweenness index, all the paths representing the shortest route to travel from all the nodes to all other nodes were given a value of 1. The spatial join tool was used to join the data of these paths with the settlement shapefile which led to the computation of the betweenness centrality index.

16.4   RESULTS AND DISCUSSION

Closeness centrality: It can be inferred that central areas have the highest closeness centrality and are reduced in the radial direction as one moves from urban to rural areas. The values are higher along the outer ring road, indicating that the development of that road has increased the accessibility of the settlements. Similarly, all the urban areas also have higher closeness values since all of them in the study region are present close to the major urban town Indore Municipal Corporation. It can be observed that the closeness centrality of the settlement in the centre is higher, as geographically it is in the centre of the region. The closeness values decrease as we move geographically away from the Indore municipal corporation. The values of closeness are higher for the settlements along the internal ring road and outer ring road or AB road (refer to Figure 16.5(a) 0. As explained earlier, centrality represents how close a particular settlement is to all other settlements and thus, it represents the settlements which are best placed in the region to influence the entire region most quickly. Since the calculation involves capturing the shortest paths to be travelled to settle, the higher values represent less time taken (or distance to cover) to reach that settlement. The closeness values would reflect the factor of travel cost for overcoming the spatial separation between the place with population and activities. Thus, the higher is the closeness value of the settlement the higher the accessibility, which in turn characterises that

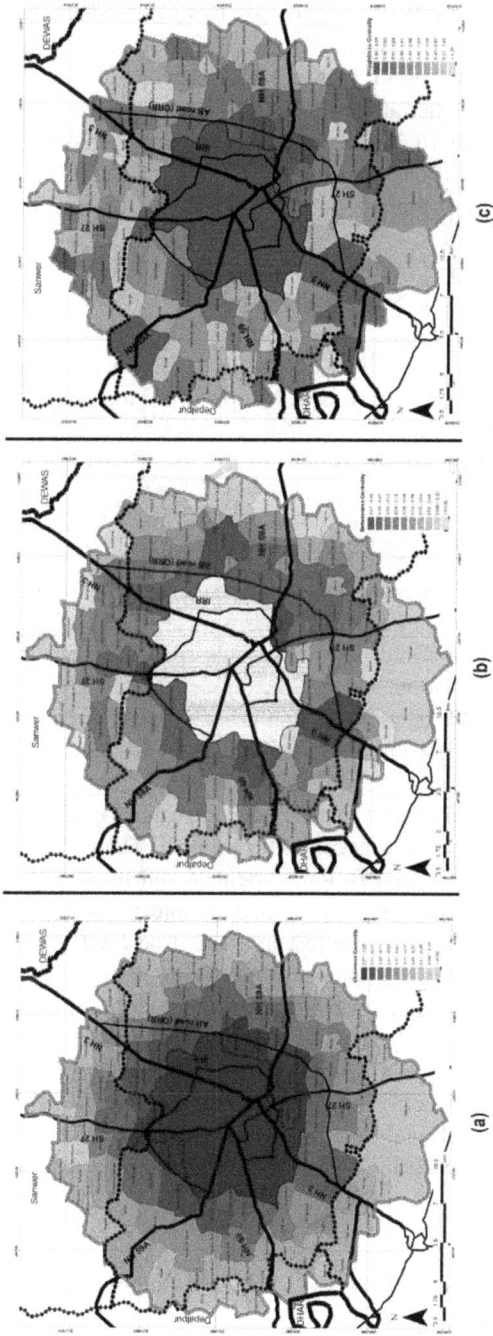

FIGURE 16.5 Computed accessibility indexes, Indore City Region: (a) closeness; (b) betweenness and (c) straightness. Choropleth map representing the settlement wise centrality values. The darker the shade, the higher is the related accessibility.

the higher is the ability of people to reach that settlement from the overall region for exploiting the services provided by it.

Betweenness centrality: When the values were computed, Indore Municipal Corporation had a value of 1, while the next highest value was 0.33, and thus while analysing class distribution, Indore M. Corp was removed (Figure 16.5(b)). Betweenness values are higher along the major radial roads from the centre and the ring roads (refer to Figure 16.5(c)). The reason for this is that these are the major roads in the hierarchy and if one wants to travel from one settlement to another, one must cross this hierarchy at a given point, and thus the settlements along the route have higher between values.

Straightness centrality: From the analysis, it was observed that the settlements along the major roads, such as NH and SH, intuitively have more straight paths than the shorter routes, i.e., the higher order roads such as freeways, expressways, and national highways obtain higher values than those of the other roads (Figure 16.5(c)). From the literature review and analysis, it was observed that longer routes and state highways are built with minimum diversions, connecting the major centres, and are thus straighter than the lower order roads.

16.5 CONCLUSIONS

With India's growing urbanisation over the years and the pressure on the land resource within the urban administration, the growth of the city region is evident. Theoretically, this concept could be used to plan road networks in the region and analyse with each link development what the impacts at the regional scale would be. Additionally, this could also be used to plan the settlement hierarchy in the region and to identify which would be the important settlements in the region based on the accessibility and according to the services/amenities that could be developed. For making the cities more liveable and fostering sustainable growth, an evidence-based planning and policy formulation approach needs to be adopted. The research concludes that accessibility can be an important factor to identify and monitor growth in the city region. Regional accessibility plays an important role in increasing the attractiveness of settlement in the region. The research could be used to identify the settlement within clusters for the development of certain higher-order services which can be easily accessed by the neighbouring settlements therein. The network centrality measures of accessibility are a powerful toolset to analyse the regional spatial structure from different perspectives. This research is a new kind in the field of regional planning and adds a constructive contribution to the literature review emerging in the application and spatial analysis of transportation networks. The regional spatial structure of the region can be interpreted by the analysis of road network-based accessibility.

16.5.1 LIMITATIONS AND SCOPE FOR FURTHER RESEARCH

This study was conducted in the city region comprising settlements up to 8 km from Indore city as per the URDPFI guidelines for mapping of peri-urban areas.

Furthermore, the census boundaries of 2011 were considered for the research. The major limitation of the study is that the centrality measure only considers the factor of the distance between the two settlements and other factors that affect the accessibility are not taken into account. The traffic potential of each settlement can be analysed, and relationships could be built, that it is co-terminus with the presence of industries and major land uses and growth. Additionally, the network centrality could be layered with various other components in addition to distance, such as road condition, right of way, volume to capacity ratios, public transport availability, etc., which would detail the calculated accessibility and provide valuable results.

DISCLAIMER

This research work is entirely that of the author/co-author and does not carry the views of the organisations mentioned herein.

REFERENCES

[1] J. Scott, Globalization and the rise of city-regions, *European Planning Studies*, vol. 9, no. 7, pp. 813–826, 01 July 2010.

[2] D. Pojani and D. Stead, Past, present and future of transit-oriented development in three European capital city-regions, *Advances in Transport Policy and Planning*, pp. 93–118, 2018.

[3] S. Armondi and S. D. G. Hurtado, *Foregrounding Urban Agendas*, Switzerland: Springer, 2020.

[4] F. J. Greene, P. Tracey and M. Cowling, Recasting the city into city-regions: Place promotion, competitiveness benchmarking and the quest for urban supremacy, *Growth and Change*, vol. 38, no. 1, pp. 1–22, 2007.

[5] Goodbody Economic Consultants, *Transport and regional development*, Dublin: Department of Urban and Regional Planning UCD, 2002.

[6] G. Dupuy, *Urban networks–Network Urbanism*, Netherlands: Techne Press, 2008.

[7] Sevtsuk and M. Mekonnen, Urban network analysis: A new toolbox for ArcGIS, *Journal of Geomatics and Spatial Analysis*, vol. 10, no. 10, pp. 1–15, 2002.

[8] G. M. Ávila, *A Contribution to Urban Transport System Analyses and Planning in Developing Countries*, Brazil: INTECH open science, 2010.

[9] D. R. Ingram, *The concept of accessibility: A serach for an operation*, 1971.

[10] R. H. M. Pereira, V. Nadalin, L. Monasterio and P. H. Albuquerque, *Urban Centrality: A Simple Index*, Brazil: Geographical Analysis, 2013.

[11] W. G. Hansen, *Accessibility and Residential growth*, MIT university Press, 1959.

[12] M. Q. Dalvi and K. M. Martin, *The measurement of accessibilty: Some preliminary results*, Transportaion UK Press, 1976.

[13] L. D. Burns, *Transportation, Temporal and Spatial Components of Accessibility*, Lexington books, 1979.

[14] J. R. Van Erk and K. T. Geurs, Accessibility evaluation of landuse and transport strategies: review and research directions, *Journal of Transport Geography*, *12*(2), 127–140, 2004.

[15] A. C. Ford, S. L. Barr, R. J. Dawson and P. James, Transport accessibility analysis using GIS: Assessing sustainable transport in London, *ISPRS International Journal of Geo-Information*, vol. 2, no. 1, pp. 124–149, 2015.

[16] O. Kotavaara, M. Pukkinen, H. Antikainen and J. Rusanen, Role of accessibility and socio-economic variables in modelling population change at varying scale, *Journal of Geographic Information System,* vol. 6, no. 1, pp. 386–403, 2014.

[17] S. Verma, Thesis: Assessing the impact of accessibility on spatial variation of land values for different land use distribution–A case study of Bhopal city, 1 ed., Bhopal: School of Planning and Architecture, Bhopal, 2017.

[18] R. Vickerman, Location, accessibility and regional development: the appraisal of trans-European networks, *Transport Policy,* vol. 2, no. 4, pp. 225–234, 1996.

[19] L. Todd, *Evaluating accessibility for transport planning: Measuring people's ability to reach desired goods and activities,* London: Victoria Transport Policy Institute, 2018.

[20] Ministry of Urban Development, *Urban and Regional Development Plans Formulation and Implementation (URDPFI) guidelines,* Delhi: Town and Country Plannig Organization, 2015.

[21] S. Hasan, X. K. Y. B. Wang and G. Foliente, "Accessibility and socio-economic development of human settlements," 06 12 2017. [Online]. Available: https://doi.org/ 10.1371/journal. [Accessed 06 03 2019].

[22] V. Oliveira, *An introduction to the study of Physical form of cities,* Springer, 2016.

[23] D. Dubois, G. Bel and M. Llibre, A set of methods in transportation network systhesis and analysis, *J. Opl Res. Soc.,* vol. 30, no. 9, pp. 797–808, 2012.

[24] R. Dirham, M. Y. Jinca and B. Hamzah, Accessibility and mobility of the road networks in supporting the regional development in Mimika Regency, *American Journal of Engineering Research (AJER),* vol. 7, no. 6, pp. 33–36, 2018.

[25] K. J. Button, S. Leitham, R. W. McQuaid and J. D. Nelson, Transport and industrial and commercial location, *The Annals of Regional Science,* vol. 29, no. 1, pp. 189–206, 1995.

17 Smart Sensors for Peak Demand Reduction in Urban Buildings

C. Rakesh, T. Vivek, and K. Balaji

CONTENTS

17.1 INTRODUCTION

Energy consumption in buildings plays a significant role in global energy requirements. Globally, building sectors consume 42% more electricity than other sectors. Increasing urbanisation in developing countries and increasing the size and number of buildings also increase the demand for electricity. Buildings consume one-third of global energy for space conditioning and generate 40% of total global CO_2 emissions. In recent years, building sector energy consumption and CO_2 emissions

DOI: 10.1201/9781003331001-17 **291**

have been increasing after levelling off between 2013 and 2016. Many factors are involved in this rise, such as increasing energy demand for space cooling and heating of buildings. The maximum energy consumption in buildings is due to the air conditioner (38%) of total building energy use, lighting (18%), major appliances (geyser, refrigerator, dryer at 18%), and remaining miscellaneous electronic appliances (36%). The primary concern is the peak demand; due to the higher heating and cooling at certain climatic conditions, all the electrical appliances will be switched on for a certain period of the day (based on the occupant requirements), which leads to peak demand (maximum electrical consumption at that time). Therefore, it is essential to understand the energy pattern on an hourly basis, which acts as the input for the smart sensors for automatically controlling the heating, ventilation, air conditioning equipment, and electronic appliances to be switched on/off based on the requirements.

Smart buildings can be defined as those that utilise information and communication technologies (ICT), such as cognitive learning and smart sensors to provide a more comfortable, sustainable, and efficient life [1]. Moreover, the Internet of Things (IoT) plays a significant role in innovative building technologies. Control of smart buildings by using IoT requires data from interconnected devices to adjust the system effectiveness, indoor comfort, safety, and to have a quicker response. Consequently, smart buildings' energy management is critical [2]. Energy conservation management consists of demand-based management, peak load reduction, and reducing carbon mitigation [3]. The industrial, commercial and residential sectors consume significant electrical energy. Therefore, identifying suitable operation strategies for effective home appliance management helps minimise energy consumption and demand during peak hours. More sources are utilised to increase the electrical power generation capacity to fulfil the increasing energy demand at peak times. Adding resources to satisfy energy demand is a costly strategy due to high demand.

17.2 SMART SENSORS FOR ENERGY MANAGEMENT IN BUILDINGS

A conventional sensor converts physical variables detected into a suitable electrical signal. The typical sensor's purpose is to send the measured information to other electronics, whatever the data captured. To be smart, the sensor should have the following properties: robust, economical, remotely operated, self-validation and identification, low power consumption, self-diagnostics and calibration, and preprocessing of data. Therefore, the need for developing smart sensors which pose various properties of inbuilt intelligence, decision-making, self-diagnostics, and validation has limited research on energy management; this study highlights the different types of smart sensors which can be used for energy savings in building applications. Figure 17.1 depicts the smart sensor building block. Smart sensors integrated with the IoT components convert the physical variable measured into a digital signal for transmission to a gateway. The inbuilt microprocessor unit is programmed, which can filter, compensate and perform any specific signal conditioning and noise reduction.

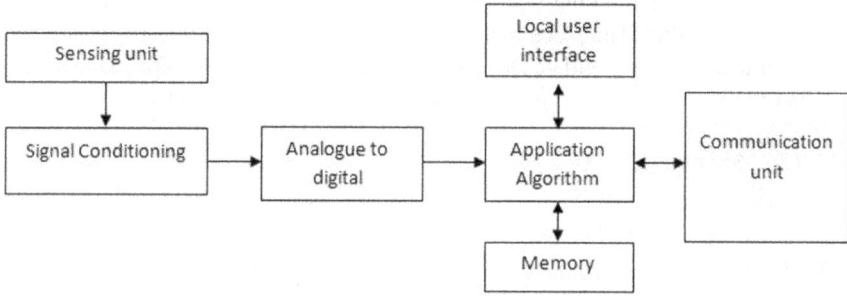

FIGURE 17.1   Smart sensor building block.

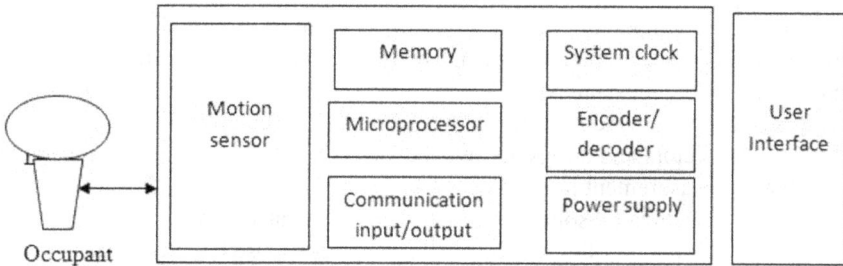

FIGURE 17.2   Block diagram of residence motion sensor.

The programmed inbuilt microprocessor reduces the load on the gateway and cloud resources, and the instructions are supplied to the microprocessor to act on the sensor automatically to perform a required operation, such as self-calibration/diagnostics/measuring the physical environment. The microprocessor has an inbuilt warning system. Any drift in measurement beyond the marginal limit will generate warning signals; hence, the user can take preventive measures before the mechanical failure of the sensors. The smart sensor measures the physical properties such as pressure, temperature, humidity, speed, and occupancy presence and converts them into electrical signals. The electrical password is converted to the user's required output (digital signal) via an inbuilt microprocessor. Multiple sensors can be coupled with different capacities depending on the functional requirements.

17.2.1 Types of Smart Sensor for Building Energy Management

17.2.1.1 Smart Motion Sensor

Motion sensors detect the presence of occupants when people enter and exit a space. From Figure 17.2, based on the number of occupants in the assigned space, the motion sensor automatically switches on/off lights, fans, or airflow from air conditioners. Most motion sensors communicate with their hubs through Z-wave or Zigbee, ranging from 100 feet to 150 feet. The detection range is from 15 to 45 feet. The sensors are sensitive to short and long fields based on room capacity. It has a wide detection angle and is versatile. The placement of the sensor should be at the centre of a wall; only specific movements will be captured. A smart motion sensor alerts via push notification or text message to the user interface when it detects any activity.

17.2.1.2 Smart Light Sensor

Light sensors automate illuminating functioning in buildings, enabling human-centric lighting, daylight harvesting, and scheduling for energy saving, efficiency, and ambience. Smart light sensors communicate with Zigbee 3.0, and the working temperature ranges from $-100^{\circ}C$ to $500^{\circ}C$, with the detection range from 0 to 83,000 lux and working humidity of 0–95%.

17.2.1.3 Smart Thermal Sensor (Infrared Detector Array)

Automatic room temperature based on occupant comfort in buildings is controlled with the help of a smart thermal sensor. Thermal sensors mainly control the internal temperature to be constant based on the user's requirement. These sensors save energy by balancing the temperature concerning climatic conditions. To save energy, the infrared detectors and arrays are intended explicitly for non-contact temperature and radiation measurement in buildings. The infrared detector array developed at the University of Michigan's solid laboratory is an integrated sensor. Polysilicon–Au thermocouples and a thin-film dielectric diaphragm were used to create the infrared-sensing element to support the thermocouples. Using silicon gate metal–oxide–semiconductor (MOS) processing, an on-chip multiplexer was created. This detector has a temperature range of 0–$100^{\circ}C$ and a response time of 10 milliseconds. A separate calibration thermopile, polysilicon resistors, diodes, and MOS transistors are also included on this chip, allowing direct measurements of the cold junction temperature and the thermoelectric power of the polysilicon lines.

17.2.1.4 Smart Wind Sensor (Accelerometer)

The wind sensor automates the shutting down of storm shutters, dust skirts, or any retractable building windows if strong winds are detected – power supply requirement 24V DC, maximum wind speed 50 m/s. The IBM research laboratory in San Jose, California, created an accelerometer that comprises a sensor device and electronics on silicon. The accelerometer is a metal-coated SiO_2 cantilever beam built on a silicon chip, with the output signal provided by the capacitance between the shaft and the substrate.

17.2.1.5 Smart Smoke Sensors (Optical Sensors)

A smart smoke sensor is an optical (photoelectric) smoke detector that enhances home safety. In the case of fire, the alarm sends real-time smoke alerts directly to a smartphone. The communication module Bluetooth 4.2 low-energy/Wi-Fi 802.11b/g/m (2.4 GHz)/ZigBee can be used, as shown in Figure 17.3. It detects smoke in real time and instantly notifies the users by flashing a light, sounding an alarm, and sending a notification to the phone.

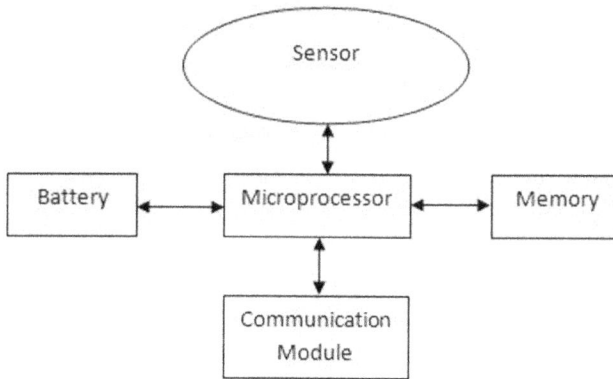

FIGURE 17.3   Architecture of smart smoke sensor.

17.2.2   BUILDING PEAK DEMAND REDUCTION WITH THE AID OF SMART SENSORS

The peak demand for buildings can be reduced by following these three strategies:

a) Scheduling peak distribution
b) Peak shrinking
c) Thermal energy storage/energy storage.

17.2.2.1   Scheduling Peak Distribution

The reason for the peak demand is multiple appliances running at the same time. The solution to this problem is smart sensors that are integrated to schedule the system to operate at off-peak hours. Hence all the equipment is not served at the same time. For example, suppose all the air conditioners in the building are conducted at the same time. In that case, the required cooling can be achieved. Still, energy savings and cost-effective cooling can be achieved with a sequential schedule instructed by the smart sensors based on occupant presence and behavior, and with a control algorithm to the sensors. Smart thermal sensors with an inbuilt microprocessor schedule program consider the forecasted weather for the day and condition the building during off-peak hours to reduce the peak demand.

The usage pattern of the appliances throughout the day will be varied and can be controlled with the help of smart sensors. Suppose there are seven appliances (refrigeration will not be considered, as this system should always be on) and $2^n - 1$ transition states, which combine 63 states. From Table 17.1, let us consider P_1 air conditioner is the highest and P_6 lighting is the lowest consumption of power. "0" represents off-state, and "1" represents the form. Let us consider 0010011 means microwave, refrigeration, and an electronic appliance are on condition, and the total power consumed at that instant is 2380 watts, if all the devices are operated at the same time peak demand will be attained. Therefore, the scheduling of the equipment has to be performed with a control algorithm, and suitable input has to be forwarded

TABLE 17.1
Schedule of the Appliance

Si. no.	Appliance	Peak power rating (watts)
1	Heater/geyser	1500
2	Iron centre	1000
3	Microwave	1200
4	Lighting	600
5	Air-conditioning	2500
6	Refrigeration	180
7	Miscellaneous	1000

to the microprocessor based on the wattage of the appliances. The conditions should compare the appliance's base load and peak demand, and sequential scheduling has to be planned. This method effectively distributes the energy used over a significant period of time, creating an energy balance and reducing peak demand.

17.2.2.2 Peak Shrinkage

Installation of energy-efficient equipment can reduce the peak demand. Equipment that can integrate with smart light, motion, and thermal sensors can save energy and shrink the entire demand curve. LED lights, 5-star ratings of home appliances, guidelines for energy conservation, and building codes for the building design are energy-efficient methods, but with an increase in the capital cost.

17.2.2.3 Thermal Storage/Energy Storage

Batteries or thermal storage systems must be dedicatedly stored during low energy consumption/off-peak conditions. During high energy consumption, this battery has to facilitate the energy requirement. The smart sensors should give instructions to the battery during the time of the day when low energy is consumed, and the microprocessor should take up the decision to charge the battery during that period.

P_g – Grid power i^{th} out the battery; P – Total power demand; P_u – Upper threshold limit; P_t – Lower threshold limit.

From the flow chart (Figure 17.4), the battery storage system is configured with two commands: charge at low demand (less than 15 kW) and discharge at peak demand (more than 500 kW). The charging command will be accepted if and only if the battery energy storage's state of charge (SOC) is less than 10%. However, the opposite requirement holds if the SOC is more than 90%, that is, the charging of the battery will be stopped if it reaches 90%.

17.2.3 Smart Building

Smart building automation includes lighting systems, control systems, and heating and cooling systems to reduce power consumption and improve building energy

```
                                                              ┌─────────┐
                                                              │  Start  │
                                                              └─────────┘
                                                                   │
                                                                   ▼
   ┌──────────────────────────┐              ◇─────────◇        ◇───────◇
   │ Charge the battery at     │◄── Yes ────│ SOC<90% │◄─ Yes ─│ Pg<Pi │
   │ moderate charging rate    │              ◇─────────◇        ◇───────◇
   └──────────────────────────┘                   │                │
                                                  No               No
                                                   │                │
                                                   ▼                │
   ┌──────────────────────────────────────────────────┐            │
   │      Charge the battery at low charge rate         │           │
   └──────────────────────────────────────────────────┘            ▼
                                                              ◇───────────◇
   ┌──────────────────────────┐                             │ Pi<Pg<Pu  │
   │  Set power output to      │◄──────── Yes ──────────────│           │
   │  be zero                  │                             ◇───────────◇
   └──────────────────────────┘                                   │
                                                                  No
                                                                   ▼
   ┌──────────────────────────┐                             ◇─────────────◇
   │  Adjust Pu to a new level │◄──────── Yes ──────────────│ Pi, Pu < 15kW │
   └──────────────────────────┘                             ◇─────────────◇
                                                                   │
                                                                  No
                                                                   ▼
   ┌──────────────────────────┐     ◇─────────◇              ◇───────────◇
   │ Injecting power according │◄─Yes│ P-Pu<15kW│◄─── Yes ───│  SOC>50%  │
   │ to the difference of P &Pu│     ◇─────────◇              ◇───────────◇
   └──────────────────────────┘          │                         │
                                         No                        No
                                          ▼                         │
                              ┌──────────────────────┐              │
                              │ Injecting full power │              ▼
                              └──────────────────────┘      ┌──────────────────┐
                                          │                 │ Stop injecting    │
                                          └─────────────────│ power             │
                                                            └──────────────────┘
```

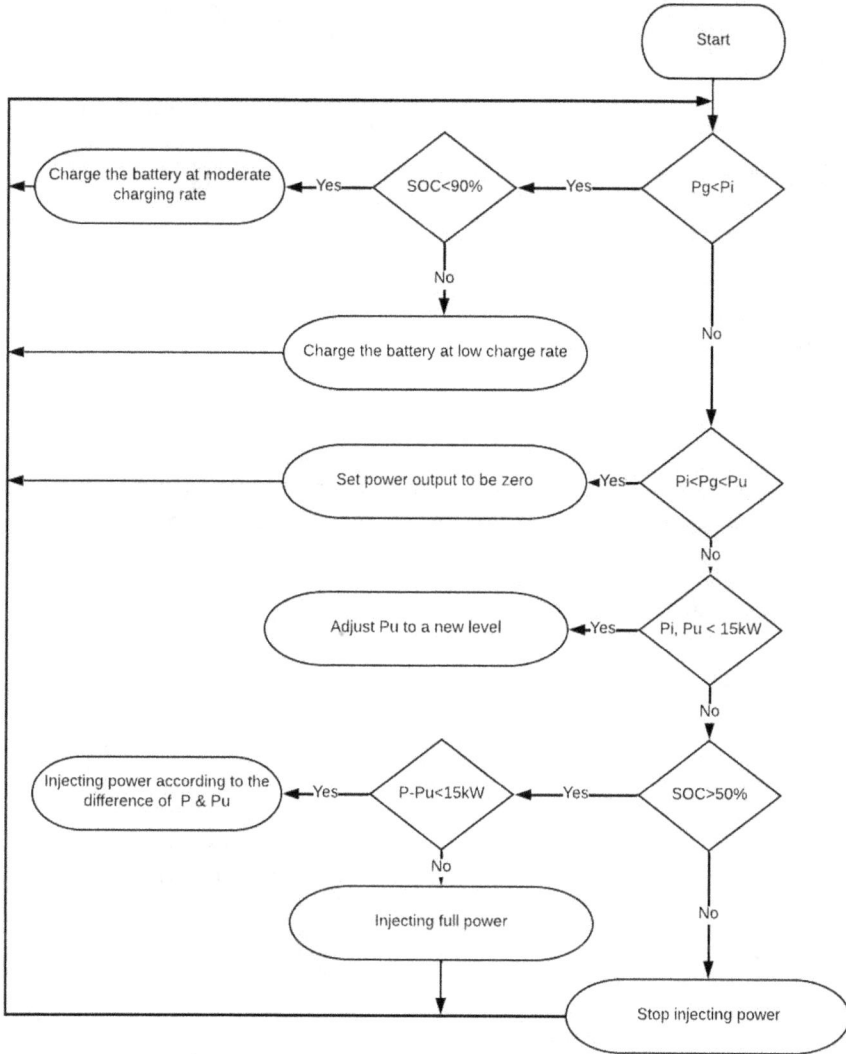

FIGURE 17.4 Flowchart of the control algorithm for the peak load shaving.

management systems [4]. Figure 17.5 represents the IoT and data mining approach for buildings.

The IoT platform uses sensors and systems such as lighting, heating, ventilation, air conditioning, and security devices connected to a standard protocol. "IoT is a fast-expanding digital ecosystem connected to devices" [5]. The aim of the IoT in retrofitting an existing building is the energy-saving potential in heating and cooling systems, predictive maintenance using machine learning techniques to determine when a service has to be performed, and condition monitoring [6].

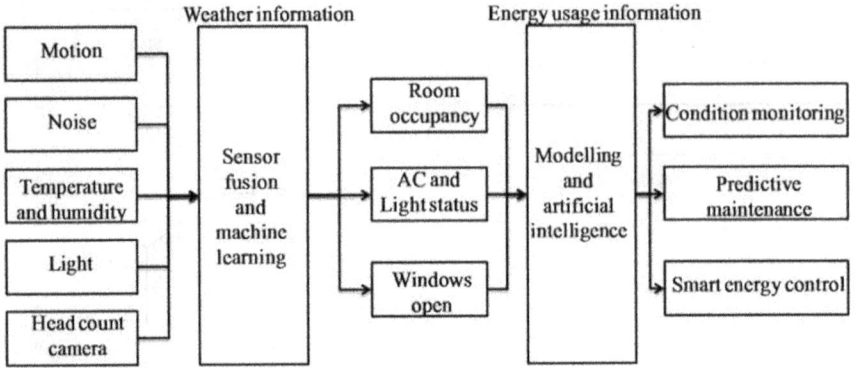

FIGURE 17.5   IoT and machine learning approach for buildings.

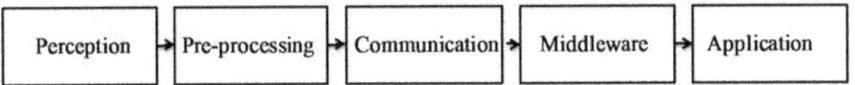

FIGURE 17.6   IoT-based architecture for smart buildings.

17.2.4   IoT-Based Architecture for Smart Buildings

Figure 17.6 represents the proposed architecture chain for IoT-based smart buildings.

17.2.4.1   Perception Layer

This layer manages information and data assortment in the substantial world [7] and is typically constituted by detection and activation. The accompanying technologies generally bolster the perception layer:

a. **Sensor networks** comprise various types of sensors that can quantify the physical world conditions, for example, position, inhabitants, motion, activities, movement, temperature, etc. Adaptability and dynamic design permit distant checking through sensor hub correspondence.
b. **Smart cameras** determine the energy use by the occupants and their activity.
c. **Radio frequency identification (RFID)** involves electromagnetic fields, which identify the object through labels. RFID can locate and track entities or individuals and communicates the data suitable for smart-built environment applications.

The perception layer involves temperature, humidity, indoor air environment sensors, cameras, RFID, and smart meters. The main challenges involved are feasibility, privacy, power consumption, security, storage requirement, sensing range limitation, cost, complex data, and stability [8].

17.2.4.2   Pre-Processing Layer

Thislayer is liable for preparing and transmitting the crude information acquired from the first layer. In addition, this layer deals with capacities, for example, figuring out the executive's form. To transfer information among systems, wired and remote communications principles are involved [9]. Since remote innovation has more points of interest over a wired connection and the subsequent IoT will be extended to the overall scale, the following portrays only some primary remote advances/media:

a. **Wi-Fi** is a correspondence innovation that utilises radio waves for the local area among devices dependent on "IEEE 802.11."
b. **Bluetooth** is another remote correspondence innovation for information trade between devices over short distances.
c. **Zigbee** is an "IEEE 802.15.4-based specification" intended for brief correspondence which involves low vitality and power utilisation. Other mainstream advances in the system layer incorporate "Z-Wave, RFID, WAVE, IrDA, and USB (wired)" [10].

17.2.4.3   Communication Layer

The transmission of information in the system layer must communicate to correspondence conventions on the side of the IoT framework. These conventions or measures are defined and proposed by different gatherings, for example, "IETF, IEEE, ETSI", and so on, and are officially acknowledged in the business for unifying the executives.

a. **IPV6 (6LoWPAN):** 6LoWPAN is structured by integrating "low-power wireless personal area networks (LoWPAN)" and "IPv6." The focal points incorporate high network and similarity with low-power utilisation.
b. **MQTT:** "Message Queue Telemetry Transport (MQTT)" is a message convention for associating distant implanted sensors and middleware. It is one of the application layers conventions.
c. **CoAP:** "Constrained Application Protocol (CoAP)" is likewise a messaging convention that depends on "REST" on the lay of "HTTP" functionalities. CoAP empowers minimal devices with reduced power, energy, computation, and capabilities to use in "RESTful interactions."

17.2.4.4   Middleware Layer

Middleware is a software layer sandwiched between the hardware and the applications that utilise the data generated by the underlying technology. It aids in the concealment of variability and facilitates application development [26].

a. **HYDRA,** renamed an innovative link project in 2014, is a middleware for an intelligently designed embedded system with a service-oriented architecture. HYDRA can be used to create applications in the smart home, health care, and agriculture sectors.

b. **AWESOME** is a web service middleware for ambient intelligence applications. It is being developed as part of the smart IHU Initiative for the implementation of smart-university deployment.

c. **The CASAS research team started CASAS smart home in a box project.** The project aims to create a large-scale data and tool infrastructure for smart environments. It is a lightweight event-based middleware that makes setting up a smart home environment simple.

d. **Context-Aware CAMUS** (Context-Aware Middleware for Ubiquitous Computing Systems) is a tuple-based middleware rendering support for innovative home application development.

e. **Smart Home Service Middleware (SHSM)** is utilised to execute administration reconciliation and insightful assistance. SHSM depends on the incorporated programming sending model. SHSM executes a gadget administration specialist for coordinating heterogeneous administrations inside the home space.

17.2.4.5 Application Layer

The application layer accommodates the working modules of the IoT framework. The application layer fills in as the front-end interface to give dynamic analytic outcomes for clients in associated businesses with augmented knowledge which is identified with the scientific apparatuses that enhance the ability to depict, anticipate, and use relativeness with different phenomena. It is a type of conceptualisation of deep learning, then again, it has enlarged insight emphasising upgrading human knowledge instead of supplanting it.

a. **Predictive analytics** This is a technique to design a complicated algorithm and model that initiates the devices to predict like humans. In industries, it is called a proactive investigation.

b. **Cloud computing** The reality of "big data" relating to processing capacity is essential to help information stockpiling and analysis. Distributed computing is a web-based figuring technique that gives shared PC assets to different devices on request. Since standard sensors don't generally have a fantastic figuring ability, distributed computing will assume the significant job of information handling for the next generation.

17.3 CONCLUSIONS

Smart sensors perform logic functions, two-way communication, and make decisions. They are a single package that performs multiple functionalities such as filtering, signal conditioning, and noise reduction; with the inbuilt microprocessor, additional hardware is not required for the system. An intelligent sensor such as an optical sensor, infrared detector array, and accelerometer has high energy-saving potential for building applications. Peak demand reduction strategies, scheduling peak distribution, peak shrinking, and thermal energy storage/energy storage have been strategically highlighted for offsetting and reducing the building peak demand. Further research on smart sensors can enable integration with the grid to reduce the grid peak demand.

ACKNOWLEDGEMENT

The authors thank the Vellore Institute of Technology (VIT), Vellore, for providing technical and financial support.

REFERENCES

[1] Yang, L., Yan, H. and Lam, J.C., 2014. Thermal comfort and building energy consumption implications – a review. *Applied Energy*, *115*, 164–173.

[2] Cao, X., Dai, X. and Liu, J., 2016. Building energy-consumption status worldwide and the state-of-the-art technologies for zero-energy buildings during the past decade. *Energy and Buildings*, *128*, 198–213.

[3] Cohen, W. S. 1997. Report of the quadrennial defense review.

[4] Nagayama, T. and Spencer Jr, B.F., 2007. *Structural Health Monitoring using Smart Sensors*. Newmark Structural Engineering Laboratory. The University of Illinois at Urbana-Champaign.

[5] Rocha, J.G., Couto, C. and Correia, J.H., 2000. Smart load cells: an industrial application. *Sensors and Actuators A: Physical*, *85*(1–3), 262–266.

[6] Leadbetter, J. and Swan, L., 2012. Battery storage system for residential electricity peak demand shaving. *Energy and Buildings*, *55*, 685–692

[7] Chua, K.H., Lim, Y.S., and Morris, S., 2016. Energy storage system for peak shaving. *International Journal of Energy Sector Management*.

[8] de Salis, R.T., Clarke, A., Wang, Z., Moyne, J., and Tilbury, D.M., 2014, July. Energy storage control for peak shaving in a single building. *2014 IEEE PES General Meeting| Conference & Exposition* (pp. 1–5). IEEE.

[9] Agamah, S.U. and Ekonomou, L., 2017. Energy storage system scheduling for peak demand reduction using evolutionary combinatorial optimization. *Sustainable Energy Technologies and Assessments*, *23*, 73–82.

[10] Suryadevara, N.K., Mukhopadhyay, S.C., Kelly, S.D.T., and Gill, S.P.S., 2014. WSN-based smart sensors and actuators for power management in intelligent buildings. *IEEE/ASME transactions on mechatronics*, *20*(2), 564–571.

18 Assessment of Flood Susceptibility in Hare River Catchment, Rift Valley, Southern Ethiopia, Using Geospatial and Drainage Morphometric Analysis

Muralitharan Jothimani, Jagadeshan Gunalan, Ephrem Getahun, Abel Abebe, and Meseret Desalegn

CONTENTS

18.1 INTRODUCTION

Floods cause one-third of all geophysical catastrophes globally [1], and they cause loss of life of humans and animals and property damage [2]. Floods are caused by various elements, including land use/land cover (LULC), terrain gradient, drainage network, weather, type of soil, etc. [3,4]. Ethiopia is a topographically diversified country with hills and lowlands. It is divided into nine major river basins, each having a drainage system that runs from the central highlands to the peripheral or surrounding lowlands.

DOI: 10.1201/9781003331001-18

303

The primary perennial rivers and tributaries' maximum discharges are during the rainy season (June–September) [5]. Due to Ethiopia's topography and elevational characteristics, flooding is a severe problem. Floods occur in various locations and multiple times, and their scale has also varied. Significant floods have been recorded in several parts of the country [5].

Flash and river floods are the two types of floods that are most probable to happen in Ethiopia and are the most destructive [5]. Following high precipitation in upstream watersheds, flash floods occur downstream, where they show themselves with remarkable intensity, velocity, and impact. As a result, floods frequently cause significant damage, which is exacerbated and lethal when they run through or along densely inhabited regions and infrastructure. Currently, the research focuses on the Hare watershed in the southern part of the Ethiopian Rift Valley. In this area, the Hare River serves as farmers' primary irrigation source. On the other hand, flooding frequently causes crop damage, and flooding has become an annual phenomenon and a source of concern for the farming area along the watershed. When the streamflow gets much more rainfall from the upper reaches, flooding occurs in the floodplain in the watershed's lowland [6].

It is necessary to examine a watershed's linear, aerial, and relief properties quantitatively to perform morphometric analysis [7]. Hard rock terrain requires the development and management of watersheds to maintain a balance between the supply and demand sides of the ecosystem. Drainage morphometric analysis helps understand the geomorphic configuration of drainage basins and their hydrologic characteristics, erosional and flood proneness, and mass movement characteristics [8,9]. Besides, it can be used for prospecting surface, groundwater, and ecological studies [10,11]. The researchers listed below have used drainage morphometric analysis to identify flood-prone locations in many parts of the world, including Ethiopia: "In the upper Jhelum basin [12], Ras En Naqb Area, South Jordan [13], Panjkora Basin, Eastern Hindu Kush, Pakistan [14], Mayurakshi River, India [15], flash flood risk estimation along the southern Sinai, Egypt [16], Agula watershed, northern Ethiopia [17], Megech River Catchment, Lake Tana Basin, North Western Ethiopia [18]".

The morphometric parameters of a watershed have traditionally been derived from topographical maps or field observations, which is a laborious and time-consuming process. Digital elevation models (DEMs) can extract these parameters more efficiently and accurately than the traditional methods used for the past two decades [19]. A number of studies have been conducted using remote sensing data and geographic information systems (GISs) to study the morphometric properties of drainage networks [20–23]. Several freely available data sets from various sensors, including the "Shuttle Radar Topography Mission (SRTM), Advanced Spaceborne Thermal Emission and Reflection Radiometer (ASTER), and Advanced Land Observing Satellite–Phased Array type L-band Synthetic Aperture Radar (ALOS–PALSAR)", have been used to perform morphometric analyses on drainage basins [12,24,25]. The precision and accuracy of the calculations of a morphometric analysis greatly depend on the pixel sizes of the DEM. It is generally accepted that DEMs with a higher resolution produce more accurate morphometric results. This study aims to quantify the morphometric drainage characteristics of the Hare River watersheds and the relationship between these characteristics and flood risk.

Identifying flood-prone locations allows decision-makers to prioritise priority areas over non-flood-prone ones. The linear and shape morphometric variables for each sub-watershed in the study area were first determined, followed by prioritisation of the sub-watersheds using compound factors. It is expected that this type of study will make a significant contribution to soil and water conservation. The current study area lacks measurements and data on hydrological function, thus requiring a quantitative morphometric assessment. There has been no study utilising the current methodology in the area under study.

18.1.1 Study Area

Approximately 161 km^2 of the Hare watershed is geographically situated between 37° 26' and 37' 36' ast and 6° 02' and 6' 17' north, as shown in Figure 18.1. There is an outlet for the River Hare that empties into Lake Abaya, one of the lakes that is part of the Abaya-Chamo lake basin in the southern Ethiopian Rift Valley. This lake is the second largest in Ethiopia after Lake Tana. In the highlands, the present research area is located at a height of 3484 meters above sea level, while in the lowlands, it is situated at 1176 meters above sea level. A slope value for the current study region ranges between 0 and 47 degrees, with steep slopes occurring in the highlands and gradual slopes in the study region's lowlands (Figure 18.2). The weather of the Hare watershed differs with the elevation. In the highlands, the average annual temperature varies from 9.6 to 22.8 degrees Celsius, whereas it ranges from 17.3 to 30.4 degrees Celsius in the lowlands. According to the Food and Agricultural

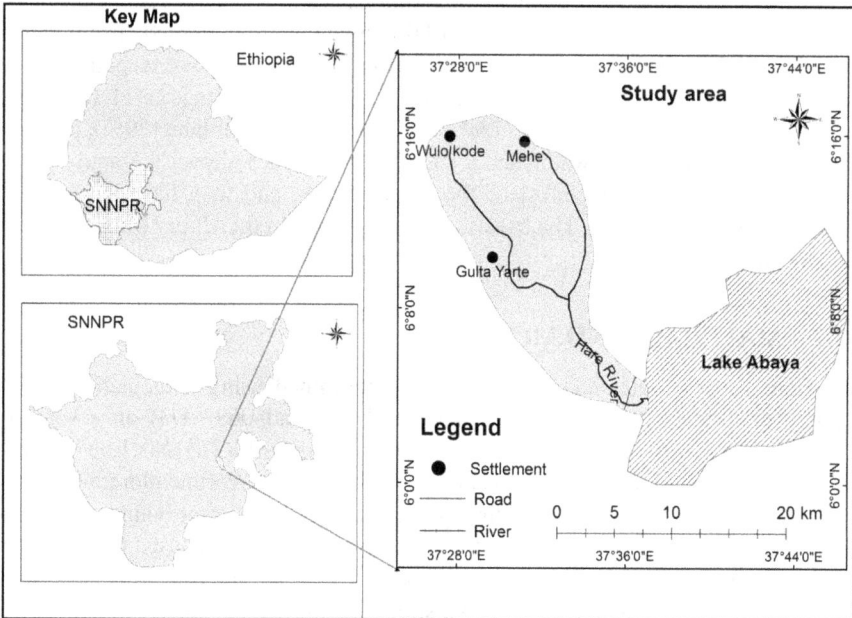

FIGURE 18.1 Study area map.

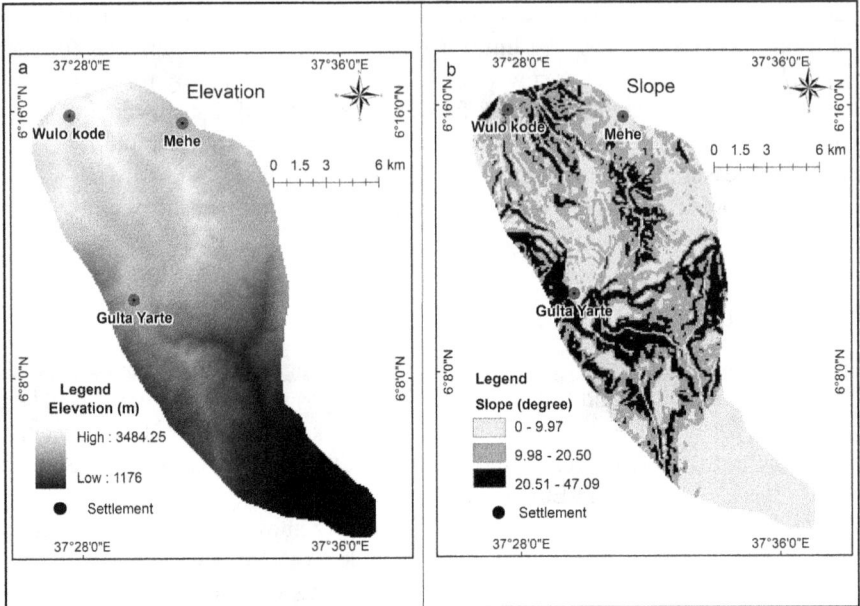

FIGURE 18.2   Elevation and slope map.

Organization, the Hare watershed's principal soil types are dystric nitisols, dystric fluvisols, orthic acrisols, and cutric fluvisols, which cover 74%, 11%, 11%, and 3% of the study area, respectively [6]. Using Landsat-8 Optical Land Image and visual interpretation of the images, the LULC map has been created. The image is available on the following website: https://earthexplorer.usgs.gov/; its path and row numbers are 169 and 56, respectively, and the date of the image is 11 December 2020. Forest (33%), water bodies (1%), farmland (11%), shrubland (29%), and bare land (26%) make up the watershed's LULC. Figure 18.3 shows the study area's LULC map. Basalts cover most of the study region, and they have a variety of weathering characteristics. The basalts range in age from the Upper Miocene to the Pleistocene.

18.2   MATERIALS AND METHODS

The Hare River's watershed boundary was determined using Ethiopian Mapping Agency toposheets (0637 C2, 0637 C4, 0637 D1, and 0637 D3) on a scale of 1:50,000. The Entity ID for the current research region is SRTM1N06E037V3, and the 30-meter resolution SRTM DEM can be accessed at http://earthexplorer.usgs.gov/. In ArcGIS, the SRTM DEM was clipped according to the current boundaries of the research area using the extract mask tool. An Arc hydro tool was used to extract the drainage network. In order to extract the drainage network, the following steps were taken: "flow direction and accumulation, stream definition, stream order, and stream to feature from the DEM". Six sub-watersheds were identified, and their boundaries

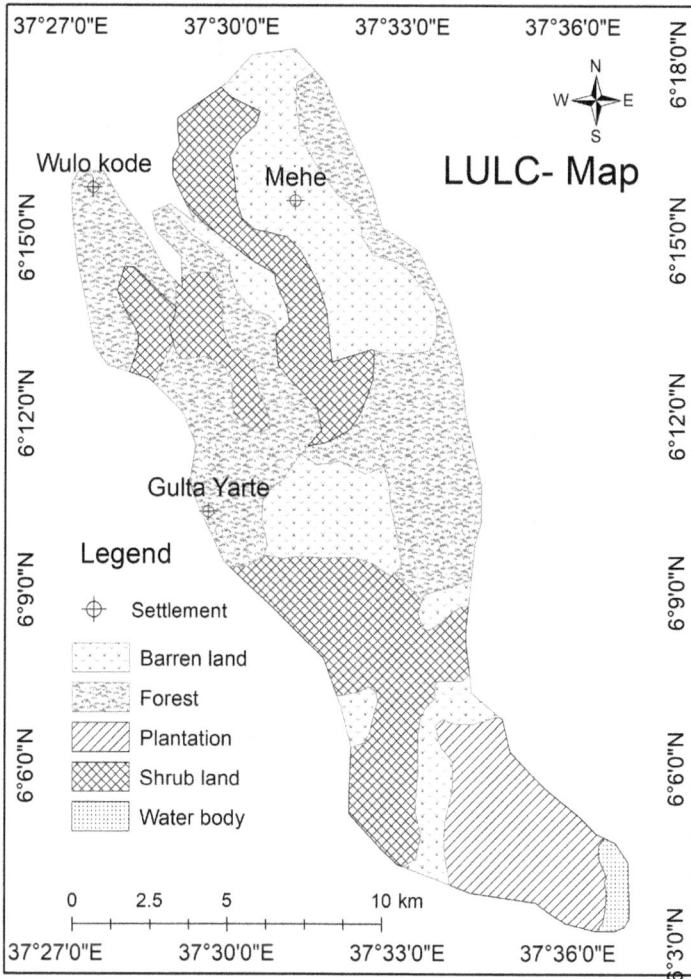

FIGURE 18.3 LULC map.

were defined based on the drainage network. The drainage system and sub-watershed boundaries of the present study region are represented in Figure 18.4.

18.2.1 THE EXTRACTION OF MORPHOMETRIC PARAMETERS

The following linear (stream length, stream order, bifurcation ratio, mean stream length, and stream length ratio, etc.) and shape (watershed area, watershed perimeter, circularity ratio, elongation ratio, form factor, drainage density, drainage frequency, compactness coefficient, drainage texture, and length of overland flow, etc.) were used. The authors of Refs. [26–30] proposed a formula for calculating morphometric parameters, which was used to determine the parameters and prioritise the sub-watersheds.

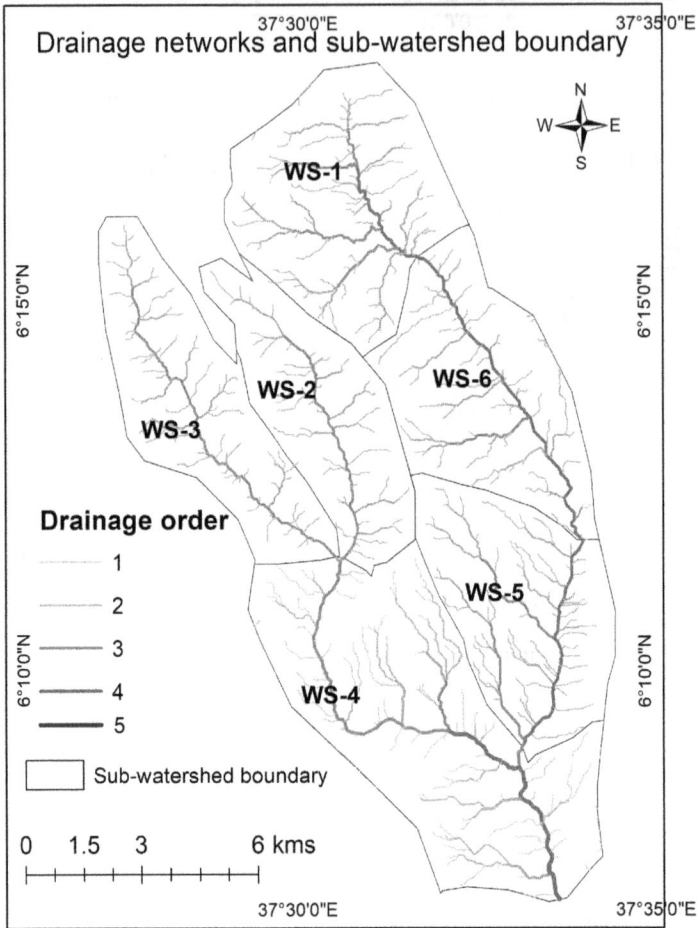

FIGURE 18.4   Drainage network and sub-watershed boundary.

18.2.1.1   Linear Morphometric Parameters

Watershed studies begin with determining stream orders, as indicated in Ref. [26]. According to Ref. [29] for stream ordering, streams were grouped according to their relative position in the stream hierarchy. As the order of the streams rises, the number of streams decreases. The number of streams in each sequence was determined using GIS. In the current research region, there were 902 stream segments and a fifth-order stream order. In Table 18.1, the stream numbers are listed in ascending order, beginning with the lowest number. Each stream's length was determined by aggregating the lengths of all streams associated with each rank. The length of the stream segments decreases significantly with increasing stream order at the current location. In Sub-watershed 1, the longest stream is 93 km long, while Sub-watershed 3 contains the shortest stream at 37 km. The current research area has a total of 351 km length of streams.

Mean drainage length parameter for each sub-watershed of the study area is found by dividing the total stream length of a stream order by the number of streams in the selected order. The average drainage system length is between 0.19 and 0.32 meters, as shown in Table 18.1. Changes in topographic slope and elevation are most likely responsible for this variance. The bifurcation ratio is directly connected to the extent of permeability and diversity of the river catchment. Bifurcation values were calculated using equation 18.1.

TABLE 18.1
Basic and Linear Morphometric Analysis Results

Morphometric parameters	WS-1	WS-2	WS-3	WS-4	WS-5	WS-6
Stream order vs. numbers						
I^{st} order	86	57	72	111	68	71
IInd order	43	24	24	41	28	26
IIIrd order	18	29	46	11	10	5
IVth order	14	0	0	30	31	34
V^{th} order	0	0	0	23	0	0
Total	161	110	142	216	137	136
Bifurcation ratio						
I^{st} / IInd	2.00	2.38	3.00	2.71	2.43	2.73
IInd / IIIrd	2.39	0.83	0.52	3.73	2.80	5.20
IIIrd/ IVth	1.29	0.00	0.00	0.37	0.32	0.15
IVth/ V^{th}	0.00	0.00	0.00	1.30	0.00	0.00
Bifurcation ratio in mean	1.13	0.64	0.70	1.62	1.11	1.62
Stream length (km)						
I^{st} order	31.59	21.48	28.10	50.86	35.78	32.45
IInd order	18.35	9.32	8.28	21.50	11.10	10.40
IIIrd order	8.03	6.61	9.72	4.74	6.34	2.43
IVth order	3.26	0	0	9.60	5.89	9.04
V^{th} order	0	0	0	6.11	0	0
Total	61.23	37.41	46.1	92.81	59.11	54.32
Mean stream length						
I^{st} order	0.37	0.38	0.39	0.46	0.53	0.46
IInd order	0.43	0.39	0.35	0.52	0.40	0.40
IIIrd order	0.45	0.23	0.21	0.43	0.63	0.49
IVth order	0.23	0.00	0.00	0.32	0.19	0.27
V^{th} order	0.00	0.00	0.00	0.27	0.00	0.00
Average	0.29	0.20	0.19	0.40	0.35	0.32
Stream length ratio						
IInd / I^{st}	0.58	0.43	0.29	0.42	0.31	0.32
IIIrd / IInd	0.44	0.71	1.17	0.22	0.57	0.23
IVth / IIIrd	0.41	0.00	0.00	2.03	0.93	3.72
V^{th} / IVth	0.00	0.00	0.00	0.64	0.00	0.00
Average	0.36	0.29	0.37	0.83	0.45	1.07

$$R_b = N_u/N_u+1 \qquad (18.1)$$

where R_b = bifurcation ratio; N_u = the total number of stream segments of the order "u"; and $N_u + 1$ = the number of stream segments of the following higher order.

When a watershed has a high Rb, its permeability decreases, and when subjected to a storm event, it is more prone to experience flash floods [26]. Table 18.1 displays the bifurcation values for the current study area, ranging from 0.64 (WS-2) to 1.62. (WS-6). The stream length ratio (R_L) is the mean length of one order to the mean length of the next lower order of a drainage segment [26]. The value fluctuates erratically with elevation [31]. The stream length ratios differ among sub-watersheds, as shown in Table 18.1. Sub-watershed WS-2 had the lowest value (0.29), while sub-watershed WS-6 had the highest value (1.07). Flooding is more likely to occur as the stream length ratio increases.

18.2.1.2 Areal Morphometric Parameters

Table 18.2 shows the area and perimeter of each sub-watershed calculated using ArcGIS software's calculated geometry function. The sub-watersheds-wise length was calculated using equation 18.2. The minimum and maximum areal extent are found in WS-2 (19.77 sq. km) and WS-4 (41.98 sq. km), and the minimum and maximum perimeters were found in WS-5 (20.07) and WS-4 (31.53), respectively.

$$L_b = 1.312*A0.568 \qquad (18.2)$$

Here, the length of the watershed is L_b, while the area of the watershed is A.

The minimum and maximum watershed lengths are 7.15 in WS-2 and 10.96 in WS-4. The length of each sub-watershed is shown in Table 18.2.

An essential aspect of drainage morphometry is the circularity ratio, and it is dimensionless. R_c reaches its maximum value of 1.0 as the watershed boundary approaches the circle [27]. The circularity ratio readings in the catchment are affected by many factors, including stream frequency, LULC, geological conditions, climatic

TABLE 18.2
Results of Areal Morphometric Parameters

Watershed area (km²)	28.22	19.77	22.79	41.98	23.52	25.28
Watershed perimeter (P) (km)	20.76	22.35	24.25	31.53	20.07	22.07
Watershed length (L_b) (km)	8.75	7.15	7.75	10.96	7.89	8.22
Elongation ratio (R_e)	0.685	0.702	0.695	0.667	0.694	0.690
Circularity ratio (R_c)	0.82	0.50	0.49	0.53	0.73	0.65
Form factor (F_f)	0.37	0.39	0.38	0.35	0.38	0.37
Stream density (D_d)	2.17	1.89	2.02	2.07	2.51	2.15
Stream frequency (D_f)	5.71	5.56	6.23	4.60	5.82	5.38
Compactness coefficient (C_c)	1.10	1.42	1.43	1.37	1.17	1.24
Stream texture (D_t)	4.14	2.55	2.97	3.52	3.39	3.22
Length of overland flow (L_g)	1.09	0.95	1.01	1.04	1.26	1.08

parameters, and slope [32]. When the Rc value is low, the basin becomes more circular, and the risk of flooding in a watershed increases as a result. Equation 18.3 was used to calculate the circularity ratio (R_c).

$$R_c = 4 * \pi * A/P^2 \tag{18.3}$$

where Rc = circularity ratio; $\pi = 3.14$, A = area of the sub-watershed, and p = perimeter of the sub-watershed.

Table 18.2 displays the Rc values for each sub-watershed in the present study region, ranging from 0.49 (WS-3) to 0.82 (WS-1). The elongation ratio (R_e) varies between 0.6 and 1 under varied climatic and geological conditions. When the number is 1, the drainage basin is considered spherical. R_e values can be categorised as less elongated (less than 0.7), oval (0.9–0.7), or round (more than 0.9) [27]. R_e was determined using equation 18.4.

$$Re = 2\sqrt{(A/\pi)}/L_b \tag{18.4}$$

where A is the watershed area, $\pi = 3.14$, and L_b is the basin's length.

In this study area, the minimum and maximum R_e values are 0.667 (WS-4) and 0.702 (WS-2). Drainage density (Dd) is influenced by rainfall intensity, rock permeability, soil infiltration, and slope of the watershed. It is proportional to the length and area of streams [26]. In order to calculate the drainage density, equation 18.5 was used.

$$D_d = \Sigma L/A \tag{18.5}$$

where Stream length is L, and the watershed area is A.

High drainage density is linked to watersheds with impermeable subsurface material, scarce vegetation cover, and high relief, which increase soil erosion and flood. Low values are linked to watersheds with permeable subsurface material, good vegetation cover, and low relief, which increase infiltration capacity [26,29]. The present area's minimum and maximum drainage density values are 1.89 (WS-2) and 2.51 (WS-5). The current study area's stream density is shown in Table 18.2.

According to Ref. [33], "stream frequency is the number of drainage sections per unit area ($F_s = N_u/A$). It's sometimes referred to as "channel frequency" or "drainage frequency". Texture and lithological features, and rainfall impact" [34]. The number of streams and frequency are closely related, with their value rising as the number of streams increases. Highly permeable lithology and low flood occurrence are connected to low stream frequency values [35–37]. Drainage frequencies within the present study area are 4.60 and 6.23 in WS-4 and WS-3, respectively. Sub-watersheds-wise drainage frequency is shown in Table 18.2.

D_t (drainage texture) is a parameter that specifies the relationship between the total number of streams for all drainage orders and the perimeter of the watershed [26]. The authors of Ref. [27] categorised drainage texture into five categories: "extremely coarse (2), coarse (2–4), moderate (4–6), fine (6–8), and extremely fine (>8)". According to the results of this study, drainage texture values vary from 2.55 (WS-2) to 4.14 (WS-1), which comes into the course of the moderate category of stream texture. Higher stream texture causes more surface runoff and flooding. Table 18.2 shows the stream texture values.

A low-valued form factor has a shorter length and a higher peak flow, while a high-valued form factor has a shorter period and a more increased peak flow. It is the ratio of watershed length squared to the watershed area. Watersheds with high form factors have a circular shape, and huge peak flows but a short concentration-time, whereas elongated watersheds with low form factors have low peak flows but a lengthy concentration-time. There is a variation in form factors between 0.37 (WS-1 and 6) and 0.39 (WS-2). Form factor values are presented in Table 18.2.

The compactness coefficient (C_c) divides the perimeter of the watershed by a circumference with a similar area. C_c affects the runoff volume and the hydrograph shape during rainfall [36]. The sub-watershed is more prone to flooding and soil erosion with lower C_c values. The minimum (1.10) and maximum (1.43) C_c values were found in WS-1 and WS-3. Water flows across the surface for a certain amount of time before gathering in a single stream segment. The half reciprocal of drainage density ($L_g = 1/Dd \times 2$) is its value. High L_g values imply a smooth gradient with less structural disturbance and runoff, while low L_g values indicate a moderate to steep slope. The present study area has L_g values ranging from 0.95 to 1.26 (Table 18.2).

18.3 RESULTS AND DISCUSSION

18.3.1 SUB-WATERSHED PRIORITISATION

Several morphometric criteria were utilised in order to prioritise all six sub-watersheds within the study area based on their results. According to Refs. [18,30,31,38], morphometric parameters like bifurcation ratio (R_b), length of overland flow (L_g), stream density (D_d), stream texture (D_t), form factor (F_f), and stream frequency (D_f) have direct effects on flooding. Hence, higher ranks were assigned to high values of the mentioned-above parameters. In contrast, elongation ratio (R_e), circularity ratio (C_r), and compactness coefficient (C_c) have an inverse relationship. A higher rank was

TABLE 18.3
Calculation of the Compound Factor

Morphometric parameters	WS-1	WS-2	WS-3	WS-4	WS-5	WS-6
Stream density	5	1	2	3	6	4
Stream frequency	4	3	6	1	5	2
Stream texture	6	1	2	5	4	3
Form factor	2	5	6	1	4	3
Bifurcation ratio	4	1	2	5	3	6
Compactness co-efficient	1	5	6	4	2	3
Elongation ratio	2	6	5	1	4	3
Circularity ratio	6	2	1	3	5	4
Length of overland flow	5	1	2	3	6	4
Compound factor	3.89	2.78	3.56	2.89	4.33	3.56

assigned to lower values of the parameters mentioned above. The first priority was given to the sub-watersheds with the lowest rating value, followed by those with the next highest rating value, and so on. The priority rating of the Hare River catchment's six sub-watersheds was determined using the compound parameter value calculation method. The calculated compound factor values were classified into three categories, namely high priority (2.78–2.89) (WS-2 and 4), medium priority (2.90–3.89) (WS-1, 3, and 6), and low priority (3.90–4.33) (WS-5). Table 18.3 shows the calculation of the compound factor values. Figure 18.5 shows the sub-watersheds priority map. The high-priority sub-watersheds were identified where high drainage density and frequency are found at higher elevations. In the medium-priority sub-watersheds, moderate slopes, D_d, D_t, and moderate to high F_f and R_c values were found.

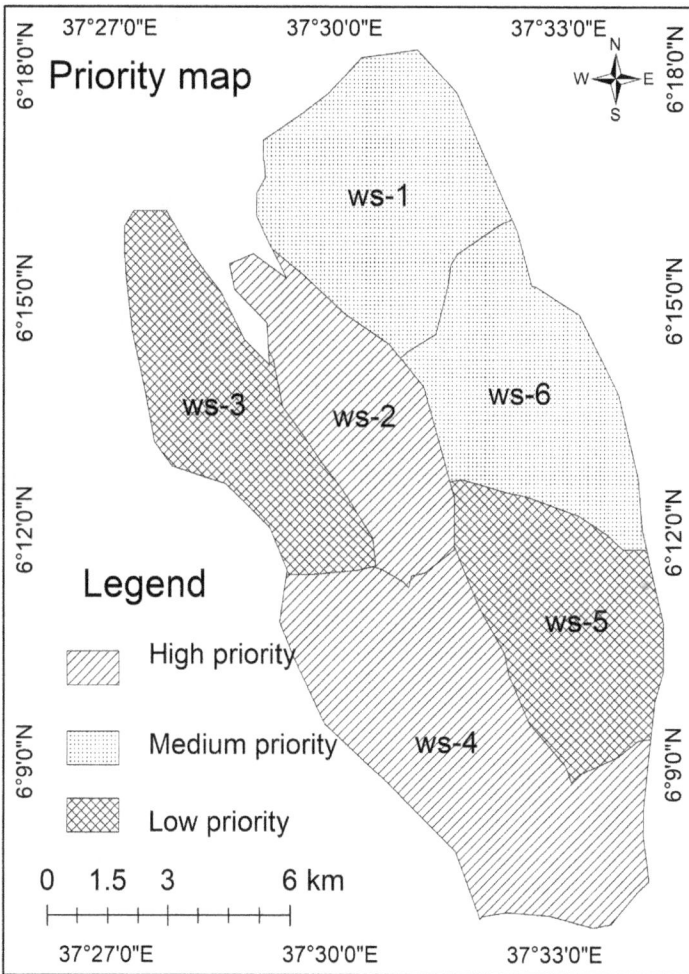

FIGURE 18.5 Sub-watershed priority map.

18.4 CONCLUSIONS

Identifying and prioritising watersheds is critical in implementing natural resource conservation measures. In the present study area, flooding is a significant environmental problem, so it demands effort to prioritise areas for flood control management. Through drainage morphometry and compound factor analysis, an attempt has been made to assess the flood-prone sub-watersheds of the Hare River catchment, Rift Valley, southern Ethiopia. The efficacy of remotely sensed data, GIS techniques, drainage morphometry, and compound factor analysis in detecting flood-prone zones at the sub-watershed level was demonstrated in this study. According to the morphometric analysis of sub-watersheds 2 and 4, flooding is more likely to occur in these watersheds. As a result, appropriate flood control steps are necessary for these watersheds to save surface water and prevent flood damage. SRTM-DEM and GIS tools for morphometric analysis are more appropriate methods, particularly flood susceptibility assessment. The technique adopted in this study is more effective and helps decision-makers in water resources conservation planning.

REFERENCES

[1] P. Adhikari, Y. Hong, K. Douglas, D. Kirschbaum, J. Gourley, R. Adler and G. Robert Brakenridge, A digitized global flood inventory (1998–2008): compilation and preliminary results, Nat. Hazards, vol. 55, no. 2, pp. 405–422, 2010. https ://doi.org/10.1007/s1106 9-010-9537-2.

[2] CEOS, The use of earth observing satellites for hazard support: assessments and scenarios. Final report of the CEOS Disaster Management Support Group (DMSG), 2003. http://hdl.handle.net/2027/mdp.39015061859966.

[3] B. Agbola, O. Ajayi, O. Taiwo and B. Wahab, The August 2011 flood in Ibadan, Nigeria: Anthropogenic causes and consequences, Int. J. Disaster Risk Sci, vol. 3, no. 4, pp. 207–217, 2012.https://doi.org/10.1007/s13753-012-0021-3.

[4] A. Youssef, S. Sefry, B. Pradhan and E. Alfadail, Analysis on causes of flash flood in Jeddah city (Kingdom of Saudi Arabia) of 2009 and 2011 using multi-sensor remote sensing data and GIS, Geomatics, Nat. Hazards Risk, vol. 7, no. 3, pp. 1018–1042, 2015. https://doi.org/10.1080/19475705.2015.1012750.

[5] W. Gashaw, D. Legesse, Flood Hazard and Risk Assessment Using GIS and Remote Sensing in Fogera Woreda, Northwest Ethiopia. In: Melesse A.M. (eds) Nile River Basin. Springer, Dordrecht, 2011.

[6] B. Menna and T. Ayalew, Modeling future flood frequency under CMIP5 Scenarios in Hare watershed, Southern Rift Valley of Ethiopia, Arab. J. Geosci., vol. 14, no. 20, 2021. doi: 10.1007/s12517-021-08479-0.

[7] J.J. Clark, "Morphometry from map, essays in geomorphology". Elsevier, New York, pp 235–274, 1966.

[8] R. Breilinger, H. Duster, R. Weingartner, Methods of catchment characterization by means of basin parameters (assisted by GIS) – empirical report from Switzerland: report, vol 120. UK Institute of Hydrology, pp 171–181, 1993.

[9] E. Bassey Eze and J. Efiong, Morphometric parameters of the Calabar River Basin: Implication for hydrologic processes, J. Geogr. Geol., vol. 2, no. 1, 2010. DOI:10.5539/jgg.v2n1p18.

[10] M. Mokarram and M. Hojati, Morphometric analysis of stream as one of resources for agricultural lands irrigation using high spatial resolution of digital elevation

model (DEM), Comput Electron Agric, vol. 142, pp. 190–200, 2017. DOI:10.1016/j.compag.2017.09.001

[11] V. Kumar, B. Chaplot, P. Omar, S. S. and H. Md. Azamathulla, Experimental study on infiltration pattern: opportunities for sustainable management in the Northern region of India, Water Sci. Technol., 2021.https://doi.org/10.2166/wst.2021.171

[12] M. Bhat, A. Alam, S. Ahmad, H. Farooq and B. Ahmad, Flood hazard assessment of upper Jhelum basin using morphometric parameters, Environ. Earth Sci, vol. 78, no. 2, 2019.https://doi.org/10.1007/s12665-019-8046-1.

[13] Y. Farhan, O. Anaba and A. Salim, Morphometric analysis and flash floods assessment for drainage basins of the Ras En Naqb area, South Jordan using GIS, J. Geosci. Environ. Prot. , vol. 04, no. 06, pp. 9–33, 2016.doi: 10.4236/gep.2016.46002.

[14] S. Mahmood and A. Rahman, Flash flood susceptibility modeling using geo-morphometric and hydrological approaches in Panjkora Basin, Eastern Hindu Kush, Pakistan, Environ. Earth Sci., vol. 78, no. 1, 2019. https://doi.org/10.1007/s12 665-018-8041-y.

[15] A. Islam and S. Deb Barman, Drainage basin morphometry and evaluating its role on flood-inducing capacity of tributary basins of Mayurakshi River, India, SN Appl. Sci., vol. 2, no. 6, 2020. https://doi.org/10.1007/s42452-020-2839-4

[16] Youssef, B. Pradhan and A. Hassan, Flash flood risk estimation along the St. Katherine road, southern Sinai, Egypt using GIS based morphometry and satellite imagery, Environ. Earth Sci., vol. 62, no. 3, pp. 611–623, 2010. https://doi.org/10.1007/s12 665-010-0551-1.

[17] Fenta, H. Yasuda, K. Shimizu, N. Haregeweyn and K. Woldearegay, Quantitative ana-lysis and implications of drainage morphometry of the Agula watershed in the semi-arid northern Ethiopia, Appl. Water Sci., vol. 7, no. 7, pp. 3825–3840, 2017. https://doi.org/10.1007/s13201-017-0534-4.

[18] M. Jothimani, Z. Dawit and W. Mulualem, Flood susceptibility modeling of Megech River catchment, Lake Tana Basin, North Western Ethiopia, using morphometric ana-lysis, Earth Syst. Environ., vol. 5, no. 2, pp. 353–364, 2020. https://doi.org/10.1007/s41748-020-00173-7.

[19] K. White, J. Bullard, I. Livingstone and L. Moran, A morphometric comparison of the Namib and southwest Kalahari dune fields using ASTER GDEM data, Aeolian Res., vol. 19, pp. 87–95, 2015. doi: https://doi.org/10.1016/j.aeolia.2015.09.006

[20] S. Tripathi, S.K. Soni, A.K. Maurya, Morphometric characterization and prioritization of sub-watershed of Seoni River in Madhya Pradesh through remote sensing and GIS technique. Int. J. Remote Sens. Geosci., Vol 2, no.3, pp.46–54. 2013

[21] S. Soni, Assessment of morphometric characteristics of Chakrar Watershed in Madhya Pradesh, India using geospatial technique, Appl. Water Sci. vol. 7, pp. 2089–2102. https://doi.org/10.1007/s13201-016-0395-2.

[22] S. Bisht, S. Chaudhry, S. Sharma and S. Soni, Assessment of flash flood vulner-ability zonation through geospatial technique in high altitude Himalayan watershed, Himachal Pradesh India, Remote Sens. Appl. Soc. Environ., vol. 12, pp. 35–47, 2018. Doi.10.1016/j.rsase.2018.09.001.

[23] F. Alqahtani and A. Qaddah, GIS digital mapping of flood hazard in Jeddah–Makkah region from morphometric analysis, Arab J. Geosci., vol. 12, no. 6, 2019. https://doi.org/10.1007/s12517-019-4338-8.

[24] M. Adnan, A. Dewan, K. Zannat and A. Abdullah, The use of watershed geomorphic data in flash flood susceptibility zoning: a case study of the Karnaphuli and Sangu river basins of Bangladesh, Nat. Hazards, vol. 99, no. 1, pp. 425–448, 2019. https://doi.org/10. 1007/s11069-019-03749-3.

[25] G. Meraj, T. Khan, S.A. Romshoo, M. Farooq, K. Rohitashw, B.A. Sheikh, An integrated geoinformatics and hydrological modelling-based approach for effective flood management in the Jhelum basin, NW Himalaya, Proceedings, vol. 7, no. 8, 2019. https://doi.org/10.3390/ECWS-3-05804.

[26] R.E. Horton, Erosional development of streams and their drainage basins: a hydrophysical approach to quantitative morphology, Geol. Soc. Am. Bull., vol. 56 pp. 275–370, 1945. https://doi.org/10.1007/BF01033300.

[27] V. Miller, A quantitative geomorphologic study of drainage basin characteristics in the clinch mountain area, In Technical Report. 3 Virginia and Tennessee. Department of Geology – Columbia University, 1953.

[28] S.A. Schumm, The evolution of drainage system and slopes in bad lands at Perth Amboy, New Jersey, Bulletin of Geological Society of America, vol. 67, pp 214–236, 1956. https://doi.org/10.1130/0016- 7606(1956)67[597:EODSAS]2.0.CO;2

[29] A.N. Strahler, Quantitative geomorphology of drainage basin and channel network, In: Chow VT (ed) Handbook of Applied Hydrology. McGraw Hill Book Company, New York, pp 39–76, 1964.

[30] K. Nooka Ratnam, Y. Srivastava, V. Venkateswara Rao, E. Amminedu and K. Murthy, Check dam positioning by prioritization of micro-watersheds using SYI model and morphometric analysis – Remote sensing and GIS perspective, J. Indian Soc. Rem. Sensing, pp. 25–38, 2005. doi:10.1007/BF02989988.

[31] P. Sreedevi, K. Subrahmanyam and S. Ahmed, The significance of morphometric analysis for obtaining groundwater potential zones in a structurally controlled terrain, Environ. Geol., vol. 47, no. 3, pp. 412–420, 2004. https://doi.org/10.1007/s00254-004-1166-1.

[32] P. Singh, J. Thakur and U. Singh, Morphometric analysis of Morar River Basin, Madhya Pradesh, India, using remote sensing and GIS techniques, Environ. Earth Sci, vol. 68, no. 7, pp. 1967–1977, 2012. https ://doi.org/10.1007/s1266 5-012-1884-8.

[33] R. Horton, Drainage-basin characteristics, Trans. Am. Geophys. Union, vol. 13, no. 1, p. 350, 1932. https://doi.org/10.1029/TR013i001p00350.

[34] J. Veeranna, K. Gouthami, P. Yadav and V. Mallikarjuna, Calculating linear and areal and relief aspect parameters using geo-spatial techniques (ArcGIS 10.2 and SWAT model) for Akkeru River Basin Warangal, Telangana, India, Int. J. Curr. Microbiol. App. Sci., vol. 6, no. 10, pp. 1803–1809, 2017.

[35] V. Joji, A. Nair and K. Baiju, Drainage basin delineation and quantitative analysis of Panamaram watershed of Kabani river basin, Kerala using remote sensing and GIS, J. Geol. Soc. India, vol. 82, no. 4, pp. 368–378, 2013. https://doi.org/10.1007/s12 594-013-0164-x.

[36] S. Maurya, P. Srivastava, M. Gupta, T. Islam and D. Han, Integrating soil hydraulic parameter and microwave precipitation with morphometric analysis for watershed pri-oritization, Water Resour. Manag., vol. 30, no. 14, pp. 5385–5405, 2016. https://doi.org/10.1007/s11269-016-1494-4.

[37] Abboud and R. Nofal, Morphometric analysis of wadi Khumal basin, western coast of Saudi Arabia, using remote sensing and GIS techniques, J. Afr. Earth Sci., vol. 126, pp. 58–74, 2017. https://doi.org/10.1016/j. jafrearsci.2016.11.024

[38] S. Gajbhiye, S. Mishra and A. Pandey, Prioritizing erosion-prone area through morphometric analysis: an RS and GIS perspective, Appl. Water Sci, vol. 4, no. 1, pp. 51–61, 2013. https://doi.org/10.1007/s13201-013-0129-7.

19 Assessment of COVID-19 Impact on the Vegetation and Urbanisation of Dehradun City Using Geospatial Techniques

Ashish Mani, Deepali Bansal,
Hanuth Saxena, Maya Kumari, Deepak Kumar,
Dharmendra Kumar, and Sulochana Shekhar

CONTENTS

19.1   INTRODUCTION

The COVID-19 pandemic has been the biggest and worst disaster of the 21st century. Bringing the whole world to a stop and claiming millions of lives globally, creating a nightmare for human civilisation. COVID-19, previously known as 2019-nCoV, belongs to the subfamily Orthocoronavirinae of the family Coronaviridae (Liu et al., 2020). It has 90% similarity with the severe acute respiratory syndrome

DOI: 10.1201/9781003331001-19

coronavirus (SARS-CoV) and hence is closer to the SARS-CoV genome in comparison to the Middle East respiratory syndrome coronavirus (MERS-CoV) genome (Kannan et al., 2020). The origin of this virus has not yet been tracked, with the hypotheses of different virologists claiming that it originated from bats and rodents. Some strains of coronavirus are common in avian species as well (Liu et al., 2020). Rumours of the virus leaking from the labs of the Wuhan Institute of Virology are widespread and are being investigated by the UN and US, although some scientists believe that there has been some sort of human influence involved with the virus for it to enter human cells (Maxmen et al., 2021). On the other hand, many scientists believe that this virus was not engineered in a lab, as we have seen it evolving and many other strands of the same family coronaviridae are present in nature and have encoded in its indentation of the cell's surface that begins the progression to infect the cells of humans and other mammals (Maxmen et al., 2021). Other than humans, coronavirus has been found in minks and multiple different species of carnivorous mammals (Maxmen et al., 2021).

As this virus has a very high rate of infection, methods like lockdowns and the use of protective gear such as masks and sanitisers have become more common these days. R_0 is the reproduction number for a virus; it is the number of cases directly caused by an infected person (Dharmaratne et al., 2020). R_0 values of coronavirus are very high and have been seen to change from country to country and strand to strand. One of the research projects showed the R_0 value to be as high as 14.8 (Billah et al., 2020). With the virus constantly evolving it is difficult for researchers to brew up a cocktail and make a vaccine, hence making it even more deadly.

Our study doesn't work on the human aspect of coronavirus, rather it works on the environmental side. We wanted to see how much the environment has benefitted from the whole lockdown situation. So, for this, we selected the smart city of Dehradun. Landsat 8 OLI satellite images were taken for pre-COVID and post-COVID times for the years 2020 and 2021. With reduced anthropogenic activities rehabilitation in nature is inevitable, but pressure on different aspects of the environment is also predominant. To be more precise, as everyone must be indoors, there is a considerable increase in electricity consumption, water, waste generation, etc. Even some people who were away from their homes due to their jobs would have returned during this period, hence increasing the overall pressure on the local natural resources. Other than that, if we look at the human aspect, massive unemployment, emotional trauma, and deaths were observed as some of the negative effects of the virus. On contemplating though, due to the decline in human interference in the environment an improvement would have been certain.

Various studies and research are focusing COVID now. They are more focused on the biology part currently. Multiple vaccines and medicines are under study, most of them are in the trial phase and hence they are not available to the public as of now. A few vaccines have been started to be administered to people. Although some of them are still in the third phases of clinical trials, the situation is so dire that the administration of vaccines has become a necessity. There are some studies related to impacts on the environment as well and these studies show some drastic changes in the processes, such as a reduction in greenhouse gases, fewer carbon emissions, reduced NO_2 and

$PM_{2.5}$ levels, less environmental noise, and cleaner beaches (Manuel et al., 2020). Some of the adverse effects of COVID on the environment are also discussed in this paper such as increased waste, reduction in waste recycling, and other indirect environmental effects (Manuel et al., 2020).

According to many studies, the conditions of many of the cities in India in terms of air pollution were improved during the time of the pandemic (Maithani et al., 2020; Siddiqui et al., 2020). Studies also showed a 27% overall improvement in air quality in cities having a population of 5 million or greater (Maithani et al., 2020; Mahato et al., 2020). In the national capital region of India, an improvement in the air quality of about 35% was evident (Chauhan et al., 2020). These major changes were all due to the lockdown introduced in India in 2020 during the first wave of the COVID-19 pandemic. Reduced industrial activities, as well as transportation, had a major contribution in terms of rehabilitation of air quality. Studies showed that the ambient air quality of the city of Dehradun itself was under the permissible limits during the COVID lockdown situation, hence showing the overall rejuvenation of the environment (Kanchan et al., 2020).

Noise pollution saw a great decline during the lockdown, it was dropped by about 20–30% in the Himalayan region (Kanchan et al., 2020). Certain studies showed that even the slightest increase in noise can increase the risk of hypertension by 3.4% (Basu et al., 2021; Kim et al., 2019; Oh et al., 2019; Eriksson et al., 2012). Not only that, but anxiety and depression have been correlated with noise for a very long time and have been considered among the leading causes of these mental illnesses (Basu et al., 2021; Belojević et al., 1997; Fyhri et al., 2010; Ongel et al., 2016). Other illnesses such as high blood pressure and different cardiovascular ailments have been also observed as the harmful effects of exposure to loud noises for a long duration of time (Basu et al., 2021; Said et al., 2016; Münzel and Sørensen, 2017). Noise pollution has been described by WHO as the third most hazardous pollution after air and water (WHO, 2005). During the lockdown, an overall decrease in noise pollution was seen. A decrease from 44.85–79.57 dB to 38.55–57.79 dB was observed, improving the ambient quality throughout the nation in terms of excessive noise (Mishra et al., 2020).

Another major improvement was seen in the water quality of different rivers. many researchers turned to remote sensing data to observe the water quality of rivers. In one of the studies, turbidities of the Ganges River were observed using a Sentinel-2 satellite. This study observed the changes in reflectance value and found reduced turbidity during the lockdown period for various stretches throughout the Ganges (Garg et al., 2020). According to the Central Pollution Control Board (CPCB), there were fewer effluent discharges due to reduced anthropogenic activities etc., and the overall quality of the rivers improved as well (CPCB, 2020). However, as a precautionary method in the beginning increased amounts of chlorine were added to the water to kill the virus, therefore making it harmful (Manuel et al., 2020; Koivusalo et al., 1997), although there is no evidence proving the survival of coronavirus in water (WHO, 2020b). Due to our limitations because of our present-day technology and the costs related to it, there isn't a lot of data available to us to give any concrete answer as to whether the virus sustains in water or not, especially wastewater (Bandala et al., 2021).

There have been a few negative impacts of COVID-19 on the environment as well. Waste generation has increased significantly as many people ordered things online from food to general goods and an overall growth in inorganic waste has been seen (Manuel et al., 2020). Not only that but growth in organic waste has been seen also (Manuel et al., 2020). The biggest rise has been seen in the medical waste category; one of the studies carried out in Wuhan, China, observed an almost 480% rise in the waste generated in comparison to pre-COVID times (Manuel et al., 2020). Other countries have seen this rise as well, especially in the case of protective clothing (Manuel et al., 2020; Calma, 2020). Another impact is seen in terms of the reduction in waste recycling, with many countries focusing on the pandemic with waste recycling becoming the least of their concerns, and hence recycling has either trickled down to very little or none (Manuel et al., 2020; Bir, 2020). Other concerning factors come from wastewater management. India is a developing nation, therefore many states do not have good wastewater treatment plants, either they dump wastewater directly into rivers, or the treatment plants are just not doing as good as a job they should be. Another major impact lies in the excessive use of disinfectants which are being sprayed on roads and other places like commercial buildings, offices etc.; these disinfectants are killing the beneficial microbes as well, which in the future might cause an ecological imbalance (Islam et al., 2016). As a result of the lockdown, massive unemployment has come into play due to which in remote areas illegal deforestation, fishing, and poaching have become widespread (Mehta, 2020).

Massive improvements have been seen in the environment since the pandemic. These positive impacts have done something which we have been trying to achieve for a long time. We should focus on keeping the positive impacts as they are and try to further move along with the movement nature has brought to us in terms of the rehabilitation of the environment. Even though there are some negative impacts correlated with the whole event, instead of worrying about them, we should focus on reducing these negative impacts. With our study as mentioned above, we want to see the positive impacts of the pandemic and lockdown in the Dehradun region of India.

19.2 STUDY AREA

Dehradun, the capital of Uttarakhand, is our region of interest. It lies between latitudes 29° 55' and 30° 30' and longitudes 77° 35' and 78° 24'. A map of the study area is shown in Figure 19.1. The total area of the city is 197.40 km^2. Dehradun city is surrounded by four major rivers: Song, Bindal, Rispana, and Asan. The origin of all four rivers is the Mussoorie mountain ranges.

19.3 METHODOLOGY

For this study, we used Landsat 8 OLI satellite images of the years 2019, 2020, and 2021. Pre-COVID and during-COVID data have been taken into consideration to visualise the changes brought about by the pandemic. ESRI's ArcGIS software has been used to conceptualise the different satellite imageries, also, all the raster operations as well as vector operations were processed on this software. Red, NIR,

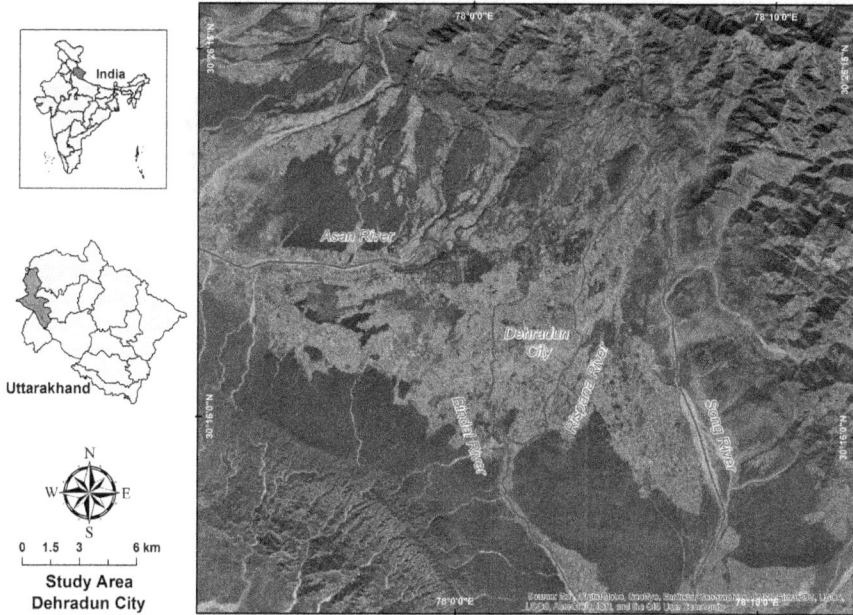

FIGURE 19.1   Study area map.

SWIR, and thermal bands of the sensors were used to create NDVI and NDBI indices and LST. Microsoft Excel was used for graphs and tables.

19.3.1 NDBI

The normalised difference built-up index, more commonly known as NDBI, is the index used for mapping and analysing built-up locations. It is calculated using the SWIR (short wave infrared) band and the NIR (near infrared) band of the satellite image. Band 5 is the NIR band which we have used and band 6 is for SWIR. The formula for calculating NDBI is:

$$NDBI = (SWIR - NIR)/(SWIR + NIR)$$

19.3.2 NDVI

Normalised difference vegetation index, more commonly known as NDVI, is the index used for mapping and analysing vegetation in a location. It is calculated using NIR (near infrared) band and the red band in the visual spectrum of a satellite image; in our case we used Landsat 8 satellite images in which band 5 is NIR and band 4 is red. The formula for calculating NDVI is:

$$NDVI = (NIR - Red)/(NIR + Red)$$

19.3.3   LST

LST, land surface temperature, is the temperature of the land at any given point of time calculated by the radiation of the land received back by a sensor. LST is calculated using thermal images from satellite imagery and the red and near-infrared bands (Franzpc, 2019). Red and near-infrared bands are used to calculate the normalised difference vegetation index, NDVI, which is explained further ahead under NDVI. To calculate LST we must go through six steps, which are discussed below:

19.3.3.1   Top of Atmospheric Spectral Radiation

Top of atmosphere spectral radiation, TOA, is the number which helps in calculating errors caused by radiance, reflection, and diffusive reflection. (UN) The formula for calculating TOA is:

$$TOA\ (L) = M_L * Q_{cal} + A_L$$

where, M_L = band-specific multiplicative rescaling factor, this is found in the meta-data of the satellite imagery and is termed as Radiance_Multi_Band_x. In our case x was 10 as we used Landsat 8 satellite imagery. The value of M_L = 3.3420E-04.

Q_{cal} = the band 10 image.

A_L = Band-specific additive rescaling factor, this is found in the metadata of the satellite imagery, and it is termed as Radiance_Add_Band_x. In our case x was 10 as we used Landsat8 satellite imagery. The value of A_L =0.10000.

Using the raster calculator input all the values to find TOA.

19.3.3.2   Brightness Temperature

Brightness temperature, BT, is the measurement of radiance from the top of the atmosphere which is travelling upwards back to the satellite. It is calculated using TOA, which we calculated in the last step; the formula for calculating BT is:

$$BT = (K_2 / (\ln (K_1 / L) + 1)) - 273.15$$

K_1 = the band-specific thermal conversion constant; it is found in the metadata of the satellite imagery, and it is termed as K_1_Constant_Band_x. In our case x was 10 as we used Landsat 8 satellite imagery. The value of K_1 = 774.8853.

K_2 = the band-specific thermal conversion constant; it is found in the metadata of the satellite imagery, and it is termed as K_2_Constant_Band_x. In our case x was 10 as we used Landsat 8 satellite imagery. The value of K_2 =1321.0789.

L = TOA, as calculated in the previous step

To calculate the temperature in Celsius, 273.15 is subtracted.

19.3.3.3   NDVI

NDVI, normalised vegetation differential index, is one of the key steps in calculating LST. It is explained below under NDVI in more detail. The formula for NDVI is:

$$NDVI=(NIR-Red)/(NIR+Red)$$

where,
NIR = near-infrared band of the satellite image, Red = red visual band of satellite image.

19.3.3.4   The Proportion of Vegetation P_v

This is another parameter which is very important for the retrieval of LST. NDVI is used to calculate P_v; although other vegetation indices can be used to calculate P_v in the case we have used NDVI. The formula for NDVI is as follows:

$$Pv = Square~((NDVI - NDVI_{min}) / (NDVImax - NDVI_{min}))$$

where,
$NDVI_{min}$ = minimum value of the NDVI
$NDVI_{max}$ = maximum value of the NDVI
$NDVI$ = data generated in the previous step

19.3.3.5   Emissivity ε

This is the power of a surface to emit heat by radiation (Merriam-Webster). Below is the formula for calculating the emissivity:

$$\varepsilon = 0.004 * Pv + 0.986$$

where,
P_v = data generated in the last step

19.3.3.6   The Final Step in Calculating LST

In the final step, LST is calculated. Previously generated brightness temperature and emissivity are used in this step. Below is the formula for calculating LST:

$$LST = (BT / (1 + (0.00115 * BT / 1.4388) * Ln(\varepsilon)))$$

where,
BT = brightness temperature generated in the previous steps.
ε = emissivity generated in the previous step.

19.3.4 Flowchart of the Following Methodology

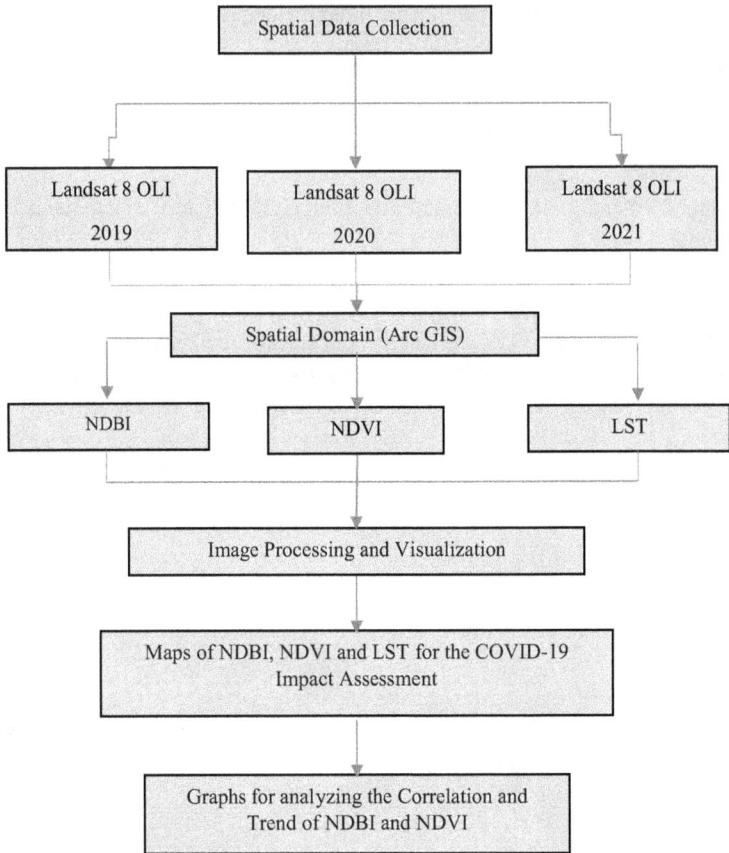

13.4 RESULT

With our study, we were able to see the changes in the vegetation and built-up area in the Dehradun region and were also able to see the temperature variations which occurred as the environment recuperated during the two lockdown periods. An alleviation in terms of the overall cooler temperature and the higher green cover area was seen. Figures 19.2–19.7 are the maps which were developed during this study and show how different changes have occurred in the year 2019, 2020, and 2021.

The maps in Figures 19.2–19.7 represent the built-up index and vegetation index in our area of interest, which is Dehradun. The scale used to prepare these maps is 1:100,000. The legend shows the distribution of vegetation and built-up index on a scale of high, medium, and low. In terms of vegetation, green represents high vegetation values, sky blue represents medium vegetation values, and deep blue represents low vegetation values, while built-up index high values are denoted by red, medium values by cream, and low values by deep blue. On the top right-hand corner the arrow

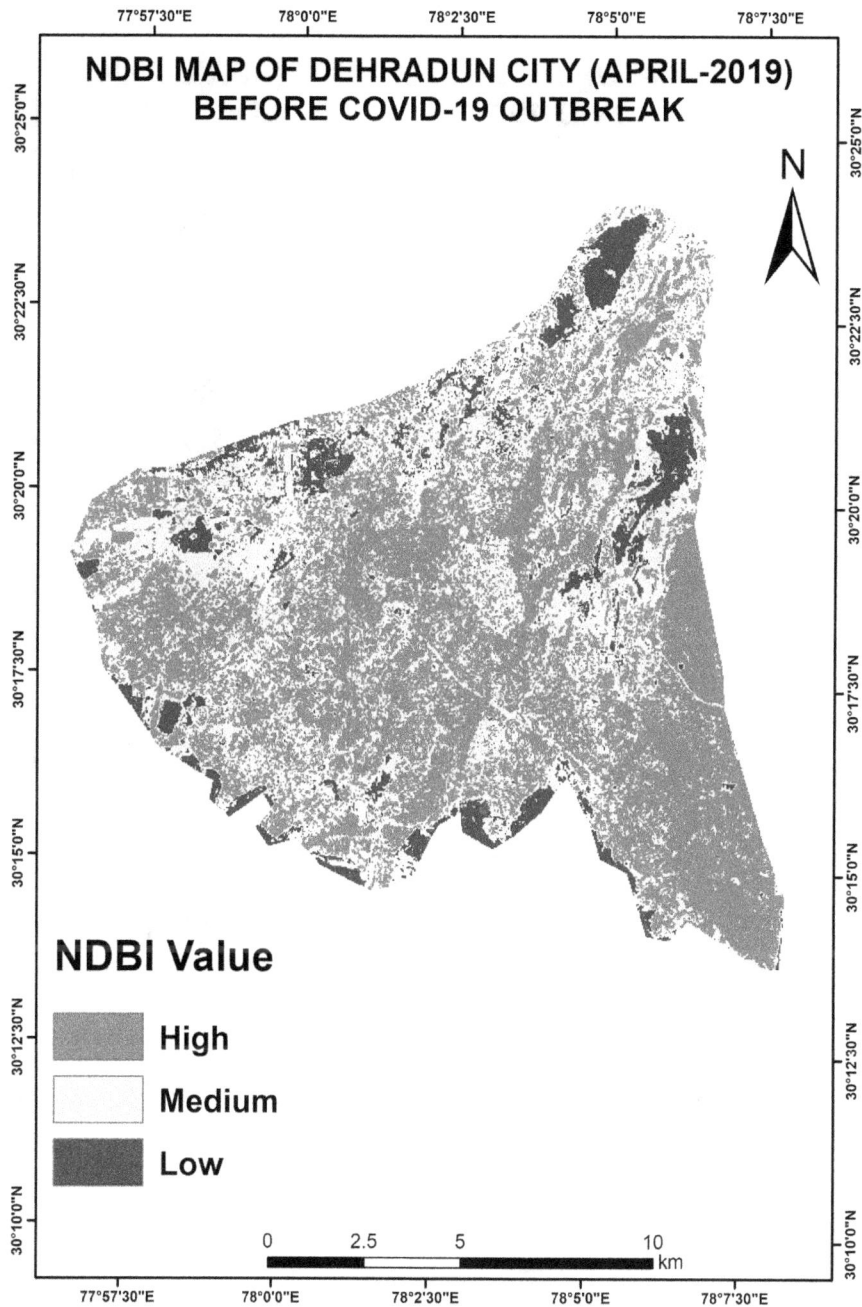

FIGURE 19.2   Map of NDBI 2019.

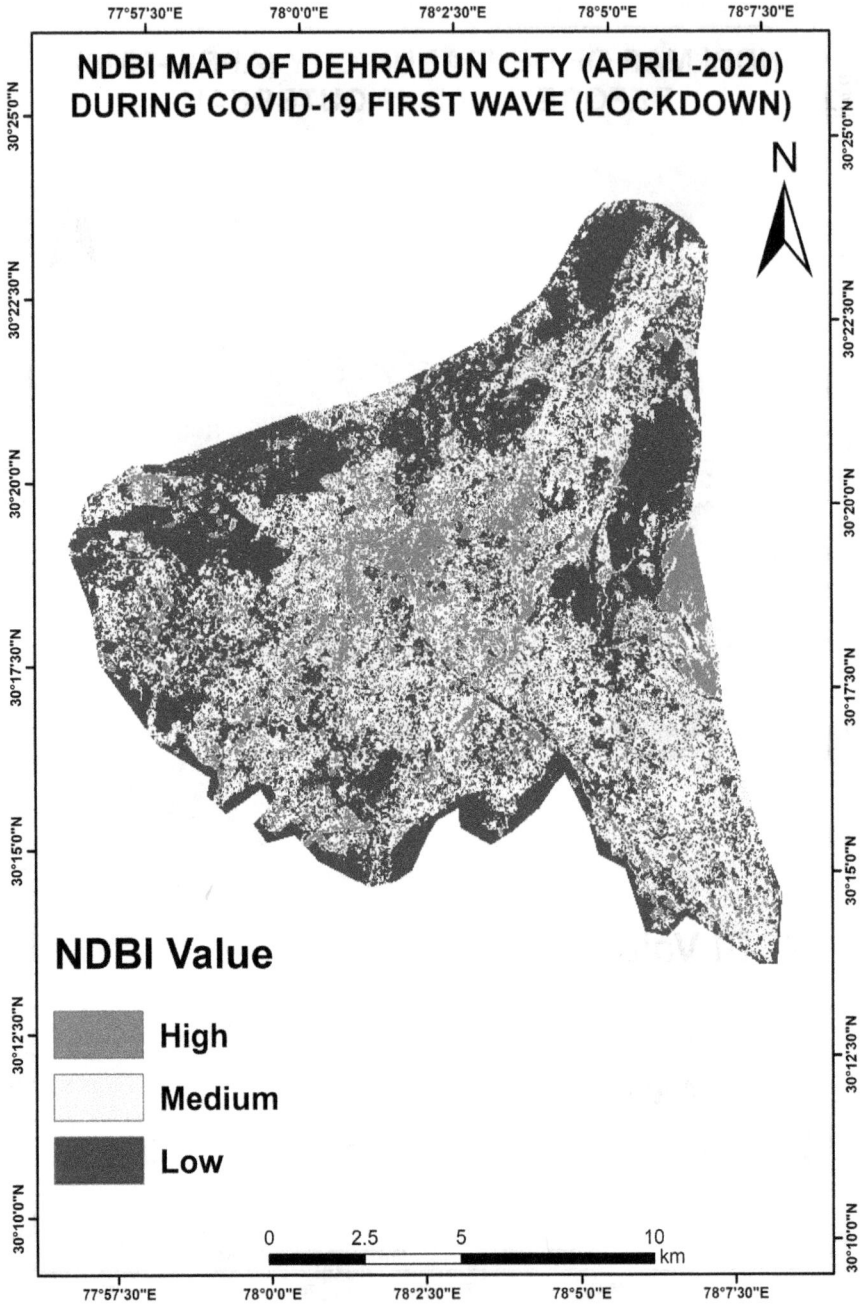

FIGURE 19.3   Map of NDBI 2020.

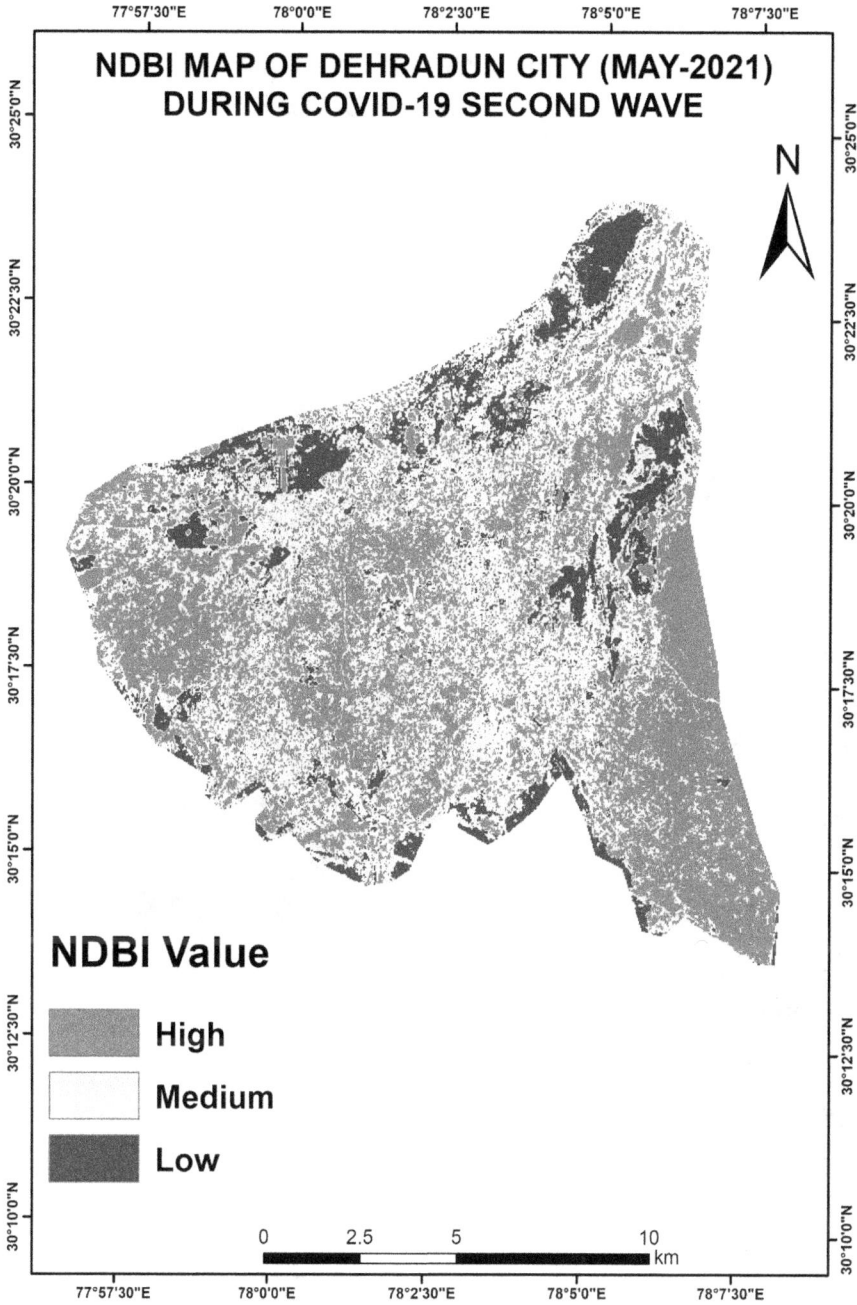

FIGURE 19.4   Map of NDBI 2021.

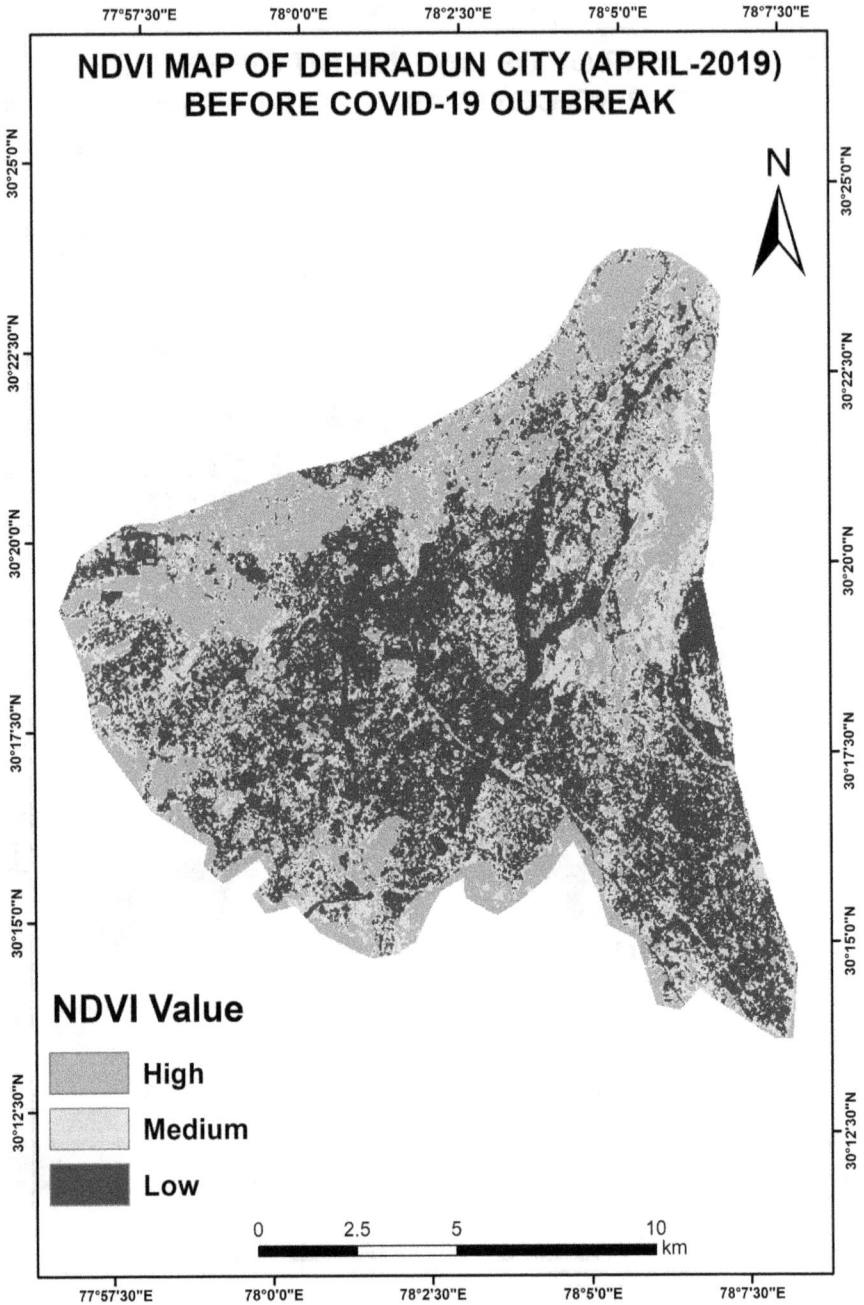

FIGURE 19.5 Map of NDVI 2019.

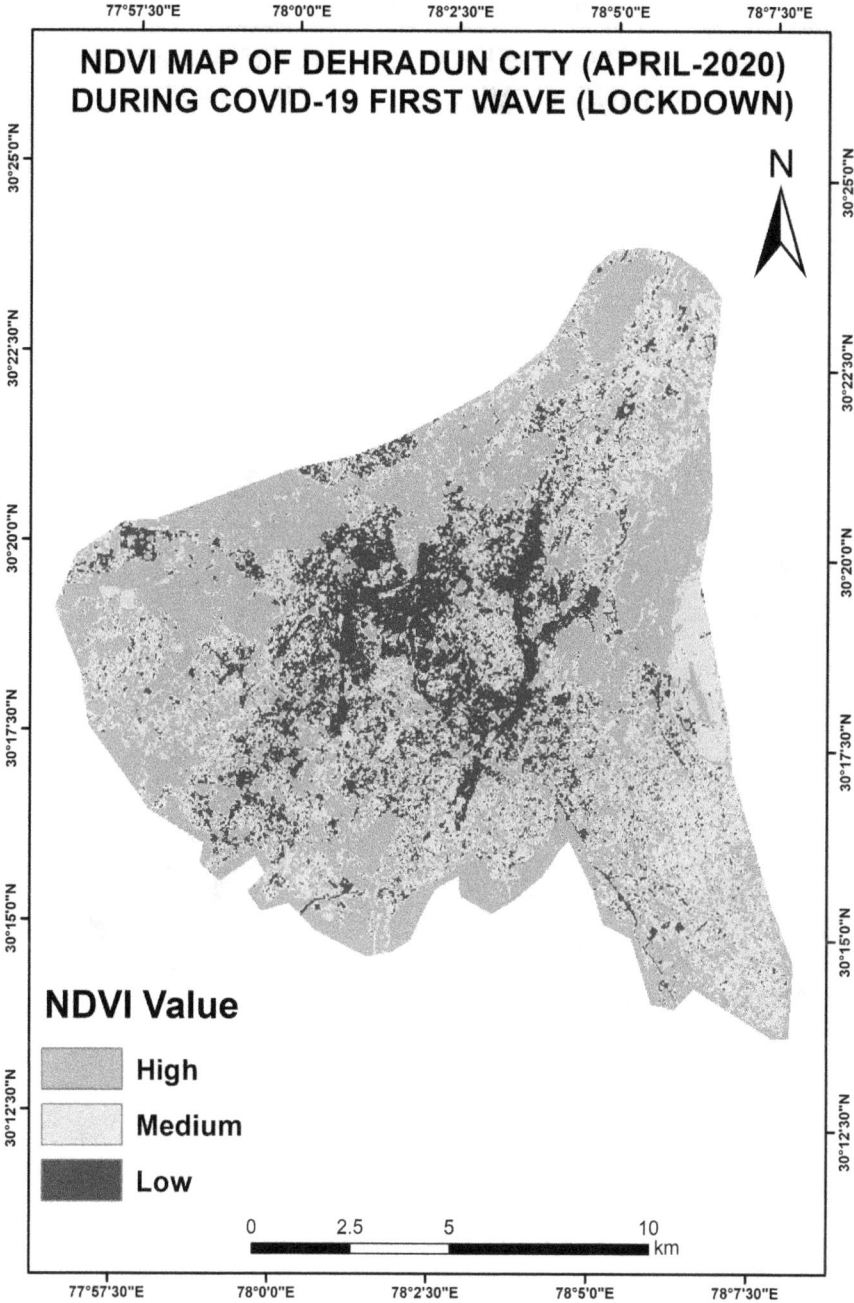

FIGURE 19.6 Map of NDVI 2020.

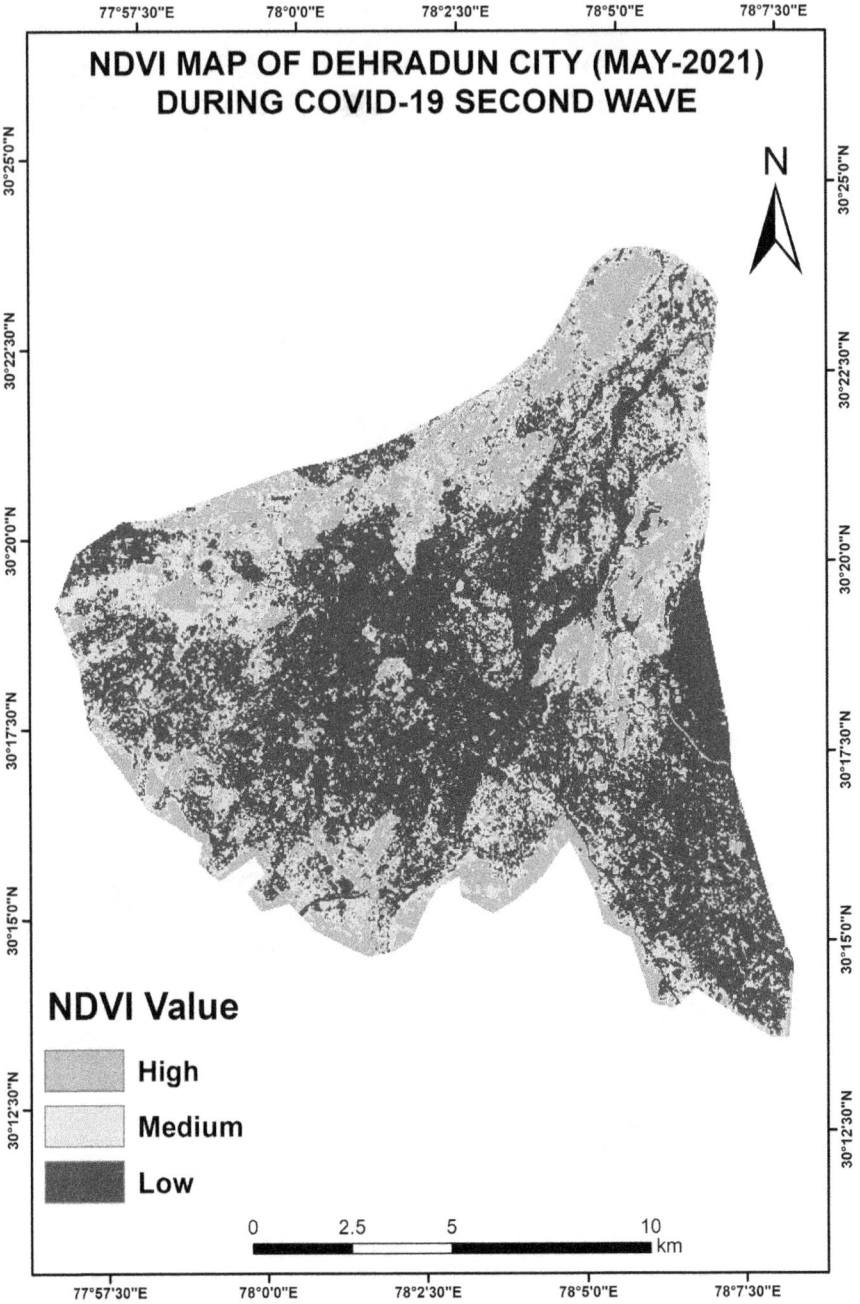

FIGURE 19.7   Map of NDVI 2021.

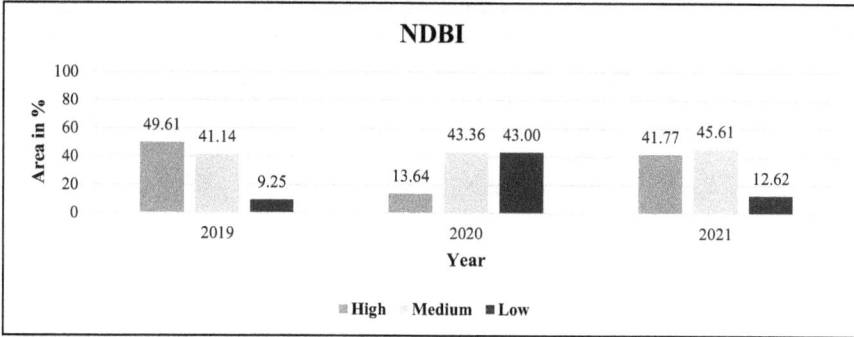

FIGURE 19.8   Graph of NDBI (2019, 2020, 2021).

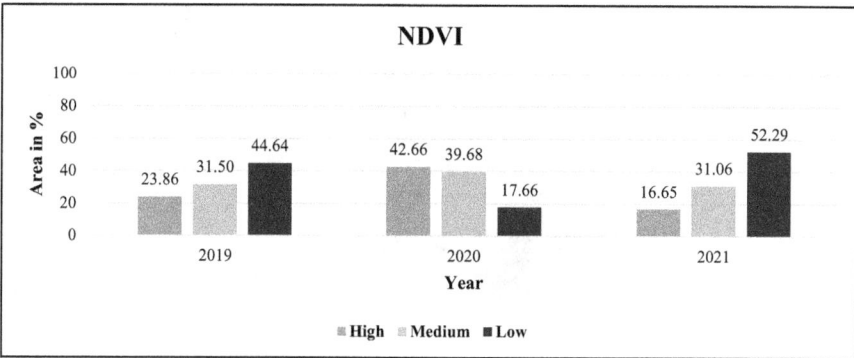

FIGURE 19.9   Graph of NDVI (2019, 2020, 2021).

denotes north and the grid on the frame shows the latitude and longitude of our region of interest.

Figures 19.8 and 19.9 present graphs representing the NDBI and NDVI area in percentage with respect to total city area.

Figures 19.8 and 19.9 represent the area in km² in percentage. Low, medium, and high values for the years 2019 and 2021 look identical as they appear in the maps. In the case of NDBI, 49.61% and 41.77% of the total area of Dehradun city were the high values for the years 2019 and 2021, whereas in 2020 the high value remained at 13.64% of the total area of the city. Medium values for all the years are as follows: 41.14%, 4.36%, and 45.61% of the total area of the city for 2019, 2020, and 2021, respectively. For low values, it is as follows: 9.25%, 43.00%, and 12.62% of the total area of the city for 2019, 2020, and 2021, respectively. In the case of NDVI, the high values are as follows: 23.86%, 42.66%, and 16.65% of the total area of Dehradun city for 2019, 2020, and 2021, respectively. For medium values they are as follows: 31.50%, 39.68%, and 31.06% of the total area of the city for 2019, 2020, and 2021, respectively. Low values are as follows: 44.64%, 17.66%, and 52.29% of the total area of the city for 2019, 2020, and 2021, respectively.

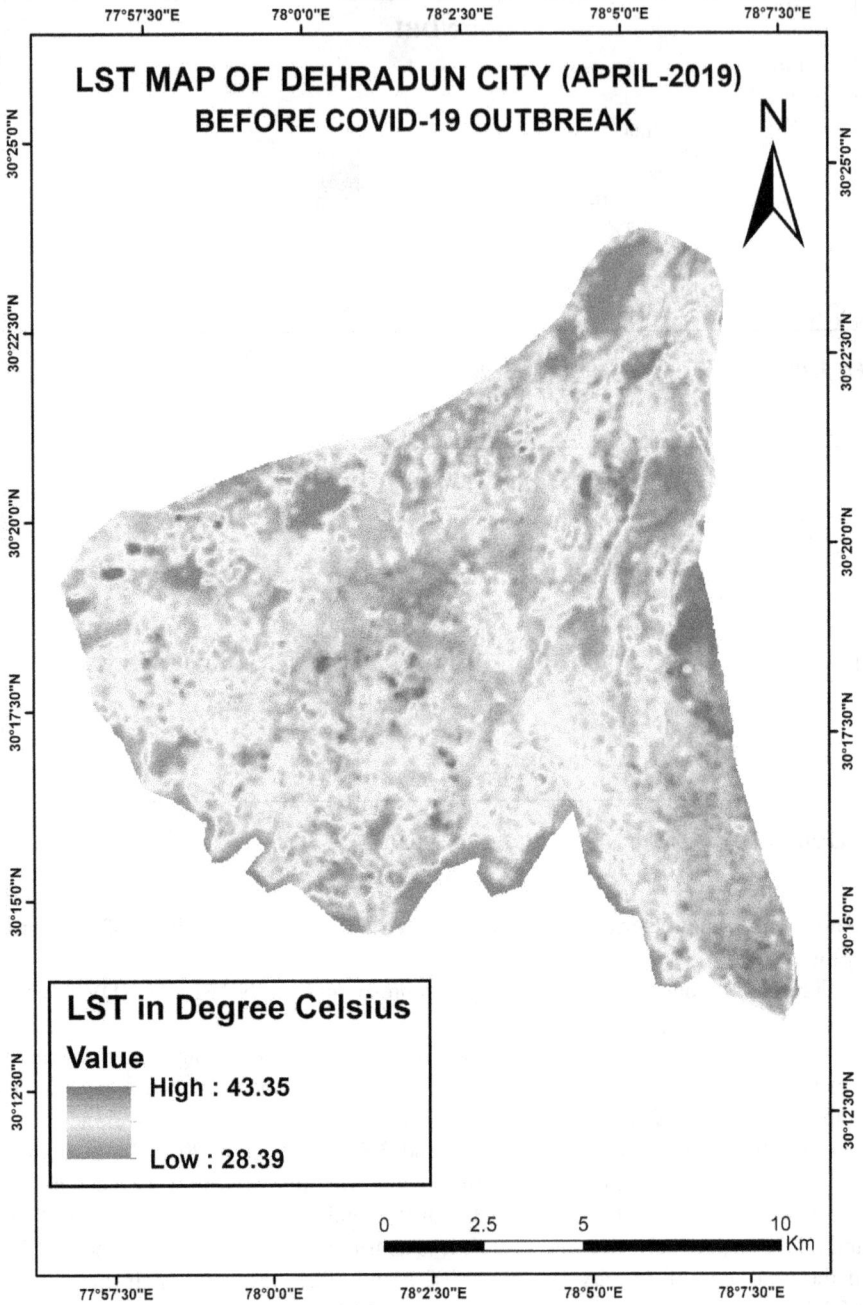

FIGURE 19.10   Map of LST 2019.

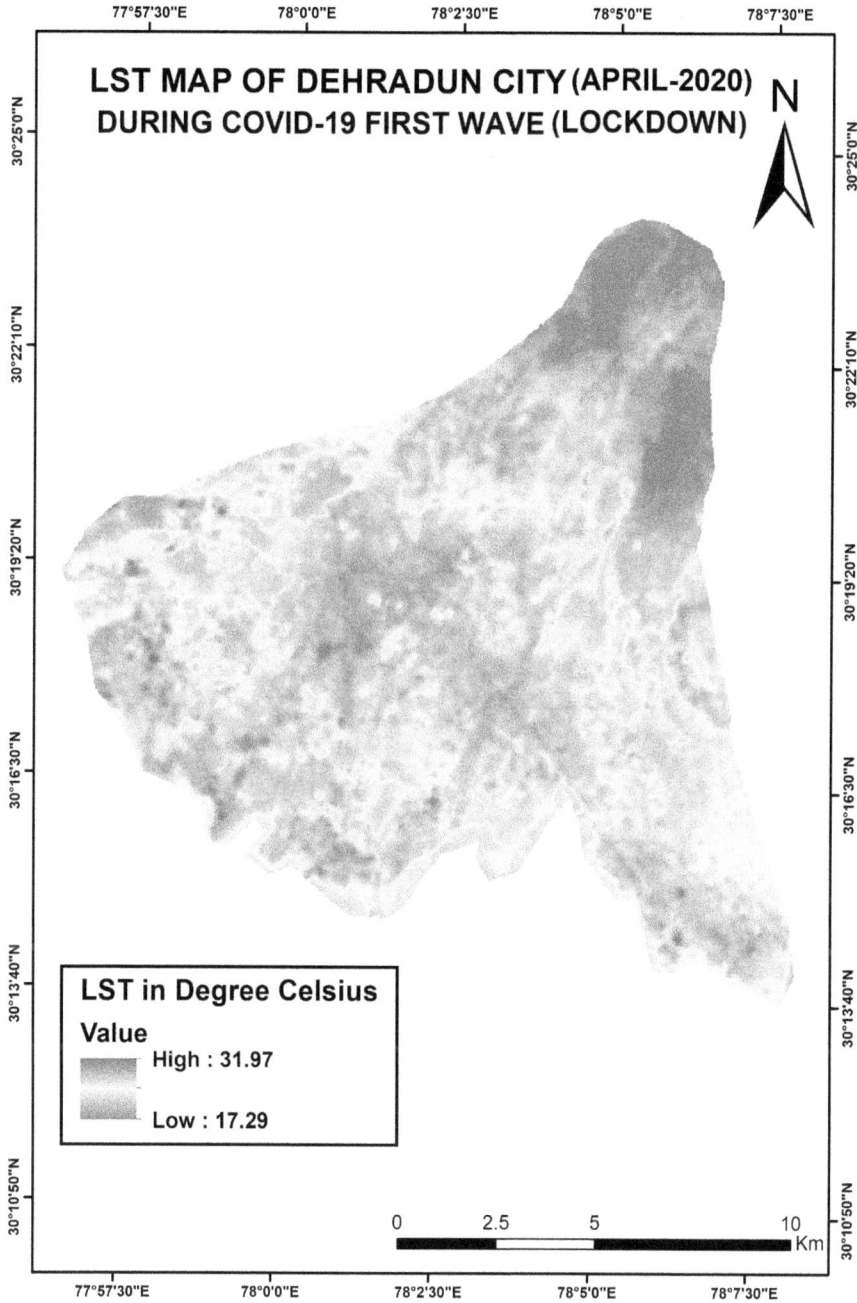

FIGURE 19.11 Map of LST 2020.

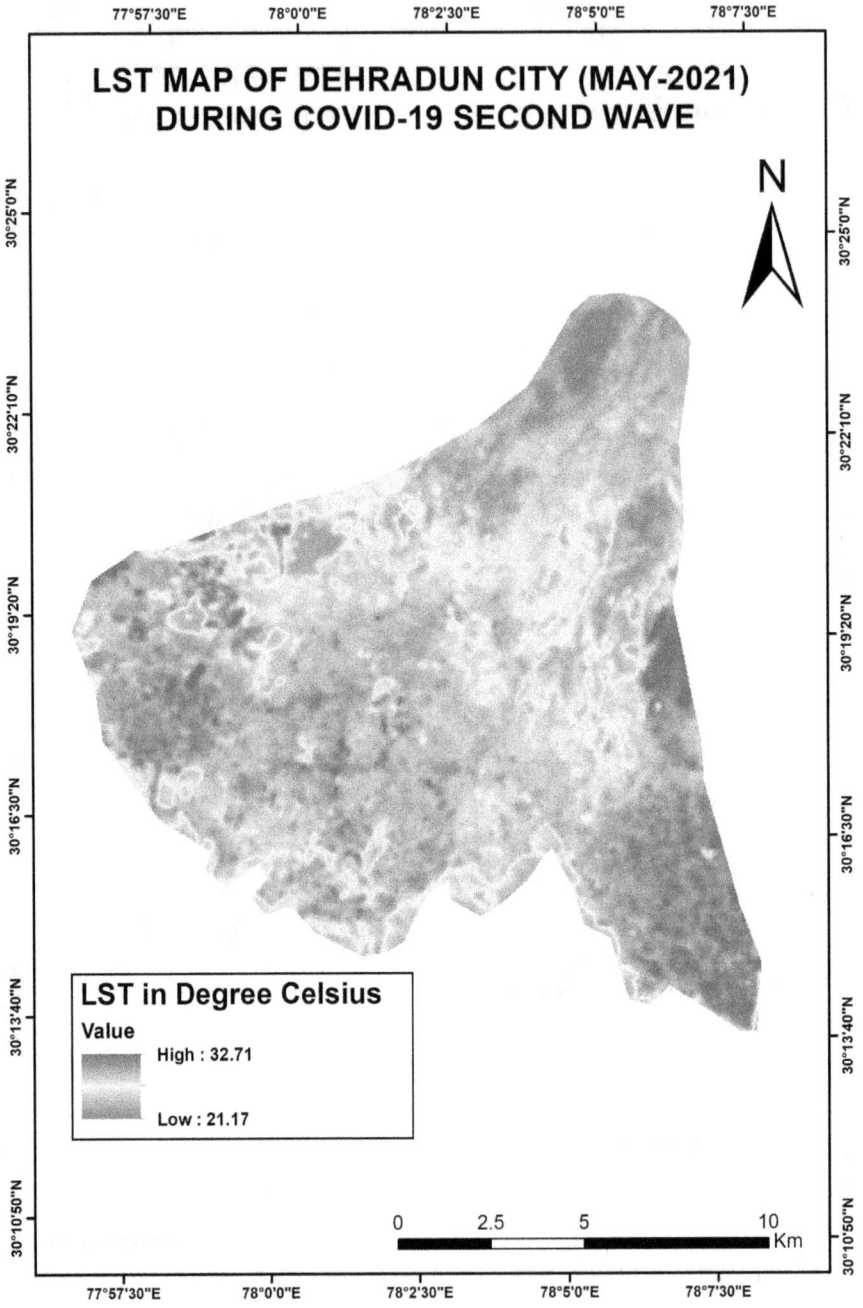

FIGURE 19.12   Map of LST 2021.

TABLE 19.1
Validation of LST Result with Other Official Sources

Year	Our LST values	Other source values
2019	Max 43.35°C, min 28.39°C	Max 38°C, min 20°C
2020	Max 31.97°C, min 17.29 °C	Max 35°C, min 23°C
2021	Max 32.71°C, min 21.17°C	Max 36°C, min 23°C

It is clear from the results that during the first lockdown, which lasted approximately 3 months, better vegetation indices were seen, showing that nature was able to recuperate and flourish during this period. An inverse relation is a general trend between the two vegetation and built-up indices. Although the second wave lockdown was of a short period, the values remained similar to those of pre-COVID times. Shorter durations have less impact on the environment and therefore, to see the fruits of lockdown in terms of the environment, a larger period is required. As the data available to us for the second wave were from May, and hence temperature and different climatic conditions would have also played a huge role in giving us a different impact factor in to the previous lockdownto check the impact of temperature we have calculated the land surface temperature of our region of interest as well.

Figures 19.10–19.12 show the three LST maps of our AOI.

These maps represent the LST data of Dehradun for the years 2019, 2020, and 2021. There is a north-pointing arrow placed at the top right corner. The grid shows the latitude and longitude of our area of interest. The scale of the map is 1 cm on the map representing 10 km on the ground. The legend gives the temperature data, where blue shows the lowest temperature and red shows the highest temperature with a gradual change in colour from blue to red. Maximum values for different years are as follows: 43.35°C, 31.97°C, and 32.71°C for 2019, 2020, and 2021, respectively. Minimum values for different years are as follows: 28.39°C, 17.29°C, and 21.17°C for 2019, 2020, and 2021, respectively.

The LST data have been validated from the official IMD site (http://data.gov.in and http://worldweatheronline.com) and the validation result was not as similar to the result of LST shown in Table 19.1.

The reason behind the temperature variation in both data sets is due to a difference in the timing of satellite imagery captured and a difference in temperature estimation techniques. From the above data, it is evident that during the lockdown overall temperatures were lower. A positive correlation with NDBI is evident, whereas a negative correlation is seen with NDVI. A 3-month lockdown helped in the recuperation of the environment as well as helping in lowering the temperature. As the second lockdown was short, not a lot of change could be seen. Small and positive changes in anthropogenic activities are seen to make a huge difference such as an increase in the green cover area and a lowering of temperature. A complete stop to the activities with only essential running has been very useful for the environment and hence small steps in helping the environment to rehabilitate would go a long way.

13.5 CONCLUSION

Dehradun city, one of the major administrative and cultural cities of India, has faced rapid urbanisation in the last few years, which has led to land-use and land-cover changes (LULC alterations), affecting the land surface temperature (LST) of the city. The core principles of mitigation and suppression strategies that were adopted to control the spread of pandemics – social distancing and community quarantine – added to the direct strengths of the global economy. As industries, transportation, and various companies were closed down due to the widespread of the COVID-19 virus, the value of LST and NDBI reduced significantly. Also, the value of NDVI increased significantly during the lockdown months. This clearly showed that this outbreak of disease and imposed lockdown provided an opportunity for our planet to heal from its wounds. Our natural environment became less polluted and ecologically healthier as compared to earlier times when, despite various attempts, we were unable to control such parameters.

However, in the second wave of COVID-19, there was undoubtedly a sense of indifference in people's minds to the coronavirus response. Due to this the restrictions and imposition of rules were comparatively flexible during 2021. This greatly affected the environmental parameters as compared to the previous year. The LST, NDBI, and NDVI values showed an inverse relationship in 2021 to what was found in 2019. This indicated the flexibility and change of perception of people to the regulations imposed by the government. A rise of 1 degree Celsius was found in this study for the year 2021. Our study is beneficial for urban planners and policymakers to bring about a change in their approach to management. Lessons can be learned from similar studies for post-COVID urban planning, design, and management.

REFERENCES

A., Manuel, et al. *Indirect Effects of COVID-19 on the Environment.* Science of the Total Environment, vol. 728, doi:https://doi.org/10.1016/j.scitotenv.2020.138813.

Bandala, Erick R., et al. *Impacts of COVID-19 Pandemic on the Wastewater Pathway into Surface Water: A Review.* Science of The Total Envrionment, vol. 774, 2021, doi:https://doi.org/10.1016/j.scitotenv.2021.145586.

Basu, Bidroha, et al. *Investigating Changes in Noise Pollution Due to the COVID-19 Lockdown: The Case of Dublin, Ireland, Sustainable Cities and Society.* Science Direct, vol. 65, 2021, doi:https://doi.org/10.1016/j.scs.2020.102597.

Belojević, G., Jakovljević, B. and Aleksić, O., 1997. *Subjective reactions to traffic noise with regard to some personality traits.* Environment International, 23(2), 221–226.

Billah, Arif, et al. *Reproductive Number of Coronavirus: A Systematic Review and Meta-Analysis Based on Global Level Evidence,* 11 Nov. 2020, doi:https://doi.org/10.1371/journal.pone.0242128.

Bir, B., 2020. www.aa.com.tr/en/health/single-use-items-not-safest-option-amid-COVID-19/1787067

Calma, J., 2020. www.theverge.com/2020/3/26/21194647/the-COVID-19-pan-demic-is-generating-tons-of-medical-waste

Chauhan, A., & Singh, R. P. (2020). *Decline in PM2.5 concentrations over major cities around the world associated with COVID-19.* Environmental Research, 187, 109634.

CPCB (Central Pollution Control Board) 2020. *Impact of lockdown on water quality of river Ganga.* CPCB, Ministry of Environment, Forest and Climate Change, Govt. of India, New Delhi. https://cpcb.nic.in/openpdffile.php?id=TGF0ZXN0RmlsZS8yOTNfMTU4Nzk3ODU3MV9tZWRpYXBob3RvMTY3MDYyucGRm.

Dharmaratne, S., Sudaraka, S., Abeyagunawardena, I. et al. *Estimation of the basic reproduction number (R0) for the novel coronavirus disease in Sri Lanka.* Virol. J. 17, 144 (2020). https://doi.org/10.1186/s12985-020-01411-0

Eriksson, C., Nilsson, M.E., Willers, S.M., Gidhagen, L., Bellander, T. and Pershagen, G., 2012. *Traffic noise and cardiovascular health in Sweden: The roadside study. Noise and Health,* 14(59), 140.

Franzpc, et al. *"How to Calculate Land Surface Temperature with Landsat 8 Satellite Images."* GIS Crack, 24 Apr. 2019, giscrack.com/how-to-calculate-land-surface-temperature-with-landsat-8-images/.

Fyhri, A. and Aasvang, G.M., 2010. *Noise, sleep and poor health: Modeling the relationship between road traffic noise and cardiovascular problems.* Science of the Total Environment, 408(21), 4935–4942.

Garg, Vaibhav, et al. *"Changes in Turbidity along Ganga River Using Sentinel-2 Satellite Data during Lockdown Associated with COVID-19."* Geomatics, Natural Hazards and Risk, vol. 11, no. 1, 2020, 1175–1195. doi:10.1080/19475705.2020.1782482.

Islam S.M.D., Bhuiyan M.A.H. *Impact scenarios of shrimp farming in coastal region of Bangladesh: an approach of an ecological model for sustainable management.* Aquacult. Int. 2016;24(4):1163–1190.

Kanchan Bahukhandi , Shilpi Agarwal & Shailey Singhal (2020): *Impact oflockdown COVID-19 pandemic on himalayan environment,* International Journal of Environmental Analytical Chemistry, DOI: 10.1080/03067319.2020.1857751

Kannan, S., et al. *"COVID-19 (Novel Coronavirus 2019)–Recent Trends."* Eur Rev Med Pharmacol Sci, vol. 24, no. 4, 2020, 2006–2011. doi:10.26355/eurrev_202002_20378.

Kim, K., Shin, J., Oh, M. and Jung, J.K., 2019. *Economic value of traffic noise reduction depending on residents' annoyance level.* Environmental Science and Pollution Research, 26(7), 7243–7255.

Koivusalo, M., Vartiainen, T., 1997. *Drinking water chlorination by-products and cancer.* Rev. Environ. Health 12, 81–90.

"Land Surface Temperature." Land Surface Temperature I Copernicus Global Land Service, 9 June 2020, land.copernicus.eu/global/products/lst#:~:text=The%20Land%20Surface%20Temperature%20(LST,direction%20of%20the%20remote%20sensor.

Liu, Yen-Chin, et al. COVID-19: *The First Documented Coronavirus Pandemic in History,* vol. 43, no. 4, 2020, pp. 328–333., doi: https://doi.org/10.1016/j.bj.2020.04.007.

Mahato, S., Pal, S. & Ghosh, K. G. (2020). *Effect of lockdown amid COVID-19 pandemic on air quality of the megacity Delhi, India.* The Science of the Total Environment, 730, Article 139086.

Mehta , Radhika. *"10 Impacts of Coronavirus on the Environment."* Earth5R, 19 Sept. 2020, earth5r.org/impacts-corona-virus-environment/.

Maithani, S., Nautiyal, G. & Sharma, A. *Investigating the Effect of Lockdown During COVID-19 on Land Surface Temperature: Study of Dehradun City, India.* J Indian Soc Remote Sens 48, 1297–1311 (2020). https://doi.org/10.1007/s12524-020-01157-w

Maxmen, Amy, and Smriti Mallapaty. *"The COVID Lab-Leak Hypothesis: What Scientists Do and Don't Know."* Nature News, Nature Publishing Group, 8 June 2021, www.nature.com/articles/d41586-021-01529-3.

Merriam-Webster, *"Emissivity."* Merriam-Webster.com Dictionary, www.merriam-webster.com/dictionary/emissivity. Accessed 21 Jul. 2021

Mishra A, Das S, Singh D, Maurya AK. *Effect of COVID-19 lockdown on noise pollution levels in an Indian city: a case study of Kanpur.* Environ Sci Pollut Res Int. 2021 Apr 22:1–13. doi: 10.1007/s11356-021-13872-z. Epub ahead of print. PMID: 33884552; PMCID: PMC8060123.

Münzel, T. and Sørensen, M., 2017. *Noise pollution and arterial hypertension.* European Cardiology Review, 12(1), 26.

Oh, M., Shin, K., Kim, K. and Shin, J., 2019. *Influence of noise exposure on cardiocerebrovascular disease in Korea.* Science of the Total Environment, 651, 1867–1876.

Ongel, A. and Sezgin, F., 2016. *Assessing the effects of noise abatement measures on health risks: A case study in Istanbul.* Environmental Impact Assessment Review, 56, 180–187.

Said, M.A. and El-Gohary, O.A., 2016. *Effect of noise stress on cardiovascular system in adult male albino rat: implication of stress hormones, endothelial dysfunction and oxidative stress.* General Physiology and Biophysics, 35(3), 371–377.

Siddiqui, A., Halder, S., Chauhan, P., & Kumar, P. (2020). *COVID-19 Pandemic and City-Level Nitrogen Dioxide (NO2) Reduction for Urban Centres of India.* Journal of the Indian Society of Remote Sensing. https://doi.org/10.1007/s12524-020-01130-7.

World Health Organization (WHO), 2005. *United Nations road safety collaboration: a handbook of partner profiles.*

WHO, 2020b. www.who.int/publications-detail/water-sanitation-hygiene-and-waste-management-for-COVID-19.

20 Framework for an Area-Based Development Approach for Predicted Urban Sprawl in Delhi City

Gaurav Kumar Mishra and Amit M. Deshmukh

CONTENTS

20.1 INTRODUCTION

The definition of urbanisation as per the National Library of Medicine is the result of migration from rural to urban areas and it is, therefore, a decrease in the population of rural areas as they adapt to this change in their social behaviour. This term is often used when new houses are built or industries are set up or other infrastructural activities take place. If we see the current scenario, we find that only 5% area of the entire terrestrial surface is urban, whereas it accommodates half of the world's population. It consumes three-fourths of the total natural resources and the same amount of pollution and waste is also generated. It involves the conversion of forest cover into agricultural land and from agricultural land to built-up land (Taubenböck et al., 2009). Therefore, it is creating a negative impact on the social, economic, and environmental aspects of life (Grimm, Grove, Pickett, & Redman, 2008; Li et al., 2013; Li & Yeh, 2000; Mage et al., 1996; Pickett et al., 1997).

India is showing the highest growth in terms of urbanisation. As per a report of the United Nations, half of the country's population will reside in cities by 2050. The main cause of this urban expansion is rural-to-urban migration. Most of the cities in India harm all aspects of living, whether it is social, economic, or environmental (Kandlikar & Ramachandran, 2000; Sudhira et al., 2007). However, it is found that

DOI: 10.1201/9781003331001-20

urbanisation fosters economic growth. From the reports of NITI AYOG by 2030, the urban GDP percentage of total GDP will be 75% and cities are thus described as the economic engines of the nation (NITI Ayog, 2012). Therefore, policies should be made that focus on only stopping unsustainable urbanisation and not preventing sustainable urbanisation (Duranton et al., 2009; Annez et al., 2009). The main challenge before city planners is rapid and uncontrolled urbanisation, as this unplanned and uncoordinated growth needs proper development of infrastructure. By 2050, approximately 500 million additional dwelling units will be needed to accommodate the future urban population. Therefore, it is a big challenge for the government to provide infrastructure and land for urbanisation. So, it is essential to assess the current status of urbanisation and to know how the formal or informal processes are contributing to urban expansion. These two components will be required to minimise the adverse effects of urbanisation on ecological and environmental aspects. An urban planner needs tools to assess all these factors which contribute to urban expansion. However, in the absence of these tools, we are unable to assess the status of the urbanisation, rate of urban expansion, and pattern and extent of the sprawl. However, this information is needed by planners to mitigate the effect of unsustainable urbanisation (Taubenböck et al., 2009).

Remote sensing allows for monitoring urban expansion through its spatial and temporal capabilities (Herold et al., 2003). It is generally used in conjunction with geographic information systems, which again strengthens its capability in spatial and temporal monitoring of urban expansion (Angel et al., 2005; Bhatta, 2009; Fan, Wang, Qiu et al., 2009; Hu, Du, & Guo, 2007; Jat et al., 2008; Masser, 2001; Seto et al., 2011; Wakode et al., 2014; Xu et al., 2007). Taking a time series analysis of urban expansion can lead to a better understanding of urban expansion processes (Geymen & Baz, 2008; Masek et al., 2000). Since urban expansion generally takes place in the form of land use land cover (LULC) change, so it becomes essential to assess the LULC change to understand the urban expansion processes.

Urban expansion quantification has been a subject of importance for the last two decades. In 2004, Sudhira, Ramachandra, and Jagdish used Shannon entropy, patchiness, and built-up density in conjunction with regression analysis for modelling the urban sprawl of the Mangalore Udipi region of India. Kumar, Pathan, and Bhanderi (2007) have used landscape metrics and Shannon entropy to monitor the urban expansion of Indore city. For Ajmer city, Jat et al., in 2008, used landscape metrics and Shannon entropy to assess the urban sprawl. Spatial metrics such as built-up density, landscape shape index, largest patch index, number of patches, total edge, and edge density were used by Taubenböck et al. in 2009 for spatial monitoring of the urban expansion of 12 high-density Indian cities.

In addition to the above quantitative measures, we need to assess the qualitative measures also. Thus, a spatiotemporal assessment of urban growth is needed so that we can easily identify the planning processes that have taken place in the expansion of the urban population. It will be easier for planners to assess the planning processes of different types of urban developments. The data coming through this type of assessment will not only provide an understanding of the regions of rapid urbanisation, but will also be indicative for implications in framing policies. It will also be indicative to assess the impact of urban growth on the environment. Along with this, it will indicate the understanding of the balance between planned and spontaneous development. This knowledge is essential for framing the policies on urban growth and minimising the negative effects of urbanisation. However, studies on both the qualitative and quantitative measures of urban growth are still very few.

This report's study area is Delhi, India. This is also the national capital of India. It has shown rapid urbanisation in the last few decades. These areas belong to either municipal corporations, municipalities, cantonment boards, or village panchayats. This chapter has focused on both measures whether quantitative or qualitative. The analysis of urban expansion shows quantitative analysis and the typology of expansion shows qualitative analysis. Further, this qualitative analysis shows the type of development that has taken place in the form of the urban core, the fringe, ribbon development, and scattered development. We have analysed the Landsat satellite images from 1989 to 2014, which is the period of rapid urbanisation of the study area. The main objectives of this study are:

a) To perform LULC change analyses which are caused by the urban expansion between the years 1989 and 2014.
b) To perform the quantitative analysis on the urban expansion of the study area.
c) Understanding of the typology of development processes which have taken place in the last 32 years.

The urban centres are becoming crowded in India. Rapid urban growth and development have resulted in an increase in the share of India's urban population from 79 million in 1961 which was about 17.92% of India's total population to 388 million in 2011, which is 31.30% of the total population (Bhat et al., 2017). Managing urban growth has been a very complex phenomenon and a big challenge this century. There are negative consequences to this urban growth. It is affecting the social, economic, and environmental aspects of sustainable urban centres. Sprawl has been loosely defined as dispersed and inefficient urban growth (Hasse & Lathrop, 2003). Sprawl opponents have blamed sprawl for weakening linkages between the residents and social capital, but there is a lack of empirical evidence to support their argument (Nguyen, 2010). The population settlements in the fast-growing urban areas need to be monitored to design a sustainable urban habitat (Deep & Saklani, 2014). Therefore, to achieve this target, it is required to model the existing scenario of urban sprawl in urban centres.

Remote sensing and GIS tools are very effective for modelling urban sprawl. The methodology most employed in analysis relies heavily on descriptive and multivariate statistics that are prone to unreliable results owing to spatial autocorrelation (Berry 1993; Fotheringham et al., 2000). The use of geospatial metrics to avoid spatial autocorrelation problems (usually a fatal roadblock when encountered in analysis) is exceptional when measuring sprawl. There is a question of sustainability for this urban sprawl as this is uncoordinated and unplanned growth of an urban centre by definition. Smart development has been identified as a sustainable worldwide solution to the existing urban planning issues, whose principles aim at providing a better quality of life and advertising livable communities; although the concept is vague to define, as no universal definition exists. Urban sprawl and uncoordinated low-density developments are creating problems for sustainable development, which need urgent attention. The area-based proposal of the urban centres can identify a peripheral district for guided compact land development as a model for future urbanisation. The strategic components for area-based development adopted are an improvement (retro-fitting), city renewal (redevelopment), city extension (greenfield development), and pan-city concept (using modern technology solutions for the existing city infrastructure); which will transform existing deteriorating areas into better-planned ones and develop new areas to accommodate the expanding populations in urban areas.

20.2   STUDY AREA

Delhi is the national capital of India. It is situated in the northern part of India, and covers 1483 square kilometres of land, with its location being between 28.33^0 and 29.00^0 north latitude and 76.83^0 and 77.33^0 east longitude. As per Census 2011, the population of Delhi City is 11,034,555, and the decadal growth was 51.45% in 1981–91, 47.02% in 1991–2001, and 11.2% in 2001–2011. The overall population density of Delhi was 9340 persons per km^2 in 2001 and 11,320 persons per km^2 in 2011, which is the highest of all Indian states/UTs. From 2016 to 2030, Delhi is projected to outperform Mumbai in terms of growth, with GDPs rising by 7.1% and 5.7% per year, respectively.

20.3   DATASETS AND METHODS

During the early 1990s, rapid urbanisation took place in India due to economic liberalisation. For analysis and mapping purposes of land use and land cover Landsat images are often used. These datasets are useful because they provide medium resolution having a long-term archive with consistent spectral and radiometric resolution. Therefore, our study deployed Landsat images of the years 1989, 1994, 1999, 2004, 2009, and 2014 to assess the urban expansion of Delhi. The toposheets for the study area were acquired from the Survey of India. These were projected to Universal Transverse Mercator (UTM) projection using WGS 1984 datum. It was further mosaicked and clipped to the study area. A detailed methodology has been provided in Figure 20.1. The methodologies show the acquisition of the Landsat satellite images which are used for the preparation of land-use land-cover maps. This is further required to be processed for simulation purposes. The actual LULC is compared with simulated LULC to train the datasets for the prediction of the LULC in 2024 (Table 20.1).

Land-cover change detection: Land-cover change is required to assess urban expansion. The use of remote sensing technology in change detection is most efficient in terms of being cost-effective, less time-consuming, and easy to operate. Meanwhile, many types of algorithms for change detection are used, e.g. image differencing, image rationing, image regression, principal component analysis, change vector analysis, decision trees, and classification methods, which are widely used for change

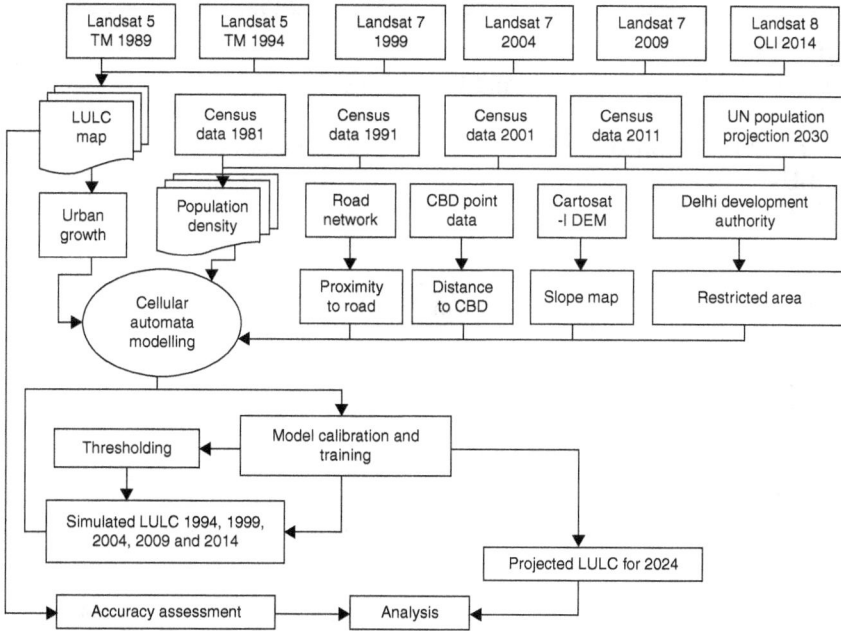

FIGURE 20.1 Flow chart showing the schematic methodology.

TABLE 20.1
Urban Expansion Metrics

Expansion metrics	Abbreviation	Formulae	Description
Expansion contribution rate	ECR	$\dfrac{B_{(i,t_2)} - B_{(i,t_1)}}{B_{t_2} - B_{t_1}} \times 100$	This describes the percentage share of urban expansion
Expansion percentage of change	EPC	$\dfrac{B_{(i,t_2)} - B_{(i,t_1)}}{B_{t_1}} \times 100$	This is a relative measure used to describe the expansion of an urban area to the previous urban extent
Annual expansion rate	AER	$\dfrac{B_{(i,t_2)} - B_{(i,t_1)}}{t_2 - t_1}$	This is a measure of urban expansion rate per annum during the study period

detection of the environment (Deng et al., 2009; Hardin et al., 2007; Hu et al., 2007; Jensen, 1996; Singh, 1989; Weng, 2002). The classification methods have been used for land use/land cover change detection. The accuracy depends on the individual classification operations. It can be calculated by taking an average of all the overall accuracies (Coppin et al., 2004; Stow et al., 1980). The transition metrics are shown in Table 20.2.

TABLE 20.2
Transition Matrix of Changed Land Cover from 1989 to 1994: (i) Percentage of the Former Land Cover 1989 (Rows) Contributed to the Current Land Cover 1994 (Columns); (ii) Percentage Change of LULC; (iii) Net Change Per Land-Use and Land-Cover Class between 1989 and 1994 as a Percentage of the Total Study Area

		Land cover 1994			
		Urban (271 km²)	Forest (193 km²)	Water (12 km²)	Other (1025 km²)
(i) Land cover 1989	Urban (161 km²)	0	1.3	5.9	69.3
	Forest (198 km²)	4.1	0	1.2	14.6
	Water (13 km²)	1.2	1.1	0	16.2
	Other (1129 km²)	94.7	97.6	92.8	0
(ii) Percentage of change		68.32%	−2.52%	−7.69%	−9.21%
(iii) Net change per LULC class between 1989 and 1994		47.86%	−2.36%	−0.47%	−49.28%
(iv)		101 km²	5 km²	1 km²	104 km²

Urban expansion metrics: Three matrices have been used for quantifying the urban expansion: (i) expansion contribution rate (ECR); (ii) expansion percentage of change (EPC); and (iii) annual expansion rate (AER). These measures were calculated for Delhi. It can be effectively used for explaining the urbanisation dynamics.

20.4   RESULTS

a) **Land-use Land-cover classification:** The accuracy assessment of classified images was done with the help of ground truth data. The result shows good overall accuracy which ranges from 89% to 90% and very good producer accuracy from approximately 73.3% to 100% for individual LULC classes.

b) **Land-cover change assessment:** After classification, change detection has been carried out in several steps for a period of 35 years from 1989 to 2024. Figure 20.2 shows the percentage of land-cover classes in 1989, 1994, 1999, 2004, 2009, 2014, 2019, and 2024.

c) **Analysis of Urban Expansion:** The urban area has expanded from 161 km² in 1989 to 785 km² in 2024.

   1) In 1989–1994, the urban area of Delhi expanded by 110 km² with an annual expansion rate of 22 km².

   2) In 1994–1999, the urban area of Delhi expanded by 80 km² with an annual expansion rate of 16 km².

   3) In 1999–2004, the urban area of Delhi expanded by 73 km² with an annual expansion rate of 12.6 km².

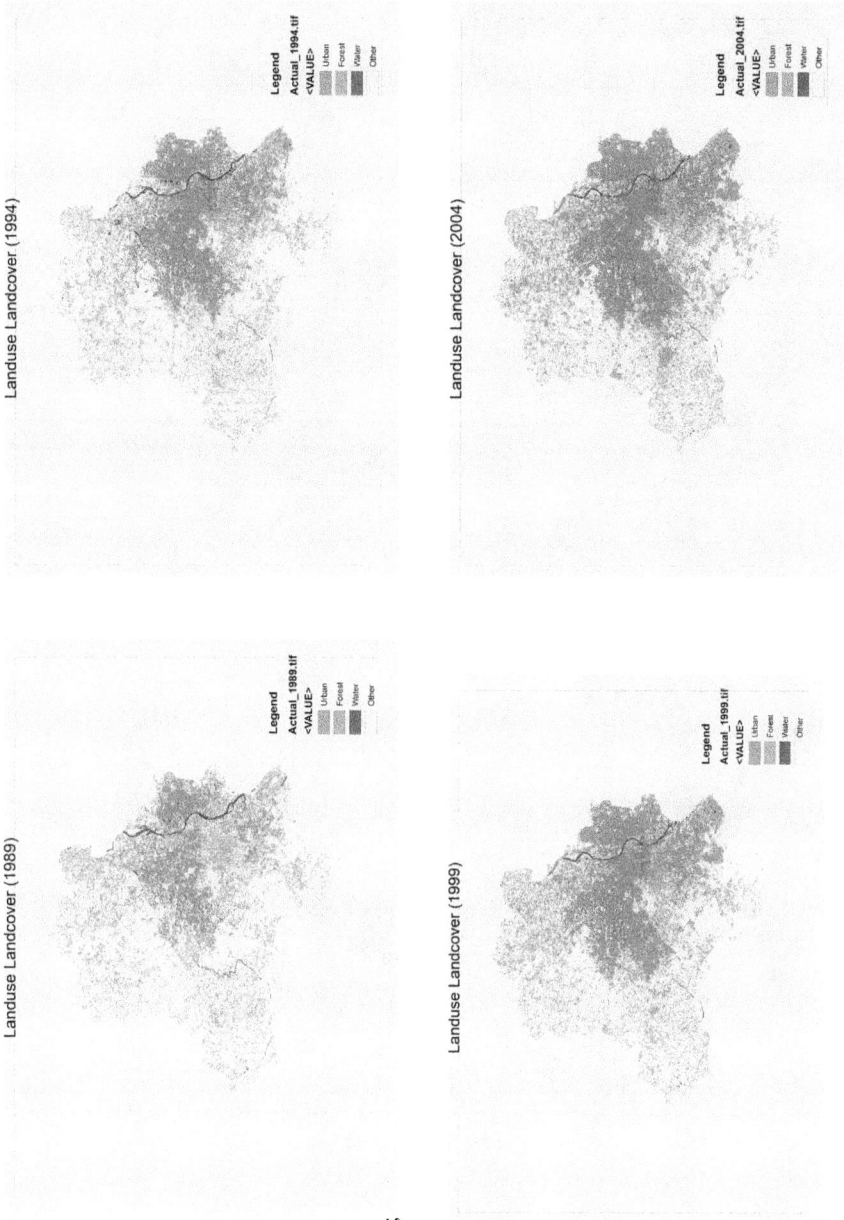

FIGURE 20.2 Data layers used in geospatial modelling.

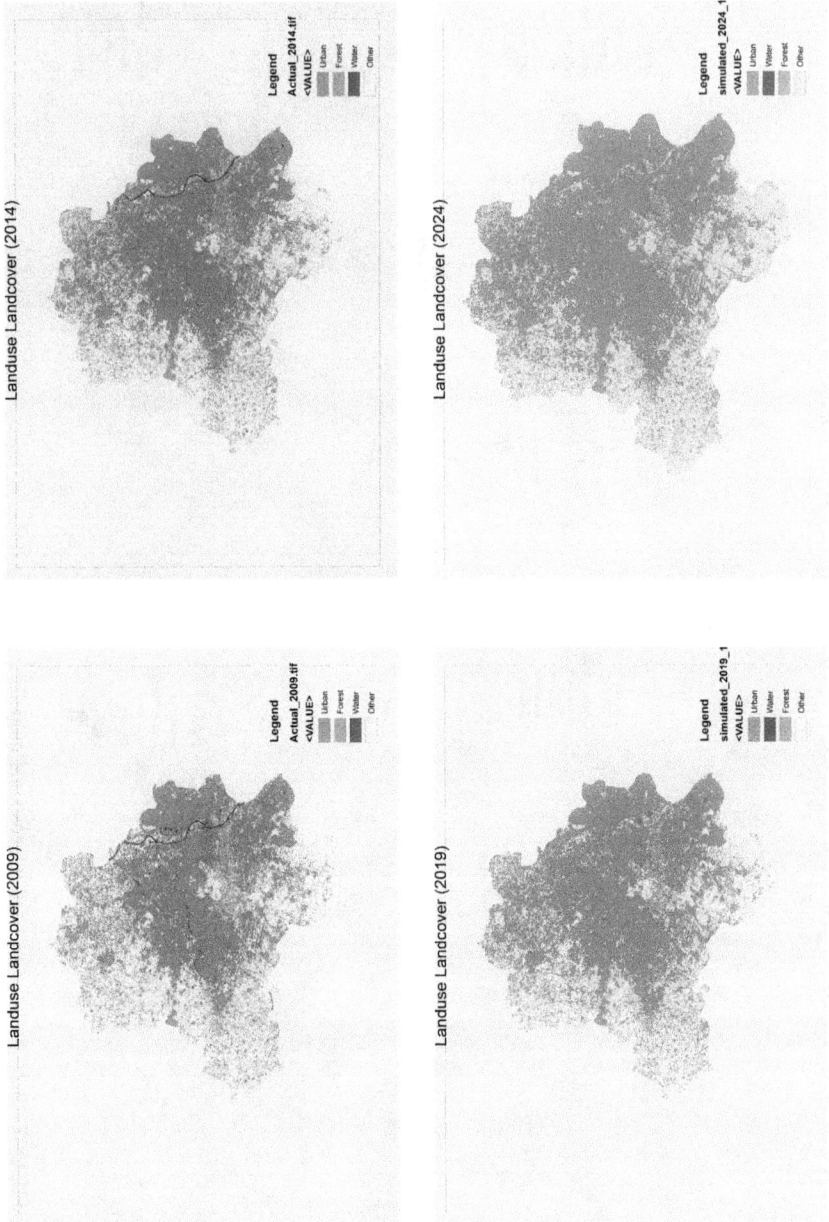

Landuse Landcover (2014)

Legend
Actual_2014.tif
<VALUE>
Urban
Forest
Water
Other

Landuse Landcover (2024)

Legend
simulated_2024_1
<VALUE>
Urban
Water
Forest
Other

Landuse Landcover (2009)

Legend
Actual_2009.tif
<VALUE>
Urban
Forest
Water
Other

Landuse Landcover (2019)

Legend
simulated_2019_1
<VALUE>
Urban
Water
Forest
Other

FIGURE 20.2 (Continued)

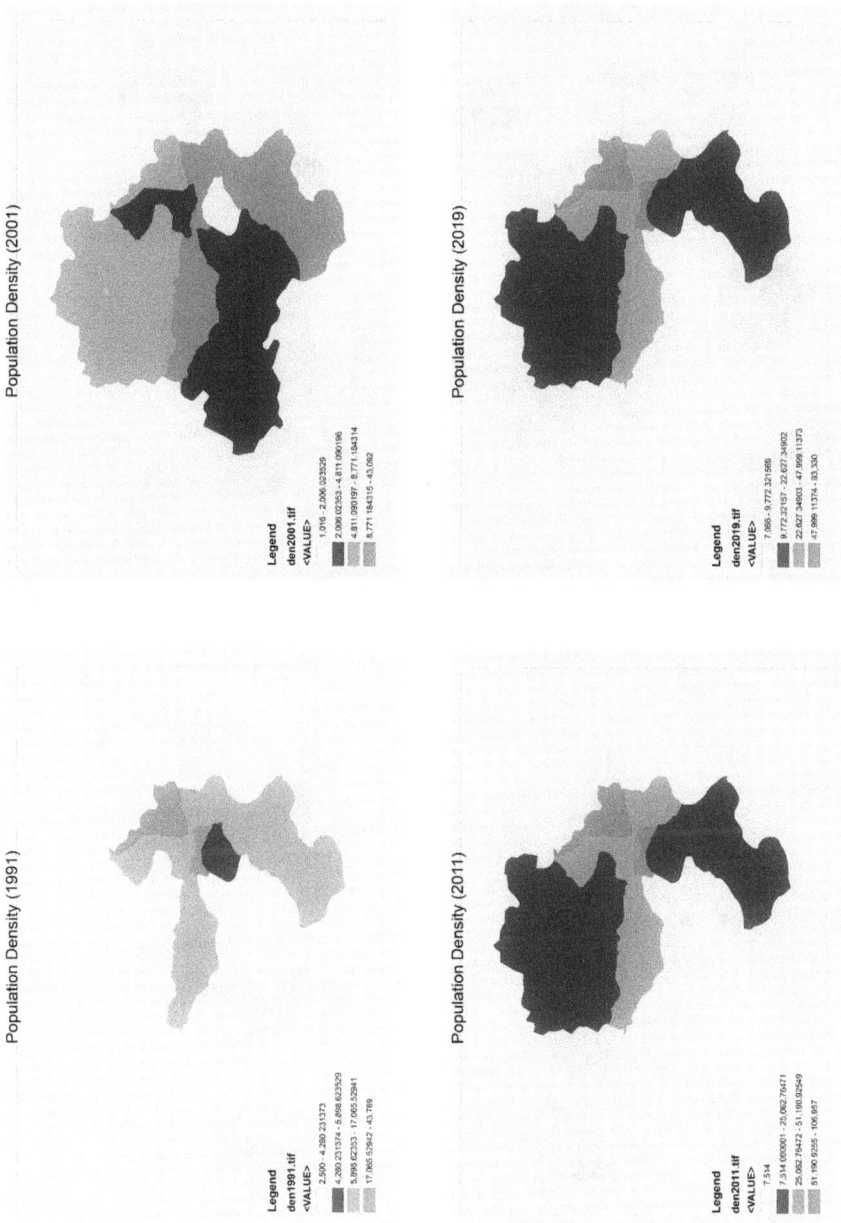

Population Density (2001)

Legend
den2001.tif
<VALUE>
1.016 - 2.006.023529
2.006.02353 - 4.811.090196
4.811.090197 - 8.771.184314
8.771.184315 - 43.082

Population Density (2019)

Legend
den2019.tif
<VALUE>
7.065 - 9.772.321986
9.772.321937 - 22.627.34902
22.627.34903 - 47.999.11373
47.999.11374 - 93.330

Population Density (1991)

Legend
den1991.tif
<VALUE>
2.500 - 4.280.231373
4.280.231374 - 5.898.623529
5.898.62353 - 17.065.52941
17.065.52942 - 43.769

Population Density (2011)

Legend
den2011.tif
<VALUE>
7.514
7.514.000001 - 25.062.78471
25.062.78472 - 51.190.92549
51.190.9255 - 106.657

FIGURE 20.2 (Continued)

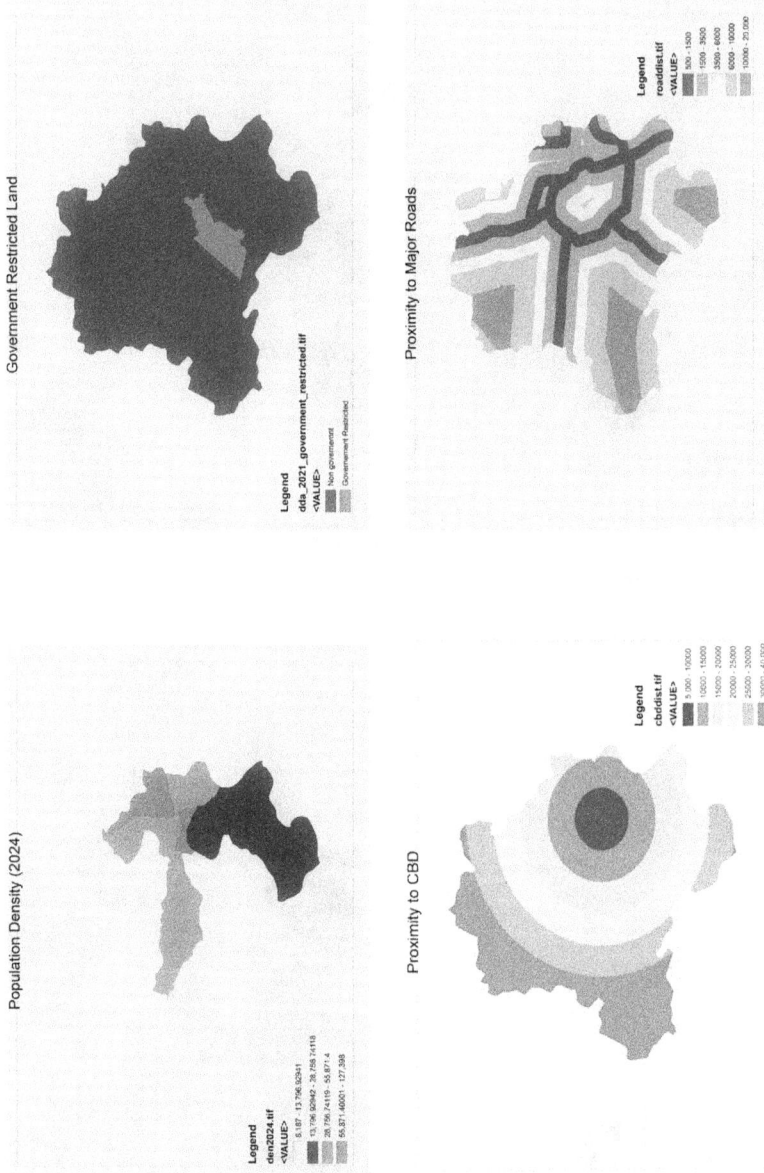

FIGURE 20.2 (Continued)

4) In 2004–2009, the urban area of Delhi expanded by 94 km² with an annual expansion rate of 18.8 km².
5) In 2009–2014, the urban area of Delhi expanded by 104 km² with an annual expansion rate of 20.8 km².
6) In 2014–2019, the urban area of Delhi expanded by 84 km² with an annual expansion rate of 16.8 km².
7) In 2019–2024, the urban area of Delhi expanded by 79 km² with an annual expansion rate of 15.8 km².

20.5   DISCUSSION

In Tables 20.3–20.9, it is found that the land use in terms of urban area is increasing, whereas forest, water, and other land use classes are decreasing. This indicates urban expansion in other land uses, e.g., forest cover, water bodies, and other land-covers, e.g., grassland, agricultural land, etc. It is making a negative impact on the ecological balance. Although in the introduction to this study it was mentioned that cities are described as the economical engines of the nation, this type of urban expansion always creates a barrier to economic growth also.

If can be seen from Table 20.3 that during the interval of 5 years from 1989 to 1994, there is net growth of 110 km² in urban area, whereas forest cover decreased by 5 km², water bodies decreased by 1 km², and other land covers decreased by 104

TABLE 20.3
Land Use Land Cover Change Matrix (1989–1994)

1994 Assessment	Urban	Forest	Water	Other	Total 1989
Urban	161	5	1	104	161
Forest	5	198	0	0	198
Water	1	0	13	0	13
Other	104	0	0	1129	1129
Total 1994	271	193	12	1025	1501
Net change	+110	−5	−1	−104	

TABLE 20.4
Land Use Land Cover Change Matrix (1994–1999)

1999 Assessment	Urban	Forest	Water	Other	Total 1994
Urban	271	2	4	74	271
Forest	2	193	0	0	193
Water	4	0	12	0	12
Other	74	0	0	1025	1025
Total 1999	351	191	8	951	1491
Net change	+80	−2	−4	−74	

TABLE 20.5
Land Use Land Cover Change Matrix (1999–2004)

2004 Assessment	Urban	Forest	Water	Other	Total 1999
Urban	351	5	0	67	351
Forest	5	191	0	0	191
Water	0	0	8	0	8
Other	67	0	0	951	951
Total 2004	424	186	8	884	1501
Net change	+73	−5	±0	−67	

TABLE 20.6
Land Use Land Cover Change Matrix (2004–2009)

2009 Assessment	Urban	Forest	Water	Other	Total 2004
Urban	424	10	0	84	424
Forest	10	186	0	0	186
Water	0	0	8	3	8
Other	84	0	3	884	884
Total 2009	518	176	11	797	1502
Net change	+94	−10	+3	−87	

TABLE 20.7
Land Use Land Cover Change Matrix (2009–2014)

2014 Assessment	Urban	Forest	Water	Other	Total 2009
Urban	518	14	4	86	518
Forest	14	176	0	0	176
Water	4	0	11	0	11
Other	86	0	0	797	797
Total 2014	622	162	7	711	1502
Net change	+104	−14	−4	−86	

TABLE 20.8
Land Use Land Cover Change Matrix (2014–2019)

2019 Assessment	Urban	Forest	Water	Other	Total 2014
Urban	622	12	0	72	622
Forest	12	162	0	0	162
Water	0	0	7	0	7
Other	72	0	0	711	711
Total 2019	706	150	7	639	1502
Net change	+84	−12	±0	−72	

TABLE 20.9
Land Use Land Cover Change Matrix (2019–2024)

2024 Assessment	Urban	Forest	Water	Other	Total 2019
Urban	706	14	0	65	706
Forest	14	150	0	0	150
Water	0	0	7	0	7
Other	65	0	0	639	639
Total 2024	785	136	7	574	1502
Net change	+79	−14	±0	−65	

km². The total study area remains the same at 1501 km² for both study periods, i.e., the years 1989 and 1994.

If can be seen from Table 20.4 that during the interval of 5 years from 1994 to 1999, there is a net growth of 80 km² in urban area, whereas forest cover decreased by 2 km², water bodies decreased by 4 km², and other land covers decreased by 74 km². The total study area remains the same at 1491 km² for both study periods, i.e., the years 1994 and 1999. This varies from the last change matrix table because of an error in rounding off the areas and production of the map. This type of minute error can be ignored for such a huge study area as the error is only 10 km².

If can be seen from Table 20.5 that during the interval of 5 years from 1999 to 2004, there is a net growth of 73 km² in urban area, whereas forest cover decreased by 5 km², water bodies remain unchanged, and other land covers are decreased by 67 km². The total study area remains the same at 1501 km² for both study periods, i.e., the years 1994 and 1999. This varies from the last change matrix table because of an error in rounding off the areas and production of the map. This type of minute error can be ignored for such a huge study area as the error is only 10 km².

If can be seen from Table 20.6 that during the interval of 5 years from 2004 to 2009, there is a net growth of 94 km² in urban area, whereas forest cover decreased by 10 km², water bodies increased by 3 km², and other land covers decreased by 87 km². The total study area remains the same at 1502 km² for both study periods, i.e., the years 2004 and 2009. This varies from the last change matrix table because of an error in rounding off the areas and production of the map. This type of minute error can be ignored for such a huge study area as the error is only 1 km².

If we look at Tables 20.7 and 20.8, it can be seen that during the interval of 5 years from 2014 to 2019, there is a net growth of 84 km² in urban area, whereas forest cover decreased by 12 km², water bodies remain unchanged, and other land covers are decreased by 72 km². The total study area remains the same at 1502 km² for both study periods, i.e., the years 2014 and 2019. This is no variance from the previous change matrix table, which this shows better accuracy in land-use land-cover classification performance.

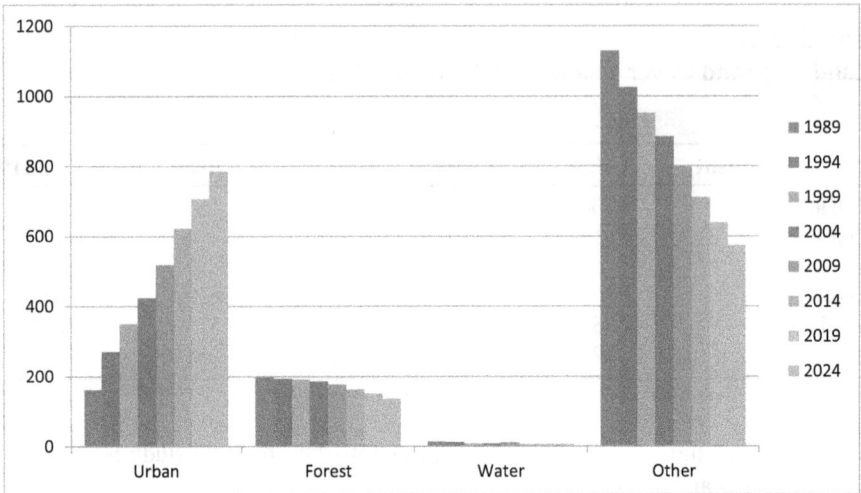

FIGURE 20.3 Land-use land-cover classes for Delhi for 1989, 1994, 1999, 2004, 2009, 2014, 2019, and 2024.

It can be seen from Table 20.9 that during the interval of 5 years from 2019 to 2024, there is a net growth of 79 km^2 in urban area, whereas forest cover decreased by 14 km^2, water bodies remain unchanged, and other land covers are decreased by 65 km^2. The total study area remains the same at 1502 km^2 for both study periods, i.e., the years 2019 and 2024. This is no variance from the previous change matrix table, which this shows better accuracy in land-use land-cover classification performance.

From these tables, we can summarise that urban land use is dominant as this is the most influencing factor in the overall change matrices and land-use land-cover maps. Most of the map area is covered by urban area. The second most prominent class is others, i.e., grassland, vacant land, agricultural land, etc. Although it is constantly decreasing in terms of area, it was also a bigger area for the initial years of study i.e., 1989 and 1994. Forest cover has had much less influence in the change of land use but yes, it is also decreasing over time. And it is a matter of concern as it is already much less in area in comparison with other land uses, whereas it plays an important role in ecological balance. The same can be said for the water bodies which also play a vital role in creating ecological balance.

Figure 20.3 shows areas of land-use land-cover classes in the years 1989, 1994, 1999, 2004, 2009, 2014, 2019, and 2024. As can be seen in this chart, urban area is increasing for the years from 1989 to 2019 and the simulated data for 2024 also show an increase in terms of urban area. Forest cover is decreasing as we do a time series analysis for the forest cover class. It is not decreasing rapidly but it is decreasing significantly. Water bodies have not shown any variation in this time series analysis, and they have retained their position with 2 square kilometres in area, which remains similar in all the years with only small fluctuations. It varies between 1–2 square kilometres. The last feature class, other, has shown a decrease in terms of area in time series analysis. Therefore, we can say that the main contributing feature class which is contributing to urban land use is another type of class as it is decreasing rapidly, and urban areas are increasing rapidly. Meanwhile the other two classes, forest and water, have much less influence on urban expansion as they are only marginally decreasing.

It was discussed in the introduction part that only that forest cover is being converted into cropland or agricultural land, and agricultural land is being converted into built-up areas. Therefore, the same thing is found here through this time series analysis, forest cover is decreasing, and it is being converted into other types of land uses and these other land uses are converted into urban areas. From the beginning of the time series analysis, i.e., 1989, we find that there is a linear increase taking place which means a constant increase in terms of area. Therefore, it is increasing constantly. Whereas forest cover has a nonlinear type of signature, it is decreasing but much less and not constantly. Also, water has a horizontal line type of signature as it is neither decreasing nor increasing, but if we analyse the pattern of other land uses, there is again a linear type of relationship among the areas which have decreased in 5 years; it is constantly decreasing linearly.

Figure 20.4 shows the percentage share of each land use in a particular year, for example, in 1989, the percentage of the area shared by the urban class was close to 10%, the forest class, was 13%; water was 2%; and another land uses were more than 75%. As we do the time series analysis, we find that in 2024 the urban area percentage will be highest in the shared area percentage and it will be close to 52% of the total area, whereas the other classes will be close to 38%, the forest will be close to 9% and water class will be 2% of the total area. In the intermediate years like 1994, 1999, 2004, 2009, 2014, and 2019, we found that the area percentage share is also constantly increasing in the case of urban areas and it increases by 7–8% in 5 years and other land uses are decreasing by 4% in the intervals of 5 years. The forest

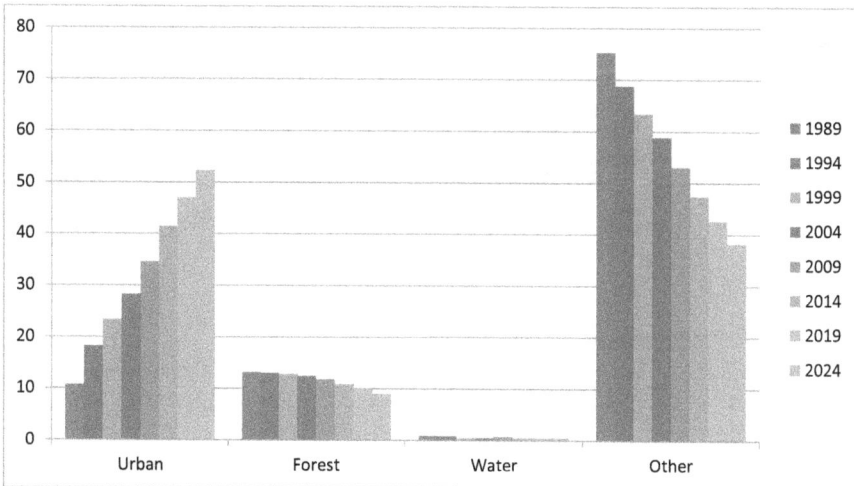

FIGURE 20.4 Percentage of land-use land-cover classes as a percentage of the total area for Delhi for 1989, 1994, 1999, 2004, 2009, 2014, 2019, and 2024.

class is decreasing very marginally by 1% to 2% in 5 years. Water is constant in this time series analysis, it is neither decreasing nor increasing in the case of percentage area share.

It is very useful for us to know that the area of study remains the same for each year as the area is of National Capitol Territory which is changing in any case. For all the years of analysis from 1989 to 2024, the area of study does not change. However, the land uses in this boundary keep changing at different time intervals. For the urban class, it is constantly increasing, for the forest it is marginally decreasing, for the water class it is constant, and for other land-use classes it is constantly decreasing.

Figure 20.5 shows the net change in area in square kilometres in a particular land-use class of Delhi. The time interval of 5 years is taken for this analysis. If we look at the interval of 1989 to 1994, we find that there is a net increase of 110 square kilometres in terms of area of urban class, there are approximately 5 square kilometres of shrinkage in forest class, 1 square kilometre reduction in water class, and approximately 104 square kilometres of reduction in other land uses. Similarly, we find the results in intervals 1994–1999, 1999–2004, 2004–2009, 2009–2014, 2014–2019, and 2019–2024. In 1994–1999, there is a reduction in other land uses, water, and forest, whereas there is an addition in urban land use. In 1999–2004, there is a reduction in other land uses, water, and forest, whereas there is an addition in urban land use. In 2004–2009, there is a reduction in other land uses, water, and forest, whereas there is an addition in urban land use. From 2009–2014, there is a reduction in other land uses, water, and forest, whereas there is an addition in urban land use. In 2014–2019, there is a reduction in other land uses, water, and forest, whereas there is an addition

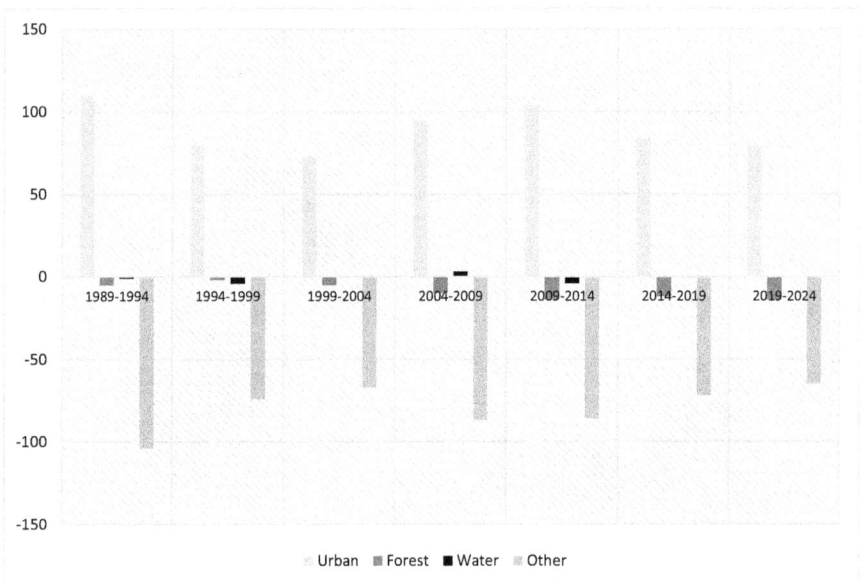

FIGURE 20.5   Bar chart showing the net change in land use land cover for Delhi in km².

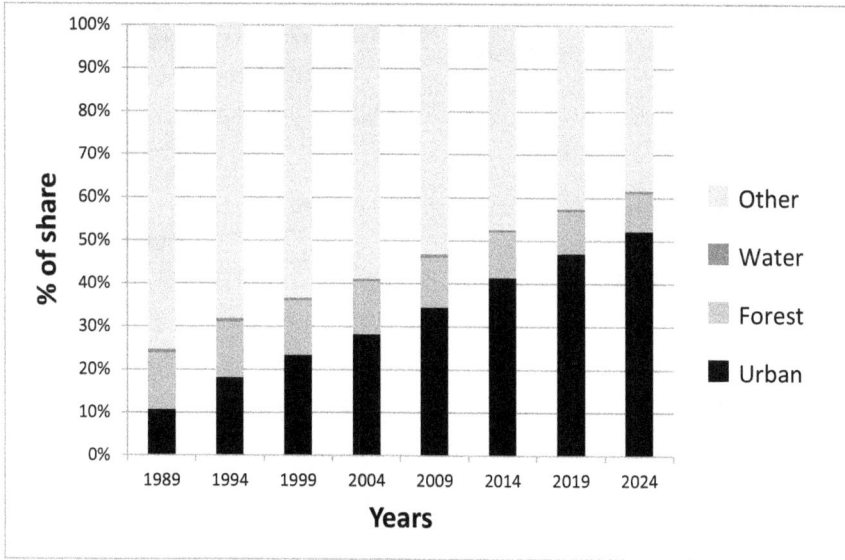

FIGURE 20.6   Bar chart showing the net change in land use land cover in 5 years.

to urban land use. In 2019–2024, there is a reduction in other land uses, water, and forest, whereas there is an addition in urban land use.

In Figure 20.6, the area has been fixed to the boundary that is the limit of the Nation Capitol Territory of Delhi for all the years of time series analysis from 1989 to the future projected year 2024. From this chart, we can see that the urban area has increased constantly and other land uses area has decreased constantly. The water land-use class and forest class have not shown very rapid change but marginal change. Here two colours are dominating, blue and violet, where blue represents the urban land-use category, whereas violet represents the other land-use category. Blue is increasing with time and violet is decreasing with time.

From this time series analysis (Figure 20.7), we can see that there is a serious problem of urban expansion, which is either taking place in a coordinated manner or unplanned manner, but it is taking place at a rapid pace. It is responsible for many issues such as environmental degradation and encroachment on agricultural land use and forest cover. It is directly or indirectly affecting humans and other creatures also. There is an imbalanced ecosystem in cities because of this rapid urbanisation. There are many reasons for urbanisation; migration is one of the reasons that plays a very important role in urban expansion. To manage this migration, many policies and planning interventions have been used such as the Ministry of MSME (Micro, Small and Medium Enterprises) has encouraged many bottom-up approaches including the establishment of small industries in rural areas, financing micro and small enterprises, and subsidies for developers those who provide EWS (Economically Weaker Section) housing for the poor. This way, there are very useful measures but still, rapid urbanisa-tion is taking place near city centres and transportation routes and in an uncoordinated

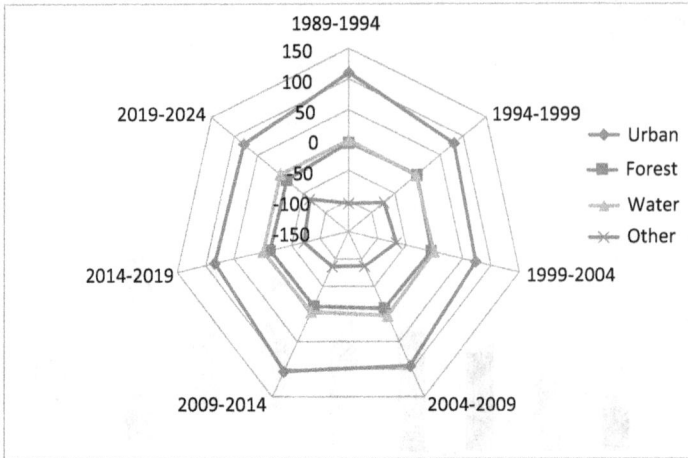

FIGURE 20.7 Radar showing classes of land use land cover.

manner. Sometimes, we refer to it as slums and this type of growth is called urban sprawl.

20.6 CONCLUSIONS

The study demonstrates the potential use of land-use land-cover analysis in the spatial and temporal assessment of urban expansion in Delhi. The study focuses on the urban dynamics of a rapidly growing city like Delhi. Further, it helps us to understand the urban expansion metrics. This matrix is easy to understand in comparison to other expansion measures such as patch, shape, edge, diversity, etc. The urban expansion metrics used in this study, such as expansion contribution rate, expansion percentage of change, and annual expansion rate, are very suitable tools for understanding the processes involved in urban expansion. Understanding the urban expansion process is essential for the planning and management of urbanisation processes.

Suggestive Framework for Policy Planning and Strategy: To minimise the effects of urban sprawl, it is not essential to overhaul the city but by improving resource management and using connectivity and intelligence within the existing infrastructure, it can be achieved. The framework should not only be focusing on making a green, clean, and efficient city, but also on making it people-centric, economically viable, and culturally conservative. It must contain solutions for the conservation of heritage monuments. It should also suggest some solutions for future challenges. To make it smart and sustainable, it needs to have all the systems working together at all levels. The framework shall be giving a smart solution considering all the social, economic, and environmental aspects. The area-based development scheme has two main factors: (i) retrofitting (approx. 500 acres) and (ii) redevelopment (approx. 50 acres) along with greenfield development (approx. 250 acres). These approaches have been selected through desk research, analysis, meeting with public representatives, and citizen engagement programs. The area-based development scheme focuses on three main areas: smart infrastructure, utility improvement, and urban transportation.

Smart Infrastructure: This will include a water supply with smart metering, sanitation including solid waste management, e-governance, and services. It will also have pedestrian-friendly/barrier-free streets and pathways, footpaths, and energy-efficient lighting. It will focus on the renovation of old and heritage buildings. In addition to the above, social infrastructure which will include gardens, parks, playgrounds, and other social gathering spaces, will be an integral part of the scheme.

Utility Improvement: The common utility duct will be provided for all basic utility infrastructures. This will reduce the lines of wires which hang between poles and create a dangerous zone. It will be underground and flow in the utility duct. This will add to the aesthetics of the space, safety measures, and ease of maintenance. It is among the key components of an area-based development scheme. In this scheme, there are proposals for self-sustained public and institutional buildings. All these buildings will have renewable energy sources, e.g., rainwater harvesting, wastewater reuse, and rooftop solar panels. All the buildings enabled with these infrastructures will reduce overall energy consumption and additional energy will be transferred to the grid for a sustainable economy.

Urban Transportation: Urban mobility nodes are one of the key components of area-based development. These will include parking, e-services, food courts, public meeting places, non-motorised vehicle availability, and a micro-level traffic interchange point.

REFERENCES

Aayog, NITI. (2012). Appraisal Document of Twelfth Five Year Plan 2012–17. *Accessed on, 27.*

Angel, S., Sheppard, S., Civco, D. L., Buckley, R., Chabaeva, A., Gitlin, L., ... & Perlin, M. (2005). *The dynamics of global urban expansion* (p. 205). Washington, DC: World Bank, Transport and Urban Development Department.

Annez, P. C., Buckley, R. M., & Spence, M. (2009). *Urbanization and growth.* Commission on Growth and Development: World Bank.

Berry, J. K. (1993). *Beyond mapping: concepts, algorithms, and issues in GIS* (No. 526.9820285 B534). Fort Collins, Colorado, USA: GIS World Books.

Bhat, P. A., ul Shafiq, M., Mir, A. A., & Ahmed, P. (2017). Urban sprawl and its impact on landuse/land cover dynamics of Dehradun City, India. *International Journal of Sustainable Built Environment, 6*(2), 513–521.

Bhatta, B. (2009). Analysis of urban growth pattern using remote sensing and GIS: a case study of Kolkata, India. *International Journal of Remote Sensing, 30*(18), 4733–4746.

Coppin, P., Jonckheere, I., Nackaerts, K., Muys, B., & Lambin, E. (2004). Review ArticleDigital change detection methods in ecosystem monitoring: a review. *International journal of remote sensing, 25*(9), 1565–1596.

Deep, S., & Saklani, A. (2014). Urban sprawl modeling using cellular automata. *The Egyptian Journal of Remote Sensing and Space Science, 17*(2), 179–187.

Deng, J. S., Wang, K., Hong, Y., & Qi, J. G. (2009). Spatio-temporal dynamics and evolution of land use change and landscape pattern in response to rapid urbanization. *Landscape and urban planning, 92*(3–4), 187–198.

Duranton, G., Rodríguez-Pose, A., & Sandall, R. (2009). Family types and the persistence of regional disparities in Europe. *Economic geography, 85*(1), 23–47.

Fotheringham, A. S., Brunsdon, C., & Charlton, M. (2000). *Quantitative geography: perspectives on spatial data analysis.* London; California: Sage.

Geymen, A., & Baz, I. (2008). Monitoring urban growth and detecting land-cover changes on the Istanbul metropolitan area. *Environmental monitoring and assessment, 136,* 449–459.

Hardin, P. J., Jackson, M. W., & Otterstrom, S. M. (2007). Mapping, measuring, and modeling urban growth. *Geo-spatial technologies in urban environments: Policy, practice, and pixels,* Berlin, Heidelberg: Springer, 141–176.

Hasse, J. E., & Lathrop, R. G. (2003). Land resource impact indicators of urban sprawl. *Applied geography, 23*(2–3), 159–175.

Herold, M., Goldstein, N. C., & Clarke, K. C. (2003). The spatiotemporal form of urban growth: measurement, analysis and modeling. *Remote sensing of Environment, 86*(3), 286–302.

Hu, Y., Xue, H., & Hu, H. (2007). A piezoelectric power harvester with adjustable frequency through axial preloads. *Smart materials and structures, 16*(5), 1961.

Hu, Z. L., Du, P. J., & Guo, D. Z. (2007). Analysis of urban expansion and driving forces in Xuzhou city based on remote sensing. *Journal of China University of Mining and Technology, 17*(2), 267–271.

Jat, M. K., Garg, P. K., & Khare, D. (2008). Monitoring and modelling of urban sprawl using remote sensing and GIS techniques. *International journal of Applied earth Observation and Geoinformation, 10*(1), 26–43.

Jensen, J. R. (1996). *Introductory digital image processing: a remote sensing perspective* (No. Ed. 2). United Kingdom: Prentice-Hall Inc.

Kandlikar, M., & Ramachandran, G. (2000). The causes and consequences of particulate air pollution in urban India: a synthesis of the science. *Annual review of energy and the environment, 25*(1), 629–684.

Kumar, J. A., Pathan, S. K., & Bhanderi, R. J. (2007). Spatio-temporal analysis for monitoring urban growth–a case study of Indore city. *Journal of the Indian Society of Remote Sensing, 35,* 11–20.

Li, X., & Yeh, A. G. O. (2000). Modelling sustainable urban development by the integration of constrained cellular automata and GIS. *International journal of geographical information science, 14*(2), 131–152.

Li, Z. L., Tang, B. H., Wu, H., Ren, H., Yan, G., Wan, Z., ... & Sobrino, J. A. (2013). Satellite-derived land surface temperature: Current status and perspectives. *Remote sensing of environment, 131,* 14–37.

Mage, D., Ozolins, G., Peterson, P., Webster, A., Orthofer, R., Vandeweerd, V., & Gwynne, M. (1996). Urban air pollution in megacities of the world. *Atmospheric environment, 30*(5), 681–686.

Masek, J. G., Lindsay, F. E., & Goward, S. N. (2000). Dynamics of urban growth in the Washington DC metropolitan area, 1973–1996, from Landsat observations. *International Journal of Remote Sensing, 21*(18), 3473–3486.

Masser, I. (2001). Managing our urban future: the role of remote sensing and geographic information systems. *Habitat international, 25*(4), 503–512.

Nguyen, D. (2010). Evidence of the impacts of urban sprawl on social capital. *Environment and planning B: planning and design, 37*(4), 61–627.

Pickett, S. T., Burch, W. R., Dalton, S. E., Foresman, T. W., Grove, J. M., & Rowntree, R. (1997). A conceptual framework for the study of human ecosystems in urban areas. *Urban ecosystems*, *1*, 185–199.

Praharaj, S., Han, J. H., & Hawken, S. (2018). Urban innovation through policy integration: Critical perspectives from 100 smart cities mission in India. *City, culture and society*, *12*, 35–43.

Qiu, Z., Wang, X., Yuan, Q., & Wang, F. (2009). Coma measurement by use of an alternating phase-shifting mask mark with a specific phase width. *Applied Optics*, *48*(2), 261–269.

Seto, K. C., Fragkias, M., Güneralp, B., & Reilly, M. K. (2011). A meta-analysis of global urban land expansion. *PloS one*, *6*(8), e23777.

Singh, A. (1989). Review article digital change detection techniques using remotely-sensed data. *International journal of remote sensing*, *10*(6), 989–1003.

Stow, D. A., LR, T., & JE, E. (1980). Deriving land use/land cover change statistics from Landsat: A study of prime agricultural land. In *Proceedings International Symposium on Remote Sensing of-Environment (USA)* (no. 14th) (pp. 1227–1235).

Sudhira, H. S., Ramachandra, T. V., & Bala Subrahmanya, M. H. (2007, July). Integrated spatial planning support systems for managing urban sprawl. In *Presentation at 10th International Conference on Computers in Urban Planning and Urban Management (CUPUM), Iguassu Falls* (pp. 11–13).

Taubenböck, H., Wegmann, M., Roth, A., Mehl, H., & Dech, S. (2009). Urbanization in India–Spatiotemporal analysis using remote sensing data. *Computers, environment and urban systems*, *33*(3), 179–188.

Wakode, H. B., Baier, K., Jha, R., Ahmed, S., & Azzam, R. (2014). Assessment of impact of urbanization on groundwater resources using GIS techniques-case study of Hyderabad, India. *International Journal of Environmental Research*, *8*(4), 1145–1158.

Weng, Q. (2002). Land use change analysis in the Zhujiang Delta of China using satellite remote sensing, GIS and stochastic modelling. *Journal of environmental management*, *64*(3), 273–284.

Index

A

Air quality, 60, 167, 179, 183
Analytic hierarchy process (AHP), 33, 37, 45
Applications, 298–300
ArcGIS, 89, 92, 98, 150

C

Classes, 81, 90, 94, 127, 128, 137, 200, 231
Climate change, 150, 174, 175, 182, 185
Communities, 58, 63, 86
COVID-19, 317, 320, 324, 336

D

Disaster, 258

E

Economic, 339, 341, 349
Energy, 291, 294, 296
    consumption, 291, 296
    expansion, 152, 155, 192
Environmental pollution, 74

G

Geographic information system (GIS), 87, 140,
    146, 149, 165, 166, 207, 314
Geospatial techniques, 105, 106, 317
GPS, 123
Groundwater resources, 260, 263, 269

H

Healthcare, 59, 161, 165

L

Land cover, 195, 196, 231
Landsat, 306, 318, 321
Land surface temperature, 25, 74, 85
Land use, 185, 190, 192, 195

LULC, 231, 244, 252
    dynamics, 27, 85
    statistics, 94, 252

M

Market, 1, 3, 16, 120
Monsoon season, 204–207

P

Planned development, 12, 15
Planning, 253, 254, 258
Population, 174, 176, 179, 182, 185

R

Random forest, 74, 82
Remote sensing, 259, 261
    and GIS, 29, 271, 341
Renewable energy, 60, 178, 357
    demand, 60, 178, 357
Resources, 314, 318, 339

S

Satellite images, 158, 159, 161
Slope, 92, 124, 126, 132
Spatial proximity, 3, 11, 13
Suitability map, 27, 48, 152, 154
Sustainable, 200, 207

T

Temperature, 72, 77, 335

U

UHI analysis, 74, 77, 78, 82, 182
Urban heat island, 64, 72, 81
Urbanisation, 3, 10
User's accuracy, 90, 126
USGS, 88, 90, 124

For Product Safety Concerns and Information please contact our EU
representative GPSR@taylorandfrancis.com
Taylor & Francis Verlag GmbH, Kaufingerstraße 24, 80331 München, Germany